CONTEMPORARY FUNDAMENTAL PHYSICS

EINSTEIN AND OTHERS

UNIFICATION

CONTEMPORARY FUNDAMENTAL PHYSICS –

VALERIY V. DVOEGLAZOV - SERIES EDITOR

UNIV. DE ZACATECAS, ZACATECAS, MEXICO

Some Contemporary Problems of Condensed Matter Physics
Stoyan J. Vlaev and L.M. Gaggero Sager (Editors)
2001. ISBN: 1-56072-889-2

The Photon
Trilochan Pradhan
2001. ISBN: 1-56072-928-7

The Problem of Electron and Superluminal Signals
V.P. Oleinik
2001. ISBN: 1-56072-938-4

Process Physics: From Information Theory to Quantum Space and Matter
Reginald T. Cahill
2005. ISBN: 1-59454-300-3

Logical Foundation of Theoretical Physics
G. Quznetsov
2006. ISBN: 1-59454-948-6

Introduction to Physics of Elementary Particles
Oleg Mikhilovich Boyarkin
2007. ISBN: 1-60021-200-X

Probabalistic Treatment of Gauge Theories
G. Quznetsov
2007. ISBN: 978-1-60021-627-5

Relativity, Gravitation, and Cosmology: New Developments
Valeriy V. Dvoeglazov (Editor)
2010. ISBN: 978-1-60692-333-7

Progress in Relativity, Gravitation, Cosmology
Valeriy V. Dvoeglazov and A. Molgado (Editors)
2012. ISBN: 978-1-61324-811-9

Einstein and Hilbert: Dark Matter
Valeriy V. Dvoeglazov (Editor)
2012. ISBN: 978-1-61324-840-9

Quantum Mechanics in Spaces of Constant Curvature
V.M. Redkov and E.M. Ovsiyuk
2012. ISBN: 978-1-61470-271-9

Revised Quantum Electrodynamics
Bo Lehnert
2012. ISBN: 978-1-62081-484-0

Maxwell Electrodynamics and Boson Fields in Spaces of Constant Curvature
E.M. Ovsiyuk, V.V. Kisel and V.M. Red'kov
2014. ISBN: 978-1-62618-891-4

Einstein and Others: Unification
Valeriy V. Dvoeglazov (Editor)
2015. ISBN: 978-1-63463-276-8

CONTEMPORARY FUNDAMENTAL PHYSICS

EINSTEIN AND OTHERS

UNIFICATION

VALERIY V. DVOEGLAZOV
EDITOR

nova
publishers
New York

LIBRARY OF CONGRESS CATALOGING-IN-PUBLICATION DATA

Einstein and others : unification / [edited by] Valeriy V. Dvoeglazov and Alberto Molgado (Univ. de Zacatecas, Zacatecas, Mexico)
pages cm. -- (Contemporary fundamental physics)
Includes bibliographical references and index.
ISBN 978-1-63463-276-8 (hardcover)
1. Quantum field theory. 2. Unified field theories. 3. Quantum gravity. I. Dvoeglazov, Valeri V., editor. II. Molgado, A., editor.
QC174.45.E56 2014
530.14'2--dc23

2014039702

Published by Nova Science Publishers, Inc. † New York

CONTENTS

EDITORIAL INTRODUCTION

Valeriy V. Dvoeglazov

We continue the Book Series "Contemporary Fundamental Physics" of the Nova Science Publishers. The thematic issue contains articles related to the problem of the unification of interactions and to other fundamental problems of the modern physics. The book includes the Editorial Introduction, 9 articles and the reply to the comment. This book may be considered as continuation of the books "Einstein and Poincaré" and "Einstein and Hilbert", published by us few years ago.

S. Capozziello et al. concern with the Extended Theory of Gravitation starting from the 5D Riemannian manifold. What is important in their presentation is that they show that the scheme may be compatible with the known observational evidence, and suggest that "eventual discrepancies could be detected in experiments, as ATLAS and CMS, today running at LHC (CERN)". They argue that the gauge groups in the general relativity must be associated with external space-time symmetries. Therefore, the gauge theory of gravity cannot be considered on the same footing with the Yang-Mills theories of the Standard Model. They also employ the tetrads and the connection forms as the fundamental fields, following Ashtekar and collaborators. Please note interesting ideas of Cartan [1] who generalized the Riemann geometry in order to include torsion in addition to curvature. Sciama [2] pointed out that intrinsic spin may be the source of torsion of the underlying spacetime manifold. Next, authors noted that Electrodynamics can be considered as a genuine geometrical theory. Apart, "after the reduction to 4D-physics, a given theory of gravity is described by the conformal-affine group of diffeomorphisms GL(4) which is characterized by $4 \times 4 = 16$ generators". By metric deformation the authors give mass to an initially massless particle in the Minkowski frame. The following statement is also interesting. "We can say that the Higgs field has the same role of the tetrads for the electroweak interactions while the electroweak bosons have the same role of the metric". And again confirm that "$f(R)$-gravity is nothing else but a scalar-tensor gravity". In the Sec. 8 the authors tried to find relations between the Dirac equation with the electroweak interaction and its interaction with the torsion field. The difference with the Standard Model is the absence of the Higgs field and the internal structure of electroweak bosons.

D. Cirilo-Lombardo, following his previous works [3], presents a comprehensive review, and he completes the analysis of his fundamental model for an Unified Field Theory. His approach is based on a hypercomplex structure coming from symmetry breaking of a group manifold in higher dimensions. This paper received a lot of critical comments by the referees. I, myself, asked him to comment about the known Weinberg-Hehl discussion in

"Physics Today" [4]. Finally, the author stresses "$-2\lambda f_{\mu\nu}$ assumes the role of a current for the torsion field, similarly to the terms proportional to the 1-form potential vector a_μ that usually act as currents of the electromagnetic field $f_{\mu\nu}$, in the field equation of the standard theory: $\nabla_\alpha f^\alpha_\mu = J_\mu$".

S. I. Kruglov considers the modified gravity. He also analyzes the origin of the Universe acceleration. So-called $f(R)$-gravity theories replace the Ricci scalar in the Einstein-Hilbert action by the function $f(R)$. The paper may also serve as a brief introduction to the modified gravitation theories and their relations to cosmology. As shown, in the Einstein frame this kind of theories is equivalent to the scalar-tensor theory of gravity and appearance of the fifth force. However, the Weak Equivalence Principle is violated.

Varlamov and Beloborodova study quantum description of extended particles. Dirac and Maxwell fields are considered on the 8-dimensional homogeneous space, which is, in fact, the direct product of the Minkowski spacetime and 2-dimensional complex sphere. Their arguments for extended objects are indeed reasonable due to the fact that in quantum mechanics the material-point idealization does not work perfectly. They use the formalism proposed by V. Ginzburg and I. Tamm long ago. This is in order to describe the extended particles of any spin (particularly, the particles of the spins 1/2 and 1) by the set of the first-order differential equations on the group manifold $\mathcal{M}_{10}$ of the Poincaré group. It is interesting that they proved convergence of 8-quantum electrodynamics on the Poincar´e group in the case of the specific homogeneous space.

Pierre presents rather a mathematical theory, namely, that of bisemifermions.

Rodrigues and de Oliveira present a pedagogical review of application of Maxwell equations in the rotating frame. They use the formalism of differential forms. Particularly, the Wilson-Wilson experiment (1913) is considered. Maxwell equations for rotating magnetic insulators in external fields provoked a lot of polemics in the literature. All conservation laws are valid in their consideration of the Wilson and Wislon experiment.

The Krasulin's chapter shows, how one can establish a relation between five-dimensional tangent vectors in space-time and spinors that would be invariant under parallel transport. The author concerns with the correct definition of the covariant derivative in 5D. The paper is written rather in a mathematical way. I would still like to note that the 8-dimensional spinors have been re-considered recently [5].

The papers of Dvoeglazov finish this editorial book. They speak for myself. They have been first presented at the 6th International Conference on the Dark Side of the Universe (DSU2010), Leon, Mexico, June 2010 (organizer: Dr. D. Delepine) and the XXXI Workshop on Geometric Methods in Physics, Bialowieza, Poland, June 2012 (organizer: Dr. P. Kielanowski). .

Again, about the history of my University in the last years [6]. Unfortunately, several distinguished professors continued to leave the Physics School: Fis. José Augusto Beltrán Mendoza (retired), Dr. Victor Manuel González Robles (he resigned), Dr. Rigoberto Juárez Maldonado (he changed the University), Dr. Jorge Luis Menchaca Arredondo (he changed the University), Dr. Alberto Molgado Ramos (he changed the University), , Dr. Adolfo Vidales Roque (retired), Dr. Alejandro Vizcarra Rendón (he was forced to resign), Dr. Stoyan Jelev Vlaev (deceased). In 2008-2012 we had the following "leaders in the Zacatecas University (UAZ)": M. Sc. Francísco Javier Domínguez Garay (Rector), Dr. Juan Francisco Álvarez (2008-10, Coordinador de Proyectos Federales y Representante Institucional ante

PROMEP), M. Sc. Rubén Ibarra Reyes (2008-10, Coodinador Operativo de PROMEP and, 2010-12, RIP ante PROMEP), Dr. José de Jesus Araiza Ibarra (Director de la UAF-UAZ). In that time, our professors achieved great results: the *Licenciatura* (B.Sc.) program has been confirmed by the Recognition of the CIEES, and the postgraduate programs of the UAF-UAZ have been approved by the PNPC (The National Program for the Quality Postgraduate [Programs]) of the CONACyT (i. e., by the federal government organizations). Nevertheless, serious cases of the violation of scientific ethics occured in the Universidad Autónoma de Zacatecas. Moreover, cheating the "facturas" was continued in those times. For instance, the authorities tried to force including some persons in the authorships of the scientific papers without any technical contributions under the menaces of administrative or financial punishments. Thus, several professors have been corrupted. These facts may diminish the credibility of the CV credentials of some of the local researchers. I hope that future generations will not confuse the Science with *soccer* or *kolhoz* games in the UAZ here.

Acknowledgment

We are very grateful to the Publisher, to our authors, referees and friends.

References

[1] E. Cartan, Ann. Ec. Norm. **42**, 17 (1925).

[2] D. W. Sciama, On the analog between charge and spin in GR, in *Recent Developments in GR, Festschrift for Leopold Infeld.* (Pergamon Press, New York, 1962), p. 415.

[3] D. J. Cirilo-Lombardo, Int. J. Theor. Phys. **49**, 1288 (2010), and references therein.

[4] S. Weinberg, F. Hehl, and others, discussion in Physics Today **58**, No. 11 (2005) 31; ibid. **59**, No. 4 (2006) 10; ibid. **60**, No. 3 (2007) 16.

[5] V. V. Dvoeglazov, Nuovo Cimento A**108**, 1467 (1995).

[6] V. V. Dvoeglazov and A. Enciso Muñoz, Hadronic J. Suppl. **17** (2002) - *Proceedings of the Zacatecas School on Theoretical Physics*; V. V. Dvoeglazov, in *Einstein and Hilbert: Dark Matter.* (Nova Science Pubs, NY, 2011).

Announcement

The Nova Science Publishers (NY, USA) continues the Project of "Relativity, Gravitation, Cosmology". The following topics are of our primary interest:

1. Dilaton gravity.
2. Quantum Mechanical Phases, Neutrino and Gravity. Photon and Gravity.
3. Spin connection and 4-potential. Axion, Torsion and Notoph.
4. Curvature as a Scalar Field over the Minkowski Space.
5. Multidimensional Gravity. De Sitter Gravity. Weyl Approach.
6. Relativistic Quantum Mechanics Approach to Gravity (a la S. Weinberg). Parity Violation.
7. Non-commutative Space-time.
8. $v_{group} > c$.
9. Quantization. Quantum Gravity.

The papers should be submitted to Prof. V. V. Dvoeglazov (valeri@fisica.uaz.edu.mx). The acceptable topics of papers are indicated in the title, and they are not restricted by those of the first special issues. We are ready to consider other candidates for the Editorial Board. Inquiries concerning submissions of papers and book requests should be sent to valeri@fisica.uaz.edu.mx and/or Nova.Main@novapublishers.com.

Valeriy V. Dvoeglazov
Profesor-Investigador Titular "C"
Unidad Académica de Física
Universidad Autónoma de Zacatecas
August 2013.

In: Einstein and Others: Unification
Editor: Valeriy V. Dvoeglazov

ISBN: 978-1-63463-276-8

Chapter 1

GRAVITATIONALLY GENERATED INTERACTIONS

***S. Capozziello*[1,2], *M. De Laurentis*[1,2], *L. Fabbri*[3,4] *and S. Vignolo*[3]**
[1]Dipartimento di Fisica, Università di Napoli "Federico II",
Compl. Univ. di Monte S. Angelo, Ed. G, Napoli, Italy
[2]INFN Sez. di Napoli, Compl. Univ. di Monte S. Angelo, Ed. G, Napoli, Italy
[3]DIPTEM Sez. Metodi e Modelli Matematici, Università di Genova,
Piazzale Kennedy, Genova, Italy
[4]INFN & Dipartimento di Fisica, Università di Bologna, Bologna, Italy

Abstract

Starting from a 5D-Riemannian manifold, we show that a reduction mechanism to 4D-spacetimes reproduces Extended Theories of Gravity (ETGs) that are direct generalizations of Einstein's gravity. In this context, the gravitational degrees of freedom can be dealt under the standard of spacetime deformations. Besides, such deformations can be related to the mass spectra of particles. The intrinsic non-linearity of ETGs gives an energy-dependent running coupling, while torsion gives rise to interactions among spinors displaying the structure of the weak forces among fermions. We discuss how this scheme is compatible with the known observational evidence and suggest that eventual discrepancies could be detected in experiments, as ATLAS and CMS, today running at LHC (CERN). We finally discuss the consequences of the present approach in view of unification of physical interactions.

PACS: 05.45-a, 52.35.Mw, 96.50.Fm

Keywords: Mass Generation; Unified Theories

1. Introduction

So far as we know, there are four fundamental forces in Nature: electromagnetic, weak, strong and gravitational interactions. The Standard Model represents well the first three, but not the gravitational interaction, described by the geometric theory of General Relativity (GR).

The Standard Model of Elementary Particles can be considered a successful relativistic quantum field theory both from particle physics and group theory points of view; technically, it is a non-Abelian gauge theory (specifically a Yang-Mills theory) associated with the tensor (Cartesian) product of the internal symmetry groups $SU(3) \times SU(2) \times U(1)$, where the $SU(3)$ color symmetry for quantum chromodynamics, is treated as exact, whereas the $SU(2) \times U(1)$ symmetry, responsible for generating the electro-weak gauge fields, is considered to be spontaneously broken. GR is built upon the Einstein equations, a non-linear set of field equations for the spacetime metric. This non-linearity is indeed a source of difficulty for the GR quantization, and with it for most of the conceptual problems that arise from having GR and quantum principles related in a unique framework.

Since the Standard Model is a gauge theory where all the fields mediating the interactions are represented by gauge potentials, the question is why the fields mediating the gravitational interaction (i.e., the gravitational potentials) are different from those of the other fundamental forces. It is reasonable to expect that there may be a gauge theory in which the gravitational fields stand on the same footing as the other gauge fields of the Standard Model [1]. As it is well-known, this expectation has prompted a re-examination of GR from the point of view of gauge theories: while the gauge groups involved in the Standard Model are all internal symmetry groups, the gauge groups in GR must be associated with external spacetime symmetries. Therefore, the gauge theory of gravity cannot be considered under the same standard of the Yang-Mills theories. It must be one in which gauge objects are not only gauge potentials but also tetrads that relate the symmetry group to the external spacetime. For this reason, we have to consider a more complex non-linear gauge theory where all the interactions should be dealt with under the same standards [2]. In GR, Einstein took the spacetime metric components as the basic set of variables representing gravity, whereas Ashtekar and collaborators employed the tetrad fields and the connection forms as the fundamental variables [3]. We also will consider the tetrads and the connection forms as the fundamental fields but with the difference that this approach gives rise to a covariant symplectic formalism capable of achieving the result of dealing with physical fields under the same standard [4, 5].

In order to frame historically our approach, let us sketch a quick summary of the various attempts where the Standard Model and GR have been considered under the same comprehensive picture. In 1956, Utiyama suggested that gravitation may be viewed as a gauge theory [6] in analogy to the Yang Mills [7] theory (1954). He identified the gauge potential due to the Lorentz group with the symmetric connection of the Riemann geometry, and reproduced the Einstein GR as a gauge theory of the Lorentz group $SO(3,1)$ with the help of tetrad fields introduced in an *ad hoc* manner. Although the tetrads were necessary components of the theory (to relate the Lorentz group) adopted as an internal gauge group to the external spacetime, they were not introduced as gauge fields. In 1961, Kibble [8] constructed a gauge theory based on the Poincaré group $P(3,1) = T(3,1) \rtimes SO(3,1)$ (the symbol $\rtimes$ represents the semi-direct product) which resulted in the Einstein-Cartan theory characterized by curvature and torsion. The translation group $T(3,1)$ is considered responsible for generating the tetrads as gauge fields. Cartan [9] generalized the Riemann geometry in order to include torsion in addition to curvature. Sciama [10], and others (Fikelstein [11], Hehl [12, 13]) pointed out that intrinsic spin may be the source of torsion of the underlying spacetime manifold. Since the form and the role of tetrad fields are very different from those of gauge potentials, it has been thought that even Kibble's attempt was not satisfactory as a full gauge theory. There have been a number of gauge theories of gravitation based on a variety of Lie groups [12–18]. It was argued that a gauge theory of gravitation corresponding to GR can be constructed with the translation group alone, in the so-called teleparallel scheme [19]. Inomata *et al.* [20] proposed that Kibble's gauge theory could be obtained, in a way closer to the Yang-Mills approach, by considering the de Sitter group $SO(4,1)$, which is reducible to the Poincaré group by a group-contraction. Unlike the Poincaré group, the de Sitter group is homogeneous and the associated gauge fields are all of gauge potential type and by the Wigner-Inönu group contraction procedure, one of the five vector potentials reduces to the tetrad. The non-linear approach to group realizations was originally introduced by Coleman, Wess and Zumino [21, 22] in the context of internal symmetry groups (1969). It was later extended to the case of spacetime symmetries by Isham, Salam, and Strathdee [23] considering the non-linear action of $GL(4, \mathbb{R})$, modulus the Lorentz subgroup. In 1974, Borisov, Ivanov and Ogievetsky [24, 25], considered the simultaneous non-linear realization (NLR) of the affine and conformal groups. They stated that GR can be viewed as a consequence of spontaneous breakdown of the affine symmetry, in the same way that chiral dynamics, in quantum chromodynamics, is a result of spontaneous breakdown of chiral symmetry. In their model, gravitons are considered as Goldstone bosons associated with the affine symmetry breaking. As we will see below, this approach can be pursued in general. In 1978, Chang and Mansouri [26] used the NLR scheme adopting $GL(4,\mathbb{R})$ as the principal group. In 1980, Stelle and West [27] investigated the NLR induced by the spontaneous breakdown of $SO(3,2)$. In 1982 Ivanov and Niederle considered non-linear gauge theories of the Poincaré, de Sitter, conformal and special conformal groups [28, 29]. In 1983, Ivanenko and Sardanashvily [30] considered gravity to be a spontaneously broken gauge theory. The tetrads fields arise, in their formulation, as a result of the reduction of the structure group of the tangent bundle from the general linear to Lorentz group. In 1987, Lord and Goswami [31, 32] developed the NLR in the fiber bundle formalism based on the bundle structure $G\ (G/H,H)$ as suggested by Ne'eman and Regge [33]. In this approach, the quotient space G/H is identified with physical spacetime. Most recently, in a series of papers, Lopez-Pinto, Julve, Tiemblo, Tresguerres and Mielke discussed non-linear gauge theories of gravity on the basis of the Poincaré, affine and conformal groups [34–39].

As it should be clear from this extensive resume, the idea of an unification theory, capable of describing all the fundamental interactions of physics under the same standard, has been one of the main issues of modern physics, starting from the early efforts of Einstein, Weyl, Kaluza and Klein [40] until the most recent approaches [41]. Nevertheless, the large number of ideas, up to now proposed, which we classify as *unified theories*, results unsuccessful due to several reasons: the technical difficulties connected with the lack of a unitary mathematical description of all the interactions; the huge number of parameters introduced to build the unified theory and the fact that most of them cannot be observed neither at laboratory nor at astrophysical (or cosmological) conditions [42]; the very wide (and several times questionable since not-testable) number of extra-dimensions requested by several approaches. Due to this situation, it seems that unification is a useful (and aesthetic) paradigm, but far to be achieved, if the trend is continuing to try to unify interactions by adding more and more ingredients, rarely with an insightful justification. A different approach could be to consider the very essential physical quantities and try to achieve unification without any *ad hoc* new ingredients. This approach can be pursued starting from straightforward considerations which lead to reconsider modern physics under a sort of economic issue: that is, let us try to unifying forces approaching new schemes but without adding new parameters, just exploiting what already is in the underlying set-up to its full extent.

So as we said, Utiyama [6] proposed that GR can be seen as a gauge theory based on the local Lorentz group in the same way that the Yang-Mills gauge theory [7] is developed on the basis of the internal isospin gauge group. In this formulation, the Riemannian connection is the gravitational counterpart of the Yang-Mills gauge fields. While $SU(2)$, in the Yang-Mills theory, is an internal symmetry group, the Lorentz symmetry represents the local nature of spacetime rather than internal degrees of freedom. In order to relate local Lorentz symmetry to the external spacetime, we need to solder the local space to the external space. The soldering tools can be the tetrad fields. Utiyama regarded the tetrads as objects given *a priori* while they can be dynamically generated [2] and the spacetime has necessarily to be endowed with torsion in order to accommodate spinor fields [43]. In other words, the gravitational interaction of spinning particles requires the modification of the Riemann spacetime of GR to be a (non-Riemannian) curved spacetime with torsion. However spinor fields can be viewed as member of Clifford algebra and exist also in flat spaces [44]. Although Sciama used the tetrad formalism for his gauge-like handling of gravitation, his theory fell shortcomings in treating tetrad fields as gauge fields. Following the Kibble approach [8], it can be demonstrated how gravitation can be formulated starting from a pure gauge viewpoint. In particular, gravity can be seen as a gauge theory which can be obtained starting from some local invariance, e.g., the local Poincaré symmetry, leading to a suitable unification prescription [2]. However, torsion is *not* the minimal ingredient we will require, as it is naturally part of the most general geometry. The *minimal* ingredient we require is a 5-dimensional extension that, in 4-dimensions, reduces to an enlargement of the Hilbert-Einstein Lagrangian (ETGs): starting from the 5D-space, we will see that the process of reduction to 4D-space *generates* the mass spectra of particles, and in this picture, we will show that deformations can be parameterized as *effective* scalar fields in a $GL(4)$-group of diffeomorphisms; alternatively, the Hilbert-Einstein action R will be taken to be in its most general form as a generic function of the Hilbert-Einstein action $f(R)$. In this circumstance, we will see that the factor $\frac{1}{f'(R)}$ will give rise to an energy-dependent running-coupling able to bring effects that are supposed to take place at the Planck scale down to the usual Fermi scale. This is a straightforward paradigm for ETGs which could involve also other curvature invariants [50, 51]. Here, we are going to eventually see that torsion gives rise to spinorial interactions which, in some situations, can be worked out to display the same structure of the weak forces.

The layout of the paper is the following: in Section 2, we shall recall the basic geometrical concepts and clarify the notations and definitions we will employ in the rest of the paper; in Section 3, we will briefly introduce the 5D-spaces, and their reduction to 4D-spacetimes; Section 4, we discuss deformations of spacetime under the standard of $GL(4)$-group of transformations. In Section 5, we discuss the reduction from 5D to 4D showing how massive states can come out. Section 6 is devoted to the generation of masses and their relation to geometrical deformations. In Section 7, we are going to deal with the concept of mass generation and symmetry breaking. The further gravitational degrees of freedom, coming from ETGs, can play a fundamental role in this picture generating natural cut-off at TeV scales: this fact could give rise to gravitationally induced electroweak interactions. In particular, as discussed in Section 8, the weak forces emerge from the interaction of torsion and spinors and a running coupling constant is directly related to the form of $f(R)$-gravity. A detailed discussion of this feature is reported. Conclusions are drawn in Section 9.

2. Geometrical Notations and Definitions

To begin with, we shall recall some concepts and introduce the notation we will follow. The fundamental objects we are going to work with are tensors, defined in terms of their transformation laws, and whose indices, upper or lower, may be converted into one another, by raising lower or lowering upper indices, by means of the tensors $g^{\alpha\beta}$ and $g_{\alpha\beta}$ respectively; in order to ensure that the raising and lowering of the indices is consistently defined, these two tensors are assumed to be non-degenerate, symmetric and one the inverse of the other $g_{\nu\rho}g^{\rho\mu} = \delta^{\mu}_{\nu}$, therefore displaying all properties that make them metric tensors.

Dynamical properties of tensors are defined through covariant derivatives D_μ which are themselves defined upon introduction of the connections $\Gamma^{\alpha}_{\mu\nu}$ defined in term of its transformation law alone, and because the most general connection is not symmetric in the two lower indices then its antisymmetric part does not vanish and writing it as

$$T^{\lambda}{}_{\mu\nu} = \Gamma^{\lambda}{}_{[\mu,\nu]} \tag{1}$$

it turns out to be a tensor called Cartan torsion tensor; in order to ensure that the raising and lowering of the indices is consistently defined even after differential properties are introduced, these two tensors are assumed vanish their covariant derivatives $D_\mu g = 0$, and therefore we will say that the metric tensors are covariantly constant or verifying metricity or again that the metric and connection are compatible: after this condition is taken into account, the most general connection can be decomposed as

$$\Gamma^{\mu}{}_{\sigma\pi} = \tfrac{1}{2}g^{\mu\rho}\left(\partial_\pi g_{\sigma\rho} + \partial_\sigma g_{\pi\rho} - \partial_\rho g_{\sigma\pi}\right) + \left(T^{\mu}{}_{\sigma\pi} + T_{\sigma\pi}{}^{\mu} + T_{\pi\sigma}{}^{\mu}\right) \tag{2}$$

where

$$K^{\mu}{}_{\sigma\pi} = T^{\mu}{}_{\sigma\pi} + T_{\sigma\pi}{}^{\mu} + T_{\pi\sigma}{}^{\mu} \tag{3}$$

is defined as the contorsion tensor, and which may turn out to be useful in some context.

Due to their symmetry properties, torsion and contorsion have only one independent component $K^{\mu} = K^{\mu\pi}{}_{\pi} = 2T_{\pi}{}^{\pi\mu} = 2T^{\mu}$ and called Cartan torsion or contorsion vectors.

The quantity be expressed as

$$R^{\rho}{}_{\sigma\mu\nu} = \partial_\nu \Gamma^{\rho}{}_{\sigma\mu} - \partial_\mu \Gamma^{\rho}{}_{\sigma\nu} + \Gamma^{\rho}{}_{\lambda\nu}\Gamma^{\lambda}{}_{\sigma\mu} - \Gamma^{\rho}{}_{\lambda\mu}\Gamma^{\lambda}{}_{\sigma\nu} \tag{4}$$

is a tensor antisymmetric in the first and second couple of indices called Riemann curvature tensor; notice that the first couple of indices is antisymmetric only after the metricity condition is accounted: this expression of the Riemann curvature tensor has been given so that, together with the expression of the Cartan torsion tensor, we may now write the commutator of two covariant derivatives as

$$[D_\mu, D_\nu]A^{\rho} = T^{\theta}{}_{\mu\nu} D_\theta A^{\rho} - \tfrac{1}{2}R^{\rho}{}_{\sigma\mu\nu} A^{\sigma} \tag{5}$$

for the case of a generic vector A^μ, and for which it is understood that for general tensor there will appear an additional contribution of the Riemann curvature tensor for any additional index. In particular for scalars, the Riemann curvature tensor would not be present at all, thus underlining the specific role of the Cartan torsion tensor in describing the failure of the commutation of the covariant derivatives even for scalar fields.

Due to their symmetry properties, curvature has only one independent component that is given by $R^\rho{}_{\sigma\rho\nu} = R_{\sigma\nu}$ called Ricci curvature tensor, with one component $R_{\sigma\nu}g^{\sigma\nu} = R$ called Ricci curvature scalar.

The commutator of three coordinate covariant derivatives in cyclic permutations gives the expressions

$$(2D_\kappa T^\rho{}_{\mu\nu} + 4T^\rho{}_{\kappa\pi}T^\pi{}_{\mu\nu} - R^\rho{}_{\kappa\mu\nu}) + (2D_\nu T^\rho{}_{\kappa\mu} + 4T^\rho{}_{\nu\pi}T^\pi{}_{\kappa\mu} - R^\rho{}_{\nu\kappa\mu}) + \\ +(2D_\mu T^\rho{}_{\nu\kappa} + 4T^\rho{}_{\mu\pi}T^\pi{}_{\nu\kappa} - R^\rho{}_{\mu\nu\kappa}) \equiv 0 \tag{6}$$

known as the Bianchi identities for torsion and

$$(D_\mu R^\nu{}_{\iota\kappa\rho} - 2R^\nu{}_{\iota\beta\mu}T^\beta{}_{\kappa\rho}) + (D_\kappa R^\nu{}_{\iota\rho\mu} - 2R^\nu{}_{\iota\beta\kappa}T^\beta{}_{\rho\mu}) + \\ +(D_\rho R^\nu{}_{\iota\mu\kappa} - 2R^\nu{}_{\iota\beta\rho}T^\beta{}_{\mu\kappa}) \equiv 0 \tag{7}$$

known as Bianchi identities for curvature.

Of course, these identities can be contracted, and in their fully contracted form they read as in the following

$$D_\rho\left(T^\rho{}_{\mu\nu} + \delta^\rho_\nu T_\mu - \delta^\rho_\mu T_\nu\right) + 2T_\rho\left(T^\rho{}_{\mu\nu} + \delta^\rho_\nu T_\mu - \delta^\rho_\mu T_\nu\right) + R_{[\mu\nu]} \equiv 0 \tag{8}$$

known as fully contracted Bianchi identities for torsion and

$$D_\nu\left(R^{\nu\rho} - \tfrac{1}{2}g^{\nu\rho}R\right) - 2R_{\nu\beta}T^{\beta\nu\rho} + T_{\beta\mu\nu}R^{\mu\nu\beta\rho} \equiv 0 \tag{9}$$

known as fully contracted Bianchi identities for curvature.

A final remark is to be made for torsion: because torsion is a tensor, then it may be separated away, so that all most general torsional quantities can be written in terms of simplest torsionless quantities plus torsional corrections. In the following, we will employ the symbols ∇ and $\mathcal{R}$ to designate the covariant derivatives and curvatures of the simplest torsionless connection.

Now, given the metric tensors $g_{\alpha\beta}$ and $g^{\alpha\beta}$ it is possible to define covariant metric concepts like lengths and angles; let us considered a pair of bases of vectors called beins V^a_β and V^β_a defined to be dual of one another $V^a_\mu V^\rho_a = \delta^\rho_\mu$ and $V^a_\mu V^\mu_r = \delta^a_r$ which we can always choose to be orthonormal $V^a_\sigma V^b_\rho g^{\sigma\rho} = \eta^{ab}$ or equivalently $V^\sigma_a V^\rho_b g_{\sigma\rho} = \eta_{ab}$ where η_{ab} and η^{ab} are unitary diagonal matrices known as Minkowskian matrices: although it is always possible to orthonormalize a basis of vectors so that it is without loss of generality that orthonormal beins are introduced, nevertheless the bein are determined up to a Lorentz transformation that can be made explicit. The main advantage of introducing the beins is that, as the metric tensors are used to move indices up or down in tensors, the bein are used to switch to spacetime Greek indices to Lorentz Latin indices, whose explicit transformation laws allow for a simpler description of the formalism; for the moment, this formalism will only give us a simpler way to deal with calculations, but in the following we will see it will be necessary for the introduction of the spinor fields.

According to the definition of bein then, the world tensors are defined in terms of their transformation law under Lorentz transformations, whose indices, upper or lower, may be converted into one another, by raising lower or lowering upper indices, by means of the Minkowskian matrices, symmetric and inverse of one another $\eta_{ab}\eta^{bm} = \delta_a^m$ as expected.

Dynamical properties of tensors are defined by covariant derivatives D_μ which are themselves defined upon introduction of the spin-connections $A^i_{j\mu}$ which is a connection under Lorentz transformation although it is a covector under general coordinate transformations; if we want that the passage between spacetime and Lorentz indices is consistent even after differential properties are introduced, then we have to assume the condition $D_\alpha V = 0$ analogously to what we have done before, and therefore the intuitive condition $D_\alpha \eta = 0$ is given automatically: these two conditions together imply that the spin-connection is given by

$$A^b{}_{j\mu} = V_j^\alpha V_\rho^b \left(\Gamma^\rho{}_{\alpha\mu} + V_\alpha^k \partial_\mu V_k^\rho\right) \tag{10}$$

in terms of the connection and antisymmetric in the two world indices.

Considering the spin-connection it is possible to define

$$R^a{}_{b\sigma\pi} = \partial_\pi A^a{}_{b\sigma} - \partial_\sigma A^a{}_{b\pi} + A^a{}_{j\pi} A^j{}_{b\sigma} - A^a{}_{j\sigma} A^j{}_{b\pi} \tag{11}$$

antisymmetric in both spacetime and world indices and such that we have the simple relationship $R^a{}_{b\sigma\pi} = R^\mu{}_{\rho\sigma\pi} V_b^\rho V_\mu^a$ in terms of the Riemann curvature tensor, as it should have been obvious since all definitions have been given so to reduce the passage from the two formalisms a mere renaming of indices; in this formalism, the antisymmetry in the two Lorentz indices is manifest. The commutators are given in an equivalent way.

As we have anticipated, the introduction of the bein are able to simplify the formalism, but their most important contribution is the fact that through them we are capable of defining new fields such as the spinors. To see that, it is enough to remark that, whereas nothing can be done for general coordinate transformations, the Lorentz transformations have an explicit structure, which can be written in different representations; of all these representations, the most special are the complex-valued representations: complex representations of the Lorentz transformations are called spinorial transformations S and they can be expanded in terms of their infinitesimal generators

$$\gamma_{ij} = \tfrac{1}{4}[\gamma_i, \gamma_j] \tag{12}$$

where the γ_a matrices belong to the Clifford algebra.

Once the spinorial transformation S is obtained, we define the complex fields that transform according to this transformation as spinor fields, classified in terms of half-integer spin; for what we will have to do, we will restrict ourselves to the case given by the fields transforming as $\psi' = S\psi$ or $\overline{\psi}' = \overline{\psi} S^{-1}$, being $\frac{1}{2}$-spin spinor fields, for which the passage between the two forms is defined by the γ_0 matrix as $\overline{\psi} \equiv \psi^\dagger \gamma_0$ or $\psi \equiv \gamma_0 \overline{\psi}^\dagger$ reciprocally.

Dynamical properties for spinor fields are defined through spinorial covariant derivatives, after the introduction of the spinorial connection Ω_μ as usual; we notice that for the way in which they have been built, the γ_j and γ_{ij} matrices are constant for the spinorial covariant derivatives automatically: consequently, we have that the most general spinorial connection can be decomposed as

$$\Omega_\mu = A^{ab}{}_\mu \gamma_{ab} + iqA_\mu \mathbb{I} \tag{13}$$

in terms of the spin-connection plus an additional term proportional to the identity matrix that will be identified with the electrodynamic potential, as we shall see in a moment.

Given the spinorial connection it is possible to define

$$R_{\sigma\pi} = \partial_\sigma \Omega_\pi - \partial_\pi \Omega_\sigma + \Omega_\sigma \Omega_\pi - \Omega_\pi \Omega_\sigma \tag{14}$$

antisymmetric and such that $R_{\sigma\pi} = R^{ij}{}_{\sigma\pi}\gamma_{ij} + iqF_{\sigma\pi}\mathbb{I}$ in terms of the Riemann and Maxwell tensors. Eventually, the commutators will equivalently be defined as above.

It is important to notice that for the decomposition of the most general spinorial connection, the additional term proportional to the identity matrix should not be surprising: in fact, because spinors are complex fields, then we have to consider an additional transformation for a complex unitary phase $\psi' = e^{iq\theta}\psi$ or $\overline{\psi}' = e^{-iq\theta}\overline{\psi}$ for a given real number q called charge, and therefore the issue of their invariance under gauge transformations is to be considered as well; the theory that is thus constructed is the widely known electrodynamic Maxwell theory. This abelian gauge potential will eventually be placed in the spinorial connection; and the fact that the spinorial connection already had enough room to host this abelian gauge field should therefore not be so surprising any longer. In this sense then, electrodynamics has the same mathematical imprint of gravitation, and likewise it should be thought as a genuine geometrical theory.

The reason for which usually it is not can be traced back to a little difference we already underlined, that is the presence of the charge; however, the charge labels a property of the matter field, and therefore what is not geometric is the individual response of a given matter field propagating in an assigned gauge field, but the electrodynamic field in itself might indeed be considered as a purely geometrical field.

This concludes the introduction of the formalism, through which we not only defined the fundamental quantities we will employ in the following, but we also defined them in order to underline the geometrical essence beneath: this does not only refer to the well-known geometrical essence of the gravitational field, but also to the geometrical essence of the electrodynamic field (at least in terms of the field itself and not the response of a matter field in it), and the geometric settings of the spinorial fields (before field equations are defined); these ideas are actually not new, and for all concepts recalled here, a general introduction as well as a deeper discussion about the idea of geometrization, can be found in [45] (additionally it is possible to find references about all different roles of torsion).

We notice that so far, the formalism we have introduced is still non-dynamical (we have not defined any field equation or action) and still valid in any dimension (we have not chosen a particular background); in the following, we will discuss generalized actions for gravity. Specifically, we will take into account a 5D-spacetime with dynamical reduction to 4D-spacetime.

3. The Geometrical Structure of 5D and 4D Spaces

In this section, we will discuss the structure of a 5D-Riemannian manifold, and its reduction to 4D-manifold: such an approach gives a useful tool to deal with the realization of effective theories of gravity in 4D and the problem of mass generation.

Let us start with the 5D-space, where we do not define *a priori* a signature for the metric and which, after a 4D-reduction procedure, must be capable of reproducing all the features of the usual spacetime [48]. Let us assume that a 5D-vector field defines a metric whose signature is given by

$$(X, X) = (x^0)^2 - \overrightarrow{X} \cdot \overrightarrow{X} + \epsilon(x_4)^2 \tag{15}$$

where $\epsilon = \pm 1$, so that, using the traditional terms, the fifth dimension can be time-like or space-like: as we shall see below, it is the 4D-dynamics that discriminates, by a bijective correspondence, the signature, giving rise to particle-like solutions for $\epsilon = -1$ or to wave-like solutions for $\epsilon = +1$. In this case, we can obtain pseudo-spheres of Lorentz signature and thus 4D-spacetimes of constant curvature (as Friedmann-Robertson-Walker ones). There are only two independent different signatures for $N = 5$: they are $(p, q) = (1, 4)$, corresponding to the case $\epsilon = -1$ and $(p, q) = (2, 3)$, corresponding to $\epsilon = +1$. The 5D-manifolds are $\mathcal{M}^{(1,4)} = \mathbf{R}^5$ and $\mathcal{M}^{(2,3)} = \mathbf{R}^5$ respectively, where $\mathbf{R}^5$ is the 5D-space, the former case is called the De Sitter space, while the latter is the Anti-De Sitter one. The fact that the standard signature of the universe is $(1, -1, -1, -1)$ can be derived from an equivalent process starting from $\mathcal{M}^{(1,4)}$ or $\mathcal{M}^{(2,3)}$. This procedure is in agreement with the Whitney Theorem [49].

An important consideration is necessary at this point. After the reduction to 4D-physics, a given theory of gravity is described by the conformal-affine group of diffeomorphisms $GL(4)$ which is characterized by $4 \times 4 = 16$ generators; a straightforward splitting is

$$\underbrace{GL(4)}_{4\times 4} \supset \underbrace{SU(3)}_{3^2-1} \otimes \underbrace{SU(2)}_{2^2-1} \otimes \underbrace{U(1)}_{1} \otimes \underbrace{GL(2)}_{2\times 2} \tag{16}$$

where the number of generators is indicated for any subgroup: the physical meaning of $SU(3)$, $SU(2)$ and $U(1)$ is clear while $GL(2)$ is a group with 4 generators.

The further generators can be recovered in the framework of ETGs being massive or ghost gravitational modes [48, 50–52]. Clearly the groups of diffeomorphisms $GL(4)$ and $GL(2)$ are not unitary and this could be considered a problem in the standard approaches to quantization. On the contrary, from our point of view, this feature are related to mechanisms suitable to induce electroweak interactions and gravitational running coupling constants. The experimental consequences of these considerations could be extremely interesting for the physics at LHC.

4. Deformations and the Physics of the GL(4)-Group

Another ingredient for our considerations is represented by the spacetime deformations that can be induced in the reduction mechanism and can be seen as scalar fields in a gravitational background [54]. Let us take into account a metric g on a spacetime manifold $\mathcal{M}$

(in N-dimensions); let us now define a new tetrad field $\tilde{\omega} = \Phi^A{}_C(x)\,\omega^C$, with $\Phi^A{}_C(x)$ a matrix of scalar fields, and finally introduce a spacetime $\widetilde{\mathcal{M}}$ with the metric $\tilde{g}$ defined as

$$\tilde{\mathbf{g}} = \eta_{AB}\Phi^A{}_C\Phi^B{}_D\,\omega^C\omega^D = \gamma_{CD}(x)\omega^C\omega^D, \tag{17}$$

where also $\gamma_{CD}(x)$ is a matrix of fields which are scalars with respect to the coordinate transformations.

If $\Phi^A{}_C(x)$ is a Lorentz matrix in any point of $\mathcal{M}$, then $\tilde{g} \equiv g$, otherwise we say that $\tilde{g}$ is a deformation of g and $\widetilde{\mathcal{M}}$ is a deformed $\mathcal{M}$. If all the functions of $\Phi^A{}_C(x)$ are continuous, then there is a one-to-one correspondence between the points of $\mathcal{M}$ and the points of $\widetilde{\mathcal{M}}$. If ξ is a Killing vector for g on $\mathcal{M}$, the corresponding vector $\tilde{\xi}$ on $\widetilde{\mathcal{M}}$ is not a Killing vector.

A particular subset of these deformation matrices is given by

$$\Phi^A_C(x) = \Omega(x)\,\delta^A{}_C. \tag{18}$$

which define conformal transformations of the metric $\tilde{g} = \Omega^2(x)g$. In this sense, the deformations defined by Eq. (17) can be regarded as a generalization of the conformal transformations.

We call the matrices $\Phi^A{}_C(x)$ First Deformation Matrices, while we can refer to

$$\gamma_{CD}(x) = \eta_{AB}\Phi^A{}_C(x)\Phi^B{}_D(x). \tag{19}$$

as the Second Deformation Matrices which, as seen above, are also matrices of scalar fields. They generalize the Minkowski matrix η_{AB} with constant elements in the definition of the metric. A further restriction on the matrices $\Phi^A{}_C$ comes from the above mentioned theorem proved by Riemann for which an N-dimensional metric has $N(N-1)/2$ degrees of freedom (see [53] for details). With this definitions in mind, let us consider the main properties of deforming matrices. Let us take into account a four dimensional spacetime with Lorentzian signature. A family of matrices $\Phi^A{}_C(x)$ such that

$$\Phi^A{}_C(x) \in GL(4)\,\forall x, \tag{20}$$

are defined on such a spacetime. These functions are not necessarily continuous and can connect spacetimes with different topologies. A singular scalar field introduces a deformed manifold $\widetilde{\mathcal{M}}$ with a spacetime singularity.

As it is well known, the Lorentz matrices $\Lambda^A{}_C$ leave the Minkowski metric invariant and then

$$\mathbf{g} = \eta_{EF}\Lambda^E{}_A\Lambda^F{}_B\Phi^A{}_C\Phi^B{}_D\,\omega^C\omega^D = \eta_{AB}\Phi^A{}_C\Phi^B{}_D\,\omega^C\omega^D \tag{21}$$

and it follows that $\Phi^A{}_C$ give rise to right cosets of the Lorentz group, *i.e.* they are the elements of the quotient group $GL(4,\mathbf{R})/SO(3,1)$. On the other hand, a right-multiplication of $\Phi^A{}_C$ by a Lorentz matrix induces a different deformation matrix.

The inverse deformed metric is

$$\tilde{g}^{\alpha\beta} = \eta^{CD}\Phi^{-1A}{}_C\Phi^{-1B}{}_D e^\alpha_A e^\beta_B \tag{22}$$

where $\Phi^{-1A}{}_C\Phi^C{}_B = \Phi^A{}_C\Phi^{-1C}{}_B = \delta^A_B$ and we can decompose the matrix $\Phi_{AB} = \eta_{AC}\,\Phi^C{}_B$ in its symmetric and antisymmetric parts

$$\Phi_{AB} = \Phi_{(AB)} + \Phi_{[AB]} = \Omega\,\eta_{AB} + \Theta_{AB} + \phi_{AB} \tag{23}$$

where $\Omega = \Phi^A{}_A$, Θ_{AB} is the traceless symmetric part and ϕ_{AB} is the antisymmetric part of the first deformation matrix. Then standard conformal transformations are nothing else but deformations with $\Theta_{AB} = \phi_{AB} = 0$ [55].

Finding the inverse matrix $\Phi^{-1A}{}_C$ in terms of Ω, Θ_{AB} and ϕ_{AB} is not immediate, but as above, it can be split in the three terms

$$\Phi^{-1A}{}_C = \alpha\delta^A{}_C + \Psi^A{}_C + \Sigma^A{}_C \tag{24}$$

where α, $\Psi^A{}_C$ and $\Sigma^A{}_C$ are respectively the trace, the traceless symmetric part and the antisymmetric part of the inverse deformation matrix above. The second deformation matrix, from the above decomposition, takes the form

$$\gamma_{AB} = \eta_{CD}(\Omega\,\delta^C_A + \Theta^C{}_A + \phi^C{}_A)(\Omega\,\delta^D_B + \Theta^D{}_B + \phi^D{}_B) \tag{25}$$

and then

$$\begin{aligned}\gamma_{AB} \;=\;& \Omega^2\,\eta_{AB} + 2\Omega\,\Theta_{AB} + \eta_{CD}\,\Theta^C{}_A\,\Theta^D{}_B + \\ &+\eta_{CD}\,(\Theta^C{}_A\,\phi^D{}_B + \phi^C{}_A\,\Theta^D{}_B) + \eta_{CD}\,\phi^C{}_A\,\phi^D{}_B\end{aligned} \tag{26}$$

as it is clear.

In general, the deformed metric can be split as

$$\tilde{g}_{\alpha\beta} = \Omega^2 g_{\alpha\beta} + \gamma_{\alpha\beta} \tag{27}$$

where

$$\begin{aligned}\gamma_{\alpha\beta} \;=\;& (2\Omega\,\Theta_{AB} + \eta_{CD}\,\Theta^C{}_A\,\Theta^D{}_B + \eta_{CD}\,(\Theta^C{}_A\,\phi^D{}_B + \\ &+\phi^C{}_A\,\Theta^D{}_B) + \eta_{CD}\,\phi^C{}_A\,\phi^D{}_B)\,\omega^A_\alpha\omega^B_\beta\end{aligned} \tag{28}$$

and in particular, if $\Theta_{AB} = 0$, the deformed metric simplifies to

$$\tilde{g}_{\alpha\beta} = \Omega^2 g_{\alpha\beta} + \eta_{CD}\,\phi^C{}_A\,\phi^D{}_B\omega^A{}_\alpha\omega^B{}_\beta \tag{29}$$

so that, for $\Omega = 1$, the deformation of a metric consists in adding to the background metric a tensor $\gamma_{\alpha\beta}$. We have to remember that all these quantities are not independent as, by the theorem mentioned in [53], they have to form at most six independent functions in a four dimensional spacetime. Similarly the controvariant deformed metric can be always decomposed in the following way

$$\tilde{g}^{\alpha\beta} = \alpha^2 g^{\alpha\beta} + \lambda^{\alpha\beta} \tag{30}$$

identically. To find the relation between $\gamma_{\alpha\beta}$ and $\lambda^{\alpha\beta}$ we may use $\widetilde{g_{\alpha\beta}g^{\beta\gamma}} = \delta^\gamma_\alpha$, and then

$$\alpha^2\Omega^2\delta^\gamma_\alpha + \alpha^2\gamma^\gamma_\alpha + \Omega^2\lambda^\gamma_\alpha + \gamma_{\alpha\beta}\lambda^{\beta\gamma} = \delta^\gamma_\alpha \tag{31}$$

and if the deformations are conformal transformations, we have $\alpha = \Omega^{-1}$, so that assuming such a condition, one obtains the following matrix equation

$$\alpha^2\gamma^\gamma_\alpha + \Omega^2\lambda^\gamma_\alpha + \gamma_{\alpha\beta}\lambda^{\beta\gamma} = 0\,, \tag{32}$$

and

$$(\delta_\alpha^\beta + \Omega^{-2}\gamma_\alpha^\beta)\lambda_\beta^\gamma = -\Omega^{-4}\gamma_\alpha^\gamma \tag{33}$$

and finally

$$\lambda_\beta^\gamma = -\Omega^{-4}(\delta + \Omega^{-2}\gamma)^{-1}{}_\beta^\alpha\gamma_\alpha^\gamma \tag{34}$$

where $(\delta + \Omega^{-2}\gamma)^{-1}$ is the inverse tensor of $(\delta_\alpha^\beta + \Omega^{-2}\gamma_\alpha^\beta)$.

To each matrix $\Phi^A{}_B$ we can associate a $(1,1)$-tensor $\phi^\alpha{}_\beta$ defined by

$$\phi^\alpha{}_\beta = \Phi^A{}_B\omega_\beta^B e_A^\alpha \tag{35}$$

such that

$$\tilde{g}_{\alpha\beta} = g_{\gamma\delta}\phi^\gamma{}_\alpha\phi^\delta{}_\beta \tag{36}$$

and viceversa from a $(1,1)$-tensor $\phi^\alpha{}_\beta$, we can define a matrix of scalar fields as

$$\phi^A{}_B = \phi^\alpha{}_\beta\omega_\alpha^A e_B^\beta \tag{37}$$

as it is to be expected.

In conclusion, the spacetime deformations are endowed with the conformal-affine structure of $GL(N)$ groups (in particular $GL(4)$) and passing from a given metric to another gives rise to induced scalar fields that, as we shall show below, can be generated by a dimensional reduction mechanism.

5. The 5D-Space and Its Reduction to 4D-Spacetime Dynamics

We are interested in a reduction mechanism capable of generating the masses of particles and a symmetry breaking capable of generating the observed interactions. To achieve these goals, we start from a theory of gravity, formulated in 5D and analyze its reduction in 4D. The results of this approach are ETGs. In the following, 5D indices will be capitalized. We can start from the matter-free case in 5D-gravity. As it turns out, the absence of the Dirac field gives rise to a torsion-free solution: the structure of the field equations thus reduce to that of the usual Einstein gravity. However, it is possible to exploit the fact that the 5D-space gives rise, after the reduction to the 4D-spacetime, to additional degrees of freedom that can be physically interpreted: these fields arise from deformations as new scalar fields.

Now, in 5D-spaces, the field equations are derived from the 5D Hilbert-Einstein action as discussed above; its variation gives the 5-dimensional field equations in the vacuum

$$G_{AB} = \mathcal{R}_{AB} - \frac{1}{2}g_{AB}\mathcal{R} = 0 \tag{38}$$

where we have defined for simplicity the tensor G_{AB} known as Einstein tensor: here the Ricci-flat space is always a solution.

The 5D energy tensor for scalar fields is defined as usual

$$T_{AB} = \nabla_A\Phi\nabla_B\Phi - \frac{1}{2}g_{AB}\nabla_C\Phi\nabla^C\Phi \tag{39}$$

where only the kinetic terms are present; as it is customary, such a tensor can be derived from a variational principle starting from a Lagrangian $\mathcal{L}_\Phi$ related to the scalar field Φ, whose physical meaning will be clear below.

Because of the definition of 5D-space itself, it is important to stress now that no self-interaction potential $V(\Phi)$ has been taken into account so that T_{AB} is a completely symmetric object and Φ is, by definition, a cyclic variable: this fact guarantees that Noether theorem holds for T_{AB} and a conservation law intrinsically exists. With these considerations in mind, the field equations can now assume the form

$$\mathcal{R}_{AB} = \chi\left(T_{AB} - \frac{1}{2}g_{AB}T\right) \tag{40}$$

where T is the trace of T_{AB}, χ is the 5D gravitational constant and $\hbar = c = 1$. The form (40) of field equations is useful in order to put in evidence the role of the scalar field Φ, if we are not simply assuming Ricci-flat 5D-spaces.

As we said, T_{AB} is a symmetric tensor, and due to the choice of the metric and to the symmetric nature of the stress-energy tensor T_{AB} and of the Einstein field equations G_{AB}, the contracted Bianchi identities $\nabla_A T^A_B = 0$ hold. Developing them gives

$$\begin{aligned}\nabla_A T^A_B &= \nabla_A\left(\partial_B\Phi\partial^A\Phi - \tfrac{1}{2}\delta^A_B\partial_C\Phi\partial^C\Phi\right) = \\ &= \left(\nabla_A\Phi_B\right)\Phi^A + \Phi_B\left(\nabla_A\Phi^A\right) - \tfrac{1}{2}\left(\nabla_B\Phi_C\right)\Phi^C - \tfrac{1}{2}\Phi_C\left(\nabla_B\Phi^C\right) = \\ &= \left(\nabla_A\Phi_B\right)\Phi^A + \Phi_B\left(\nabla_A\Phi^A\right) - \Phi_C\left(\nabla_B\Phi^C\right)\end{aligned} \tag{41}$$

and since our 5D-space is a Riemannian manifold, then $\nabla_A\Phi_B = \nabla_B\Phi_A$ so that in this case, partial and covariant derivatives coincide for the scalar field Φ. Finally we get

$$\nabla_A T^A_B = \Phi_B\,{}^{(5)}\Box\Phi \tag{42}$$

where ${}^{(5)}\Box$ is the 5D d'Alembert operator defined as $\nabla_A\Phi^A \equiv g^{AB}\nabla_A\nabla_B\Phi \equiv {}^{(5)}\Box\Phi$.

The general result is that the conservation of the energy tensor T_{AB} implies the Klein-Gordon equation, which assigns the dynamics of Φ, that is $\nabla_A T^B_A = 0$ implies that ${}^{(5)}\Box\Phi = 0$ assuming of course $\Phi_B \neq 0$ since we are dealing with a non-trivial physical field. As we shall see below, the conservation law of the 5D energy of the scalar field gives a physical meaning to the fifth dimension.

The reduction to the 4D-dynamics can be accomplished by taking into account the Campbell theorem [56]: this theorem states that it is always possible to consider a 4D Riemannian manifold, defined by the line element $ds^2 = g_{\alpha\beta}dx^\alpha dx^\beta$, in a 5D one with $dS^2 = g_{AB}dx^A dx^B$. We have $g_{AB} = g_{AB}(x^\alpha, x^4)$ with x^4 the yet unspecified extra coordinate; as we discussed above, g_{AB} is covariant under the group of 5D coordinate transformations but not under the restricted group of 4D transformations. This fact has the relevant consequence that the choice of 5D coordinates results as the *gauge* necessary to specify the 4D physics also in its non-standard aspects. Viceversa, in specifying the 4D physics, the bijective embedding process in 5D gives physical meaning to the fifth coordinate x^4.

In other words, the fifth coordinate, in 4D can assume a physical meaning, and we are going to see how this might be connected to the mass of elementary particles.

The reduction to the 4D-spacetime is encoded by the Lagrangian

$$\mathcal{L} = \sqrt{-g^{(5)}}\left[{}^{(5)}\mathcal{R} + \lambda(g_{44} - \epsilon\Phi^2)\right] \tag{43}$$

with λ as a Lagrange multiplier, Φ a scalar field and $\epsilon = \pm 1$. This approach is completely general and used in theoretical physics when we want to put in evidence some specific feature [57]; in this case, we need it to derive the physical gauge for the 5D-metric.

We can write down the metric as

$$dS^2 = g_{AB}dx^A dx^B = g_{\alpha\beta}dx^\alpha dx^\beta + g_{44}(dx^4)^2 = g_{\alpha\beta}dx^\alpha dx^\beta + \epsilon\Phi^2(dx^4)^2\,, \quad (44)$$

from which we obtain directly particle-like solutions for $\epsilon = -1$ or wave-like solutions for $\epsilon = +1$ in the 4D-reduction procedure. The standard signature of 4D-component of the metric is $(+1, -1, -1, -1)$ and $\alpha, \beta = 0, 1, 2, 3$.

Furthermore, the 5D-metric can be written in a Kaluza-Klein fashion as the matrix

$$g_{AB} = \begin{pmatrix} g_{\alpha\beta} & 0 \\ 0 & \epsilon\Phi^2 \end{pmatrix}\,, \quad (45)$$

and the 5D-curvature Ricci tensor can be derived. After the projection from 5D to 4D, $g_{\alpha\beta}$, derived from g_{AB}, no longer explicitly depends on x^4, and a useful expression for the Ricci scalar can be derived:

$$^{(5)}\mathcal{R} = \mathcal{R} - \frac{1}{\Phi}\Box\Phi\,, \quad (46)$$

where the dependence on ϵ is explicitly disappeared and $\Box$ is the 4D-d'Alembert operator, which gives $\Box\Phi \equiv g^{\mu\nu}\nabla_\mu\nabla_\nu\Phi$.

The action above can therefore be written in a 4D-reduced Brans-Dicke action of the following form

$$\mathcal{L} = \frac{1}{16\pi G_N}\sqrt{-g}\,[\Phi\mathcal{R} + \mathcal{L}_\Phi] \quad (47)$$

where the Newton constant G_N is given in terms of the 5D gravitational constant and a characteristic length associated to the Compton length as $G_N = \frac{^{(5)}G}{2\pi l}$. Defining the function of the 4D-scalar field ϕ as

$$-\frac{\Phi}{16\pi G_N} = \phi \quad (48)$$

we get in 4D a general action in which gravity is non-minimally coupled to a scalar field as

$$\mathcal{A} = \int_\mathcal{M} d^4x\sqrt{-g}\left[\phi\mathcal{R} + \tfrac{1}{2}g^{\mu\nu}\nabla_\mu\phi\nabla_\nu\phi - V(\phi)\right] + \int_{\partial\mathcal{M}} d^3x\sqrt{-b}K \quad (49)$$

where the form and the role of $V(\phi)$ are still general. The second integral is a boundary term where $K \equiv h^{ij}K_{ij}$ is the trace of the extrinsic curvature tensor K_{ij} of the hypersurface $\partial\mathcal{M}$ embedded in the 4D-manifold $\mathcal{M}$ and b is the metric determinant of the 3D-manifold.

The Einstein field equations are derived by varying with respect to the 4D-metric $g_{\mu\nu}$

$$\mathcal{R}_{\mu\nu} - \tfrac{1}{2}g_{\mu\nu}\mathcal{R} = T_{\mu\nu} \quad (50)$$

where

$$T_{\mu\nu} = \tfrac{1}{\phi}\left[-\tfrac{1}{2}\nabla_\mu\phi\nabla_\nu\phi + \tfrac{1}{4}g_{\mu\nu}\nabla_\alpha\phi\nabla^\alpha\phi - \tfrac{1}{2}g_{\mu\nu}V(\phi) - g_{\mu\nu}\Box\phi + \nabla_\mu\nabla_\nu\phi\right] \quad (51)$$

is the *effective* energy tensor containing the non-minimal coupling contributions, the kinetic terms and the potential of the scalar field ϕ.

By varying with respect to ϕ, we get the 4D-Klein-Gordon equation, which are nothing else but the contracted Bianchi identity, as we have already shown; this implies that the effective energy tensor in the right-hand side of (50) is a divergenceless tensor, which is compatible with Einstein theory of gravity also if we started from a 5D-space, as long as the theory is torsionless, that is in the vacuum of spinor fields.

Finally in order to physically identify the fifth dimension, we may write the Klein-Gordon equation in the form

$$\left(\Box + m_{eff}^2\right)\phi = 0 \tag{52}$$

where

$$m_{eff}^2 = \frac{1}{\phi}\left[V'(\phi) - \mathcal{R}\right] \tag{53}$$

is the effective mass given in terms of scalar field self-interactions and external gravitational contributions. In any quantum field theory formulated on curved spacetimes, these contributions, at one-loop level, have the same weight, and indeed, Extended Theories of Gravity are renormalizable at one loop-level [58].

What we want to show next is that a natural way to generate particle masses can be achieved starting from a 5D picture. In other words, the concept of mass can be derived from a geometric viewpoint.

6. The Generation of Masses

6.1. Particle Masses as Eigenstates from 5D-Spaces

The above considerations allow a straightforward mechanism for the generation of masses that can be related to the projection from 5D to 4D-manifolds.

In a 5D-space, the 5D d'Alembert operator for particle-like solutions can be split selecting the value $\epsilon = -1$ in the metric as ${}^{(5)}\Box = \Box - {\partial_4}^2$, so that for a given scalar field Φ, we have

$${}^{(5)}\Box\Phi = \left[\Box - {\partial_4}^2\right]\Phi = 0 \tag{54}$$

identically; because the (45) is diagonal in the fifth component, we may separate the variable as $\Phi = \phi(t, \vec{x})\psi(x_4)$ and thus

$$\frac{\Box\phi}{\phi} = \frac{1}{\psi}\left[\frac{d^2\psi}{dx_4^2}\right] = -k_n^2 \tag{55}$$

where k_n must be a constant. From Eq.(55), we obtain the two equations of motion

$$\left(\Box + k_n^2\right)\phi = 0 \tag{56}$$

$$\frac{d^2\psi}{dx_4^2} + k_n^2\psi = 0\,. \tag{57}$$

the former being the evolution equation for the scalar field with a mass term.

The constant k_n has the physical dimension of the inverse of a length and, assigning boundary conditions, we can derive the eigenvalue relation $k_n = \frac{2\pi}{l}n$ where n is an integer.

As a result, in standard units, we recover the Compton length $\lambda_n = \frac{\hbar}{2\pi m_n c} = \frac{1}{k_n}$ which assigns the mass of a particle. It has to be stressed that, the eigenvalues of Eq.(57) are the masses of particles which are generated by the reduction process from 5D to 4D. On the other hand, different values of n fix the families of particles, while, for any given value of n.

6.2. 4D-Spacetimes and Dynamics of Massive Scalar Fields

Dynamics of 4D-component of the induced scalar field Φ can be related to the spacetime deformations discussed above; in this way, the role of $GL(4)$-group of diffeomorphism will result of fundamental importance. Let us start by showing how particles can acquire mass by deformations and let us relate the procedure with the above reduction mechanism.

As usual, a particle with zero mass is characterized by the invariant relation $\eta^{\alpha\beta}p_\alpha p_\beta = 0$ in the Minkowski spacetime. Deforming the spacetime and considering the above operators, one has

$$\eta^{AB}\Phi^C_A\Phi^D_B e^\alpha_C e^\beta_D p_\alpha p_\beta = g^{\alpha\beta}p_\alpha p_\beta = 0\,, \tag{58}$$

so we have defined two frames, one, the Minkowski spacetime, defined by the metric $\eta^{\alpha\beta}$ and the other defined by the metric $g^{\alpha\beta}$, generated by the projection from 5D. The two frames are related by the matrices of deformation functions $\Phi^C_A(x)$. In both frames the massless particle follow a null path, but we observe that using the decomposition (30) the particle does not appear massless with respect to the first frame. As matter of fact, Eq. (58) becomes

$$\Omega^{-2}\eta^{\alpha\beta}p_\alpha p_\beta + \chi^{\alpha\beta}p_\alpha p_\beta = 0\,, \tag{59}$$

so that

$$\eta^{\alpha\beta}p_\alpha p_\beta = -\Omega^2\chi^{\alpha\beta}p_\alpha p_\beta \neq 0. \tag{60}$$

and now in the first frame we are able to define a rest reference system for the particle.

For a massless particle it is not possible to define a rest reference system since considering the invariant relation,

$$g^{00}p_0^2 + 2g^{0i}p_0p_i + g^{ij}p_ip_j = 0\,, \tag{61}$$

and defining as rest frame the system in which $p_j = 0$, the solution exists only for $p_0 = 0$ *i.e.* only the trivial solution $p_\alpha = 0$ satisfies the " rest reference frame" condition.

On the other hand, if we consider massive particle, then

$$g^{00}p_0^2 + 2g^{0i}p_0p_i + g^{ij}p_ip_j = m^2\,, \tag{62}$$

the rest frame is characterized by the conditions $p_j = 0$ and consequently $g^{00}p_0^2 = m^2$. This means that the (squared) mass is proportional to the (squared) energy and the solution is no more trivial. To overcome this problem, let us fix the deformation such that

$$-\Omega^2\chi^{\alpha\beta}p_\alpha p_\beta = m^2\,, \tag{63}$$

m being the mass "attribute" to the particle. In the second frame we have

$$p_0^2 - \vec{p}^{\,2} + \Omega^2\chi^{00}p_0^2 + 2\chi^{0i}p_0p_i + \chi^{ij}p_ip_j = 0\,, \tag{64}$$

if $\vec{p} = 0$ then $p_0^2(1 + \Omega^2\chi^{00}) = 1$ which implies, besides the trivial solution, also the condition $\Omega^2\chi^{00} = -1$ which, together with Eq. (63), gives $p_0^2 = m^2$, the rest system condition in the first frame.

It is also possible to show that there is an equivalence between deforming a metric and giving mass to a massless particle. As we have shown, we have described the spacetime deformation by using matrices of scalar fields. The same problem can be addressed in terms of spacetime tensors

$$\eta_{AB}\Phi^A_C\Phi^B_D e^C_\mu e^D_\nu = g_{\alpha\beta}\Phi^\alpha_\mu\Phi^\beta_\nu = \tilde{g}_{\mu\nu}\,. \tag{65}$$

It should be observed that the Φ^α_μ do not represent coordinates transformations as far as they cannot be reduced to Jacobian matrices. This means that Φ^α_μ have a fundamental physical meaning. Starting from $\tilde{g}$ we observe that if $g_{\mu\nu}p^\mu p^\nu = 0$ then $\tilde{g}_{\mu\nu}p^\mu p^\nu \neq 0$ when $\tilde{g}$ is a general deformation that can be related to the conformal transformations of g. Eq. (65) tells us that, equivalently, we can read the deformation as a transformation of the 4-momentum of the particle

$$\tilde{g}_{\mu\nu}p^\mu p^\nu = g_{\alpha\beta}\Phi^\alpha_\mu\Phi^\beta_\nu p^\mu p^\nu \equiv g_{\alpha\beta}\tilde{p}^\alpha\tilde{p}^\beta\,, \tag{66}$$

in such a way there exists deformations of spacetime which give mass to massless particles. In other words, deformations, that are elements of the conformally-invariant $GL(4)$-group can be related, in principle, to the generation of the masses of particles.

6.3. Mass Generation from Deformations

So far we have considered the geometrical definition of mass from the point of view of relativistic mechanics. Now we would like to extend this result to classical field theory with the aim to extend it to quantum field theory.

Let us consider a scalar free massless particle. It can be described by the Lagrangian

$$\mathcal{L} = \frac{1}{2}\sqrt{-g}g^{\mu\nu}(\partial_\mu\phi)(\partial_\nu\phi) = \frac{1}{2}\sqrt{-g}\,(\eta^{\mu\nu} + \chi^{\mu\nu})\,(\partial_\mu\phi)(\partial_\nu\phi). \tag{67}$$

It is well-known that the free propagator has a pole in $p^2 = 0$. In order to introduce a mass term in the Lagrangian and in the field equations, we have to eliminate a divergence from it, that is considering

$$\begin{aligned}\partial_\mu\left[\partial_\nu\sqrt{-g}\,\chi^{\mu\nu}\phi^2\right] &= \partial_\mu\partial_\nu\left(\sqrt{-g}\,\chi^{\mu\nu}\right)\phi^2 + \\ &+4\,\partial_\nu\left(\sqrt{-g}\,\chi^{\mu\nu}\right)\phi\,\partial_\mu\phi + 2\sqrt{-g}\,\chi^{\mu\nu}\partial_\mu\phi\,\partial_\nu\phi + 2\sqrt{-g}\,\chi^{\mu\nu}\phi\,\partial_\mu\partial_\nu\phi\,,\end{aligned} \tag{68}$$

the Lagrangian takes the form

$$\begin{aligned}\widetilde{\mathcal{L}} &= \frac{1}{2}\sqrt{-g}\,(\eta^{\mu\nu})\,(\partial_\mu\phi)(\partial_\nu\phi) - \frac{1}{4}\partial_\mu\partial_\nu\left(\sqrt{-g}\,\chi^{\mu\nu}\right)\phi^2 - \\ &+\partial_\nu\left(\sqrt{-g}\,\chi^{\mu\nu}\right)\phi\,(\partial_\mu\phi) - \frac{1}{2}\sqrt{-g}\,\chi^{\mu\nu}\phi\,\partial_\mu\partial_\nu\phi.\end{aligned} \tag{69}$$

In this way a "mass" term $m^2 = \frac{1}{2}\partial_\mu\partial_\nu\left(\sqrt{-g}\,\chi^{\mu\nu}\right)$ can be defined in a new Lagrangian. On the other hand, considering an action $\mathcal{A} = \int\sqrt{-g}\widetilde{\mathcal{L}}$ and a variational principle $\frac{\delta\mathcal{A}}{\delta\phi} = 0$

implies the equation

$$\frac{\partial\widetilde{\mathcal{L}}}{\partial\phi} - \partial_\alpha \frac{\partial\widetilde{\mathcal{L}}}{\partial(\partial_\alpha\phi)} + \partial_\alpha\partial_\beta \frac{\partial\widetilde{\mathcal{L}}}{\partial(\partial_\alpha\partial_\beta\phi)} = 0\,, \tag{70}$$

which gives

$$\Box\phi = \eta^{\mu\nu}\partial_\mu\partial_\nu\phi + \Omega^2\chi^{\mu\nu}\partial_\mu\partial_\nu\phi = 0\,. \tag{71}$$

It is well-known that this equation cannot give, in general, a potential or a mass term, except when we take as a solution a plane wave $\phi = \exp ik_\mu x^\mu$. In this case the χ part of the equation can be interpreted as a mass term according to

$$\chi^{\mu\nu}\partial_\mu\partial_\nu\phi = -\Omega^2\chi^{\mu\nu}k_\mu k_\nu \exp ik_\mu x^\mu = m^2\phi\,. \tag{72}$$

This equation can be compared with Eq. (63), previously derived for a relativistic particle. We can extend this result to each function which can be expressed by a Fourier transform,

$$\phi(x) = \int \exp(ik_\mu x^\mu)\tilde{f}(k)d^4k\,, \tag{73}$$

where the spacetime dependent mass term is defined by the equation

$$m^2(x) = -\int \chi^{\mu\nu}k_\mu k_\nu \exp(ik_\mu x^\mu)\tilde{f}(k)d^4k. \tag{74}$$

With these restrictions, the Klein-Gordon equation for a massless particle in the g-frame is seen in the η-frame as a *massive particle* as soon as the plane wave solutions are considered. This result makes more sense when the mass is interpreted as a quantum effect. In the case a curved spacetime, the above arguments can be implemented by the substitutions of operators $\eta \rightarrow g$ and $\partial \rightarrow \nabla$. It is important to note that the scalar field equation is not conformally invariant. In order to have a conformally invariant scalar field equation, it is necessary to introduce, in the Lagrangian density a non-minimal coupling between geometry and field. A possible choice is

$$\mathcal{L} = \frac{1}{2}\sqrt{-g}g^{\mu\nu}(\partial_\mu\phi)(\partial_\nu\phi) - \xi\mathcal{R}\phi^2\,, \tag{75}$$

By varying with respect to ϕ, the Klein-Gordon equation

$$\Box\phi + \xi\mathcal{R}\phi = 0\,, \tag{76}$$

is recovered. Minimal coupling is obtained for $\xi = 0$. By analogy with the previous considerations, a deformation with constant curvature $\mathcal{R} = m^2$ gives a mass term in the original frame. However, we need also to interpret the other terms appearing in the new frame. Expanding Eq. (76),

$$\eta^{\mu\nu}\partial_\mu\partial_\nu\phi + \chi^{\mu\nu}\partial_\mu\partial_\nu\phi + (\eta^{\mu\nu} + \chi^{\mu\nu})\,\Gamma^\lambda_{\mu\nu}\partial_\lambda\phi + \xi\mathcal{R}\phi = 0\,, \tag{77}$$

we need that the connection is compatible with the metric tensor $(\eta^{\mu\nu} + \chi^{\mu\nu})$, that is

$$\nabla\,(\eta^{\mu\nu} + \chi^{\mu\nu}) = 0. \tag{78}$$

In order to have a deformation defining a mass $m^2 = \mathcal{R}$ in the original frame, two conditions must be satisfied, $\mathcal{R} = \mathcal{R}_0 > 0$ where $\mathcal{R}_0$ is a constant and

$$\chi^{\mu\nu}\partial_\mu\partial_\nu\phi + (\eta^{\mu\nu} + \chi^{\mu\nu})\,\Gamma^\lambda_{\mu\nu}\partial_\lambda\phi = 0\,, \tag{79}$$

which is a restriction on the deformation tensor $\chi^{\mu\nu}$. This condition allows to determine the mass by a conformal transformation. In fact, in the new frame, the equation is

$$g^{\alpha\beta}\nabla_\alpha\partial_\beta\phi + \xi\mathcal{R}\phi = 0\,, \tag{80}$$

which, in the case of a conformal transformation $g_{\alpha\beta} = \Omega^2\eta_{\alpha\beta}$ becomes

$$\Omega^{-2}\eta^{\alpha\beta}\left(\partial_\alpha\partial_\beta\phi + 2\Omega^{-1}\partial^\gamma\Omega\partial_\gamma\phi\right) - 6\xi\eta^{\alpha\beta}\Omega^{-3}\partial_\alpha\partial_\beta\Omega = 0\,. \tag{81}$$

It is equivalent to a massive scalar field equation (in the first frame), if the condition

$$2\Omega^{-1}\partial^\gamma\Omega\partial_\gamma\phi - 6\xi\eta^{\alpha\beta}\Omega^{-1}\partial_\alpha\partial_\beta\Omega = m^2\,, \tag{82}$$

holds. Since Eq. (81) is linear in ϕ, we can define $\phi = e^{i\lambda_\alpha x^\alpha}$. Eq. (82) becomes

$$2\Omega^{-1}\partial^\gamma\Omega\lambda_\gamma - 6\xi\eta^{\alpha\beta}\Omega^{-1}\partial_\alpha\partial_\beta\Omega = m^2\,, \tag{83}$$

then a solution is $\Omega = e^{ik_\alpha x^\alpha}$ and the mass depends on the combination

$$m^2 = 6\xi\eta^{\alpha\beta}k_\alpha k_\beta - 2\eta^{\alpha\beta}k_\alpha\lambda_\beta. \tag{84}$$

In this example, the deformation is a specific complex function, but we can extend this result to any function by taking

$$\Omega(x) = \int \tilde{\Omega}(k)e^{ikx}d^4k\,, \tag{85}$$

and also non-constant masses can be is obtained being

$$m^2 = \frac{\int \tilde{\Omega}(k)(6\xi\eta^{\alpha\beta}k_\alpha k_\beta - 2\eta^{\alpha\beta}k_\alpha\lambda_\beta)e^{ikx}d^4k}{\int \tilde{\Omega}(k)e^{ikx}d^4k}\,. \tag{86}$$

These results mean that a geometrical definition of mass is always possible. It can be induced by the conformal transformation Ω a restriction of the deformations group $GL(4)$.

Summarizing we have shown that the fifth dimension of a reduction mechanism from 5D to 4D-manifolds can be interpreted as a mass generator, where the degrees of freedom coming from ETGs have a physical meaning and cannot be simply gauged away, and as soon as particles acquire masses $GL(4)$ can induce symmetry breaking. In other words, the further degrees of freedom related to the ETGs may be related to a Higgs-like mechanism as we are going to discuss below.

7. Gravitational Massive States and Induced Symmetry Breaking

7.1. A Gravitational Cut-off at TeV Scales

The above results could be interesting to investigate quantum gravity effects and symmetry breaking in the range between GeV and TeV scales. Such scales are actually investigated by the experiments at LHC. It is important to stress that any ultra-violet model of gravity (e.g., at TeV scales) have to explain also the observed weakness of gravitational effects at largest (infra-red) scales. This means that massless modes have to be considered in any case.

The above 5D-action is an example of higher dimensional action where the effective gravitational energy scale (Planck scale) can be rescaled according to Eq. (48). In terms of mass, being $M_p^2 = \frac{c\hbar}{G_N}$ the constraint coming from the ultra-violet limit of the theory (10^{19} GeV) , we can set $M_p^2 = M_\sharp^{D-2} V_{D-4}$, where V_{D-4} is the volume coming from the extra dimension. It is easy to see that V_{D-4}, in the 5D case, is related to the fifth component of Φ. $M_\sharp$ is the above discussed cut-off mass that becomes relevant as soon as the Lorentz invariance is violated. Such a scale, in the context discussed here, could be of the order TeV.

As we have shown above, it is quite natural to obtain effective theories containing scalar fields of gravitational origin. In this sense, $M_\sharp$ is the result of dimensional reduction. To be more explicit, the 4D dynamics is led by the effective potential $V(\phi)$ and the non-minimal coupling $F(\phi)$. Such functions could be tested since they are related to massive states.

In particular, the effective Extended Gravity, produced by the reduction mechanism from 5D to 4D, can be chosen as

$$\mathcal{A} = \int d^4x \sqrt{-g} \left[-\frac{\phi^2}{2}\mathcal{R} + \frac{1}{2} g^{\mu\nu} \partial_\mu \phi \partial_\nu \phi - V \right] \tag{87}$$

plus contributions of ordinary matter terms. The potential for ϕ can be assumed as

$$V(\phi) = \frac{M_\sharp^2}{2}\phi^2 + \frac{\lambda}{4}\phi^4 , \tag{88}$$

where a massive term and the self-interaction term are present. This is the standard choice of quantum field theory which perfectly fits with the arguments of dimensional reduction. Let us recall again that the scalar field ϕ is not put *by hand* into dynamics but it is given by the extra degrees of freedom of gravitational field generated by the reduction process in 4D. It is easy to derive the vacuum expectation value of ϕ, being $M_\sharp^2 = 2\lambda M_p^2$, which is a fundamental scale of the theory.

Some considerations are in order at this point. Such a scale has to be confronted with Higgs vacuum expectation value which is 246 GeV and then with the *hierarchy problem.* If $M_\sharp$ is larger than Higgs mass, the problem is obviously circumvented. It is important to recall that hierarchy problem occurs when couplings and masses of effective theories are very different than the parameters measured by experiments. This happen since measured parameters are related to the fundamental parameters by renormalization and fine cancellations between the fundamental quantities and the quantum corrections are necessary. The hierarchy problem is essentially a fine-tuning problem.

In particle physics, the question is why the weak force is stronger and stronger than gravity. Both of these forces involve constants of nature, Fermi's constant for the weak force and Newton's constant for gravity. From the Standard Model, it appears that Fermi's constant is unnaturally large and should be closer to Newton's constant, unless there is a fine cancellation between the bare value of Fermi's constant and the quantum corrections to it.

More technically, the question is why the Higgs boson is so much lighter than the Planck mass (or the grand unification energy). In fact, researchers are searching for Higgs masses ranging from 115 up to 350 GeV with different selected decay channels from $b\bar{b}$ to $t\bar{t}$ (see for example [59] and references therein). One would expect that the large quantum contributions to the square of the Higgs boson mass would inevitably make the mass huge, comparable to the scale at which new physics appears, unless there is an incredible fine-tuning cancellation between the quadratic radiative corrections and the bare mass. With this state of art, the problem cannot be formulated in the context of the Standard Model where the Higgs mass cannot be calculated. In a sense, the problem is solvable if, in a given effective theory of particles, where the Higgs boson mass is calculable, there are no fine-tunings. If one accepts the *big-desert* assumption and the existence of a hierarchy problem, some new mechanism (at Higgs scale) becomes necessary to avoid fine-tunings.

The model which we are discussing contains a "running" scale that could avoid to set precisely the Higgs scale. If the mass of the field ϕ is in TeV region, there is no hierarchy problem being ϕ a gravitational scale. In this case, the Standard Model holds up plus an extended gravitational sector derived from the fifth dimension.

In other words, the Planck scale can be dynamically derived from the vacuum expectation value of ϕ. In some sense, our model, in its low energy realization, works like the model proposed by Antoniadis et al. [60]. The Planck scale can be recovered, as soon as the coupling λ is of the order 10^{-31}. Action (87) (and the other Extended Gravity Theories discussed above) as effective models valid up to a cutoff scale of a few $M_\sharp \sim$ TeV (see also [61–63]). The tiny value of λ, coming from the extra dimension, is in agreement with the above considerations allowing the presence of physical (quasi-) massless gravitons with very large interaction lengths.

As in [60], gravity is weak although the gravitational scale is low. Also the string theory limit corresponds to a large scalar field vacuum expectation value at TeV. It is important to stress that, by a conformal transformation from the Jordan frame to the Einstein frame, the Planck scale is decoupled from the vacuum expectation of the scalar field ϕ. However the scalar field redefinition has to preserve the vacuum of the theory. Besides, the gauge couplings and masses depend on the vacuum expectation value of ϕ and are dynamically determined. This means that Standard Model and Einstein Gravity (in the above conformal-affine sense) could be recovered *without the hierarchy problem*. As discussed in [64,65], it is possible to show that the operators generated by the self-interaction of the scalar field are of the form

$$\frac{1}{M_\sharp^{N-4}}\lambda^{\frac{N}{2}}\phi^N \tag{89}$$

and they are always suppressed by the small parameter λ and do not destabilize the potential of the theory. This result holds also for perturbative corrections from quantum gravity.

Considering again the problem of mass generation, one can assume that particles of

Standard Model have sizes related to the cut off, that is $M_\sharp^{-1}$, and their collisions could lead to the formation of bound states as in [60,66]. Potentially, such a phenomenon could mimic the decay of semi-classical quantum black holes and, at lower energies, it could be useful to investigate substructures of Standard Model. This means that we should expect some strong scattering effects in the TeV region involving the coupling of ϕ to the Standard Model fields. The "signature" of this phenomenon could lead to polarization effects of the particle beam as discussed in the previous section. Furthermore the strong dynamics derived from the phenomenon could resemble compositeness as discussed in [68]. Furthermore, bounds on the production of mini-black holes can be derived from astroparticle physics [69–73]. In [70] a bound on the cross-section is smaller than 0.5 TeV^{-2}. In our case, assuming the cross-section $M_\sharp^{-2}$, we get a bound of TeV-order. If the fundamental scale of our theory is of this order, strong scattering processes at LHC would have the cross-section 10^7fb and would dominate the cross-sections expected from the Standard Model.

7.2. Gravitational and Electroweak Interactions

The Higgs mechanism is an approach that allows: $i)$ to generate the masses of electroweak gauge bosons; $ii)$ to preserve the perturbative unitarity of the S-matrix; $iii)$ to preserve the renormalizability of the theory. The masses of the electroweak bosons can be written in a gauge invariant form using either the non-linear sigma model [22] or a gauge invariant formulation of the electroweak bosons. However if there is no propagating Higgs boson, quantum field amplitudes describing modes of the electroweak bosons grow too fast violating the unitarity around TeV scales [74–77]. There are several ways in which unitarity could be restored but the Standard Model without a Higgs boson is non-renormalizable.

A possibility is that the weak interactions become strongly coupled at TeV scales and then the related gauge theory becomes unitary at non-perturbative level. Another possibility for models without a Higgs boson consists in introducing weakly coupled new particles to delay the unitarity problem into the multi TeV regime where the UV limit of the Standard Model is expected to become relevant. Dvali et al. [78] proposed that, as black holes in gravitational scattering, classical objects could form in the scattering of longitudinal W-bosons leading to unitary scattering amplitude.

These ideas are very intriguing and show several features of electroweak interactions. First of all, the Higgs mechanism is strictly necessary to generate masses for the electroweak bosons. Beside, some mechanisms can be unitary but not renormalizable or vice-versa. In summary, the paradigm is that three different criteria should be fulfilled: $i)$ a gauge invariant generation of masses of electroweak bosons, $ii)$ perturbative unitarity; $iii)$ renormalizability.

Here we have proposed an alternative approach, based on Extended Theories of Gravity deduced from a 5D-manifold where the Standard Model is fully recovered enlarging the gravitational sector and avoiding the hierarchy problem.

It is important to point out that, in both the non-linear sigma model and in gauge invariant formulation of Standard Model, it is possible to define an action in terms of an expansion in the scale of the electroweak interactions v. The action can be written as

$$\mathcal{A} = \mathcal{A}_{SMw/oHiggs} + \int d^4x \sum_i \frac{C_i}{v^N} O_i^{4+N}\,, \tag{90}$$

where O_i^{4+N} are operators compatible with the symmetries of the model. The electroweak bosons are gauge invariant fields defined by

$$\underline{W^i_\mu} \quad = \quad \frac{i}{2g}\mathrm{Tr}\,\Omega^\dagger \overleftrightarrow{D}_\mu \,\Omega\tau^i\,, \tag{91}$$

with $D_\mu = \partial_\mu - igB_\mu(x)$ and

$$\Omega = \frac{1}{\sqrt{\phi^\dagger\phi}}\begin{pmatrix} \phi_2^* & \phi_1 \\ -\phi_1^* & \phi_2 \end{pmatrix}, \qquad \text{where } \phi = \begin{pmatrix} \phi_1 \\ \phi_2 \end{pmatrix} \tag{92}$$

is a $SU(2)_L$ doublet scalar field which is considered to be a dressing field and does not need to propagate. The same approach can be applied to fermions [79–81].

The analogy between the effective action for the electroweak interactions (90) and that of Extended Gravity is striking. Considering only the leading terms, the above theory can be written as a Taylor series of the form

$$f(R) \simeq \Lambda + \sum_k \frac{1}{k!}\left.\frac{d^k f(R)}{dR^k}\right|_0 R^k \tag{93}$$

where the coefficients are the derivatives of $f(R)$ calculated at a certain value of R. Clearly, as shown in previous sections,, the extra gravitational degrees of freedom can be suitably transformed in a scalar field ϕ which allows to avoid the hierarchy problem. Both electroweak theory and Extended Gravity have a dimensional energy scale which defines the strength of the interactions. The Planck mass sets the strength of gravitational interactions while the weak scale λ determines the range and the strength of the electroweak interactions. As shown in the previous subsection, these scales can be compared at TeV energies.

In other words, the electroweak bosons are not gauge bosons in standard sense but they can be "derived" from the above further gravitational degrees of freedom. The local $SU(2)_L$ gauge symmetry is imposed at the level of the quantum fields. However there is a residual global $SU(2)$ symmetry, *i.e.* the custodial symmetry. In the case of gravitational theories formulated as the $GL(4)$-group of diffeomorphisms, tetrads are an unavoidable feature necessary to construct the theory. They are gauge fields which transform under the local Lorentz transformations $SO(3,1)$ and under general coordinate transformations, the metric $g_{\mu\nu} = e^a_\mu e^b_\nu \eta_{ab}$ which is the field that is being quantized, transforms under general coordinate transformations which is the equivalent of the global $SU(2)$ symmetry for the weak interactions (in our case the residual $GL(2) \supset SU(2)$) . Such an analogy between the tetrad fields and the Higgs field is extremely relevant. As shown above for deformations, we can say that the Higgs field has the same role of the tetrads for the electroweak interactions while the electroweak bosons have the same role of the metric. Dynamics would be given by deformations.

A gravitational action like (93) is, in principle, non-perturbatively renormalizable if, as shown by Weinberg, there is a non-trivial fixed point which makes the gravity asymptotically free [82]. This scenario implies that only a finite number of the Wilson coefficients in the effective action would need to be measured and the theory would thus be predictive and probed at LHC.

Measuring the strength of the electroweak interactions in the electroweak W-boson scattering could easily reveal a non-trivial running of the electroweak scale v. If an electroweak fixed point exists, an increase in the strength of the electroweak interactions could be found, as in the strongly interacting W-bosons scenario, before the electroweak interactions become very weak and eventually irrelevant in the fixed point regime. In analogy to the non-perturbative running of the non-perturbative Planck mass, it is possible to introduce an effective weak scale

$$v_{eff}^2 = v^2\left(1 + \frac{\omega}{8\pi}\frac{\mu^2}{v^2}\right), \tag{94}$$

where μ is an arbitrary mass scale, ω a non-perturbative parameter which determines the running of the effective weak scale and v is the weak scale measured at low energies. If ω is positive, the electroweak interactions would become weaker with increasing energy. This asymptotically-free weak interaction would be renormalizable at non-perturbative level.

The asymptotically free weak interaction scenario could also solve the unitarity problem of the standard model. In the standard model without taking into account the Higgs boson, there are five amplitudes contributing at tree-level to the scattering of two longitudinally polarized electroweak W-bosons. Summing these five amplitudes, one finds at order s/M_W^2

$$\mathcal{A}(W_L^+ + W_L^- \to W_L^+ + W_L^-) = \frac{s}{v_{eff}^2}\left(\frac{1}{2} + \frac{1}{2}\cos\theta\right), \tag{95}$$

where s is the center of mass energy squared and θ is the scattering angle. Clearly if v_{eff} grows fast enough with energy, the ultra-violet behaviour of these amplitudes can be compensated and the summed amplitude can remain below the unitary bound. A similar proposal has been made to solve problems with unitarity in extra-dimensional models [83].

It is important to stress that our approach does not require new physics but to take only into account the whole budget of gravitational degrees of freedom. The monitoring of the strength of the electroweak interactions in the W-bosons scattering at LHC could establish the existence of a fixed-point in the weak interactions. Using the one-loop renormalization group of the weak scale could help in formalizing this picture [84]. To be more precise, let us consider the scale of electroweak interactions

$$v(\mu) = v_0\left(\frac{\mu}{\mu_0}\right)^{\frac{\gamma}{16\pi^2}}, \quad \text{with } \gamma = \frac{9}{4}\left(\frac{1}{5}g_1^2 + g_2^2\right) - \mathrm{Tr}(3\mathrm{Y}_u^\dagger \mathrm{Y}_u + 3\mathrm{Y}_d^\dagger \mathrm{Y}_d + \mathrm{Y}_e^\dagger \mathrm{Y}_e), \tag{96}$$

where Y_i are the respective Yukawa matrices. If the theory is in the perturbative regime e.g., at m_W, the Yukawa coupling of the top dominates since at this scale $g_1 = 0.31$ and $g_2 = 0.65$ and γ is negative. In this case, the scale of the weak interactions become smaller. If the weak interactions become strongly coupled at TeV region, g_2 becomes large and γ is expected to become positive. We obtain the expected running and the weak scale becomes larger. This is not possible in the framework of a perturbative approach. This result could represent a "signature" for the approach presented here. However, we stress once again that there are indications of a non-trivial fixed point for the non-linear sigma model using exact renormalization group techniques [85,86]. In conclusion, the unitarity problem of the weak interactions could be fixed by a non-trivial fixed point in the renormalization group of the weak scale. A similar mechanism could also fix the unitarity problem for fermions masses [87–91] if their masses are not generated by the standard Higgs mechanism but in the same way considered here (let us remind that also $SU(3)$ could be generated by the splitting of $GL(4)$ group). In the case of electroweak interactions this approach could be soon checked at LHC but good indications are also available for QCD [92].

8. $f(R)$-Gravity with Torsion and Spinor Fields

Considering torsion in the discussion means to take into account also spin fields [13]. In particular, matter fields can be given by Dirac fields: Dirac fermions are defined in terms of gamma matrices satisfying the Clifford algebra. Such fields can be achieved in the process of 4D-reduction considering the whole geometric budget where generic connections are assumed (not only symmetric ones). The fifth gamma matrix is a parity-odd gamma matrix whose index has no geometrical meaning: it is $\gamma^5 = i\gamma^0\gamma^1\gamma^2\gamma^3$ in terms of the other independent gamma matrices. The dynamics of the gravitational field can be taken in a more general instance: that is instead of considering the usual Hilbert-Einstein Lagrangian R, we will take a general function of the Hilbert-Einstein Lagrangian $f(R)$ derived from the reduction process [48]. The special case given by the Einstein-Hilbert Lagrangian is immediately recovered for $f(R) = R$, but in this case the peculiar form of the Lagrangian is such that yields an Hessian determinant null, and therefore it is degenerated with respect to all other more general $f(R)$-theories. In the following we will focus on these most general $f(R)$-theories of gravitation which can be always conformally related to the presence of scalar fields [50, 51]. Dirac fermions possess in general both energy and spin density, and so their coupling to both metric and torsion is very natural: for them, it is possible to write an effective Lagrangian as

$$\mathcal{L} = \sqrt{-g}\left[f(R) + \tfrac{i}{2}\left(\overline{\psi}\gamma^\alpha D_\alpha\psi - \overline{\psi}\gamma^\alpha D_\alpha\psi\right) - m\overline{\psi}\psi\right] \tag{97}$$

that is the Dirac action plus a gravitational action given by a general function $f(R)$ of the Hilbert-Einstein action of gravity. Note that now we are including torsion degrees of freedom. By varying this action with respect to the fields involved, we get the field equations in the form

$$i\gamma^\alpha D_\alpha\psi + i\gamma^\alpha T_\alpha\psi - m\psi = 0 \tag{98}$$

where m is the mass of the Dirac matter field and T_α is the contorsion vector (see Section 2). Energy and spin density are

$$T_{\alpha\beta} = \tfrac{i}{4}\left(\bar{\psi}\gamma_\alpha D_\beta\psi - D_\beta\bar{\psi}\gamma_\alpha\psi\right) \tag{99a}$$

$$S_{\mu\alpha\beta} = \tfrac{i}{4}\bar{\psi}\left\{\gamma_\mu, [\gamma_\alpha, \gamma_\beta]\right\}\psi \tag{99b}$$

the spin density tensor being completely antisymmetric; the field equations for gravity are

$$f'(R)R_{\alpha\beta} - \tfrac{1}{2}f(R)g_{\alpha\beta} = T_{\alpha\beta} \tag{100a}$$

$$f'(R)T_{\mu\alpha\beta} - \tfrac{1}{4}\left(g_{\alpha\mu}\partial_\beta f'(R) - g_{\beta\mu}\partial_\alpha f'(R)\right) = \tfrac{1}{2}\left(S_{\alpha\beta\mu} + \tfrac{1}{2}g_{\alpha\mu}S_{\beta\rho}{}^{\rho} - \tfrac{1}{2}g_{\beta\mu}S_{\alpha\rho}{}^{\rho}\right) \tag{100b}$$

After substitution of energy and spin density into these field equations, we get

$$f'(R)R_{\alpha\beta} - \tfrac{1}{2}f(R)g_{\alpha\beta} = \tfrac{i}{4}\left(\bar{\psi}\gamma_{\alpha}D_{\beta}\psi - D_{\beta}\bar{\psi}\gamma_{\alpha}\psi\right) \tag{101a}$$

$$f'(R)T_{\mu\alpha\beta} - \tfrac{1}{4}\left(g_{\alpha\mu}\partial_{\beta}f'(R) - g_{\beta\mu}\partial_{\alpha}f'(R)\right) = \tfrac{i}{8}\bar{\psi}\left\{\gamma_{\mu}, [\gamma_{\alpha}, \gamma_{\beta}]\right\}\psi \tag{101b}$$

linking the Dirac matter to the spacetime geometry. Here $f'(R)$ means the ordinary derivative with respect to R. We see that when the Dirac matter field equations are used for the energy and spin density, the geometric field equations reduce to the Bianchi identities, showing that the model is consistently defined. It is important to stress that we are again in the realm of ETGs and that $f(R)$-gravity is nothing else but a scalar-tensor gravity where the 4D-scalar field ϕ has been suitably recast in terms of the curvature scalar R [specifically $\phi \to f'(R)$] [67].

It is possible to see that from (100b) there could be torsion even in absence of the spin density tensor: therefore we may say that while the energy is the source of the curvature of the spacetime, both the spin and the non-linearity of $f(R)$ gravity are the sources of the spacetime torsion, in particular the non-linearity being the source of the torsion trace vector; however in the case of Dirac field, because of the complete antisymmetry of the spin, matter is the source of the completely antisymmetric part of torsion alone.

In the following, we suppose that the trace of the field energy-curvature coupling equation (100a) given by $f'(R)R - 2f(R) = T^{\alpha}_{\alpha} = T$ is an invertible relation between the Ricci curvature scalar R and the trace of the stress-energy tensor; since $f(R) = kR^2$ is only compatible with the traceless energy condition, then we assume that $f(R) \neq kR^2$.

Clearly, it is now possible to split all torsional quantities into torsionless quantities plus torsional contributions. The Dirac theory is algebraically related to the spin, therefore we can substitute torsion with the spin everywhere, to get the Dirac field equations

$$i\gamma^{\alpha}\nabla_{\alpha}\psi - \tfrac{3}{16\varphi}\bar{\psi}\gamma^{\alpha}\psi\gamma_{\alpha}\psi - m\psi = 0 \tag{102}$$

where ∇ designates the purely metric derivatives and the contributions of torsion are written in terms of the spin of the spinor field itself. Here have introduced the field $\varphi = f'(R)$: this means that Dirac field equations[1] , in the most general case with torsion, are dynamically equivalent to those with no torsion but with additional self-interaction potentials (see consideration in the previous sections). By employing the Fierz rearrangements they can be written in the form of Nambu–Jona–Lasinio potentials; moreover this means that Dirac field equations in the most general $f(R)$-gravity are dynamically equivalent to those in the simplest case $f(R) = R$ in which the additional potentials of self-interaction are scaled by a factor φ, which may be either positive or negative yielding repulsive or attractive Nambu-Jona–Lasinio potentials respectively [46,47].

In the 4D case, and in absence of spinor fields, it is possible to work out the field equations that we would obtain by varying with respect to the metric only, as in the purely metric case, that is

$$f'(\mathcal{R})\mathcal{R}_{\alpha\beta} - \frac{1}{2}f(\mathcal{R})g_{\alpha\beta} = \left(g_{\alpha\mu}g_{\beta\nu} - g_{\alpha\beta}g_{\mu\nu}\right)\nabla^{\mu}\nabla^{\nu}f'(\mathcal{R}), \tag{103}$$

[1]Here we are using the symbol φ instead of ϕ since torsional degrees of freedom are present in $f(R)$.

which are fourth-order equations; after a suitable manipulation, the above equation can be rewritten as

$$\mathcal{R}_{\alpha\beta} - \frac{1}{2}\mathcal{R}g_{\alpha\beta} = \frac{1}{\phi}\left[\frac{1}{2}g_{\alpha\beta}\left(f(\mathcal{R}) - \mathcal{R}f'(\mathcal{R})\right) + \nabla_\alpha\nabla_\beta f'(\mathcal{R}) + g_{\alpha\beta}\Box f'(\mathcal{R})\right] \quad (104)$$

where the right-hand side can be interpreted as an effective energy tensor constructed by the extra curvature terms [50,51].

8.1. Torsion and Weak Interactions

Having established a general framework in which the gravitational background can be related to the electroweak phenomenology, we shall now focus on a specific example in which such an approach can be realized. It is easy to see that the field $\frac{1}{\varphi}$ plays the role of an energy-dependent scaling. Let us assume that the action (93) is truncated at the fourth-order power: the Taylor coefficients can be written as

$$f(R) = R + \tfrac{1}{4}\varepsilon R^2 + \tfrac{1}{27}\eta^2\varepsilon^2 R^3 \quad (105)$$

in terms of the two parameters ε and η that have to be determined. The Λ term has been discarded since it is unessential for the following considerations. In this situation, the trace of the energy-curvature coupling is a cubic equation in R with three solutions, of which we are going to take into account the only real one; this is also the only solution for which R vanishes as the trace of the energy-momentum tensor Σ vanishes. In this sense, this is the correct infrared limit. Explicitly φ is

$$\varphi = 3\left(1 - \tfrac{3}{8\eta^2}\right) + \left(\tfrac{3}{4\eta} + \sqrt[3]{s + \sqrt[2]{s^2 - 1}}\right)^2 + \left(\tfrac{3}{4\eta} + \sqrt[3]{s - \sqrt[2]{s^2 - 1}}\right)^2 \quad (106)$$

function of the energy-momentum trace $\frac{1}{2}\eta\varepsilon\Sigma = s$ in the only parameter η. This $\frac{1}{\varphi}$ energy-dependent coupling starts from the unity in the infrared regime, and then it increases up to its maximum value reached at some scale that may be fine-tuned eventually, before decreasing to vanish asymptotically in the ultraviolet regime. Henceforth any Nambu-Jona–Lasinio interaction will be negligible in the infrared, then it would get larger up to a given energy, before ensuring asymptotic freedom in the ultraviolet regime.

Having established how the non-linearity of $f(R)$-gravity behaves as an energy-dependent running coupling for a given Nambu-Jona–Lasinio interaction, we may now leave the treatment of the single-field in self-interaction to study the couple of fields in interaction. In this case, the role of torsion is drastically important for the following reason: each spinor has a dynamics governed by a covariant derivative that contains torsion related to the spin density of the total system. If we now have only two spinor fields in a region of spacetime getting close to one another, it is clear that either of them will have dynamical behavior influenced by the presence of the other one, according to whether the other spinor is actually close enough to make relevant contributions to the spin density or far enough to consider those contributions to the spin density negligible. This means that the presence of torsion for a system of two or many spinors has effects for their dynamics, which make them look like interacting even if no other interaction has been defined.

The presence of the running coupling may fine-tune the scale of the interaction to any value. Consequently it is straightforward to wonder if this torsionally induces interaction might be similar to the nuclear interactions, somehow. These torsional potentials are similar to the forces in the Fermi model: this hypothesis was conjectured since the 1970s in some seminal works [12, 13, 45, 93–95] although one of the problems people had to face was precisely that of the energy-dependent scale: since this problem appears to be addressable by simply setting the non-linearity of the present gravitational action, we may ask ourselves what are the possible effects of torsional interactions and, in particular, if they may look like the Fermi nuclear interactions.

8.2. The Massless Spinor-Semispinor System

Let us consider now massless fields, so to split the left-handed and right-handed projections. As a first example, we shall deal with the case in which one of the two spinors is a semispinor having only the left-handed projection [96]. The aim is to compare this case with that of the electroweak interactions for leptons before the symmetry breaking. In the case of massless spinors of which one is a spinor and the other is a semispinor (having only the left-handed projection) we set $\psi^1_R \equiv 0$. The field equations are

$$i\gamma^\mu\nabla_\mu\psi^1_L + \tfrac{3}{16}\left(\bar{\psi}^1_L\gamma_\mu\psi^1_L\gamma^\mu\psi^1_L + \bar{\psi}^2_L\gamma_\mu\psi^2_L\gamma^\mu\psi^1_L - \bar{\psi}^2_R\gamma_\mu\psi^2_R\gamma^\mu\psi^1_L\right) = 0 \quad (107)$$

$$i\gamma^\mu\nabla_\mu\psi^2_L + \tfrac{3}{16}\left(\bar{\psi}^1_L\gamma_\mu\psi^1_L\gamma^\mu\psi^2_L + \bar{\psi}^2_L\gamma_\mu\psi^2_L\gamma^\mu\psi^2_L - \bar{\psi}^2_R\gamma_\mu\psi^2_R\gamma^\mu\psi^2_L\right) = 0 \quad (108)$$

$$i\gamma^\mu\nabla_\mu\psi^2_R - \tfrac{3}{16}\left(\bar{\psi}^1_L\gamma_\mu\psi^1_L\gamma^\mu\psi^2_R + \bar{\psi}^2_L\gamma_\mu\psi^2_L\gamma^\mu\psi^2_R - \bar{\psi}^2_R\gamma_\mu\psi^2_R\gamma^\mu\psi^2_R\right) = 0 \quad (109)$$

which can be written in the form

$$i\gamma^\mu\nabla_\mu L - \tfrac{1}{2}g\vec{A}_\mu\cdot\vec{\sigma}\gamma^\mu L + \tfrac{1}{2}g'B_\mu\gamma^\mu L - G_Y\phi R = 0 \quad (110)$$

$$i\gamma^\mu\nabla_\mu R + g'B_\mu\gamma^\mu R - G_Y\phi^\dagger L = 0 \quad (111)$$

by defining

$$\left(\psi^2_R\right) = R \qquad \begin{pmatrix} \psi^1_L \\ \psi^2_L \end{pmatrix} = L \quad (112)$$

with

$$\tfrac{3}{8}\bar{R}L = \phi \quad (113)$$

and

$$\tfrac{3}{8}\left(\bar{L}\gamma_\mu\tfrac{\mathbb{I}}{2}L + \bar{R}\gamma_\mu R\right) = -\left(\tfrac{g'}{1-G_Y}\right)B_\mu \quad (114)$$

$$\tfrac{3}{8}\bar{L}\gamma_\mu\tfrac{\vec{\sigma}}{2}L = \left(\tfrac{3g}{3-G_Y}\right)\vec{A}_\mu \quad (115)$$

as new fields. We see that field equations (110) and (111) are formally identical to the field equations for the lepton fields before the symmetry breaking in the Standard Model.

We notice that from the fundamental lepton fields, it is possible to build up a complex singlet and a complex doublet: from these, we also see that the composite scalar field is a complex doublet; and the composite vector fields are a real singlet and a real triplet. In fact, consider that the massless fundamental leptons R and L are complex and therefore they can be subject to two independent $U(1)$ transformations: the massless composite scalar field, given by the definition $\frac{3}{8}\bar{R}L$, is consequently complex and thus it transforms according to the combined $U(1)$ transformations. Finally the massless composite vector fields are real and do not transform at all. Moreover we have that the massless fundamental leptons R and L are such that R is the only right-handed spinor and it does not transform into anything else whereas L is formed by a doublet of left-handed spinors which, in principle, are indistinguishable and therefore they can transform into one another according to an $SU(2)_L$ transformation: the massless composite scalar field $\frac{3}{8}\bar{R}L$ transforms according to the $SU(2)_L$ transformation above; furthermore, the massless composite vector fields are such that the vector field, given by the right-handed projections $\bar{R}\gamma^\mu R$, does not transform whereas, on the other hand, the vector fields, given by the left-handed projections, are such that the vector field, given by $\bar{L}\gamma_\mu L$, transforms into itself. The three vector fields $\bar{L}\gamma_\mu\vec{\sigma}L$ transform into each other and so the vector fields given by $2\bar{R}\gamma^\mu R + \bar{L}\gamma_\mu L$ and $\bar{L}\gamma_\mu\vec{\sigma}L$ transform as the singlet and the triplet of the adjoint representation of the same $SU(2)_L$ transformation that has been given above. This establishes the transformation laws of the lepton, the scalar and vector fields before the symmetry breaking in the Standard Model.

8.3. The Massless Spinor-Spinor System

As a second example, we shall deal with the case in which two spinors have both projections, but for which the two right-handed projections have different Abelian charges so to be forbidden to mix [97]. The aim is to compare this case with that of the electroweak interactions for quarks before the symmetry breaking.

In the case of massless spinors having both projections, there is nothing that would forbid the two right-hand projections to mix forming a doublet, contrary to what is observed; to avoid this situation maintaining the two right-handed projections as two singlets, we have to postulate in addition that Abelian fields are introduced and the charges are assigned to the two right-handed fields with two different values. The field equations then are

$$i\gamma^\mu\nabla_\mu\psi_L^1 + \tfrac{3}{16}\left(\bar{\psi}_L^1\gamma_\mu\psi_L^1\gamma^\mu\psi_L^1 + \bar{\psi}_L^2\gamma_\mu\psi_L^2\gamma^\mu\psi_L^1 - \bar{\psi}_R^1\gamma_\mu\psi_R^1\gamma^\mu\psi_L^1 - \bar{\psi}_R^2\gamma_\mu\psi_R^2\gamma^\mu\psi_L^1\right) = 0 \quad (116)$$

$$i\gamma^\mu\nabla_\mu\psi_R^1 - \tfrac{3}{16}\left(\bar{\psi}_L^1\gamma_\mu\psi_L^1\gamma^\mu\psi_R^1 + \bar{\psi}_L^2\gamma_\mu\psi_L^2\gamma^\mu\psi_R^1 - \bar{\psi}_R^1\gamma_\mu\psi_R^1\gamma^\mu\psi_R^1 - \bar{\psi}_R^2\gamma_\mu\psi_R^2\gamma^\mu\psi_R^1\right) = 0 \quad (117)$$

$$i\gamma^\mu\nabla_\mu\psi_L^2 + \tfrac{3}{16}\left(\bar{\psi}_L^1\gamma_\mu\psi_L^1\gamma^\mu\psi_L^2 + \bar{\psi}_L^2\gamma_\mu\psi_L^2\gamma^\mu\psi_L^2 - \bar{\psi}_R^1\gamma_\mu\psi_R^1\gamma^\mu\psi_L^2 - \bar{\psi}_R^2\gamma_\mu\psi_R^2\gamma^\mu\psi_L^2\right) = 0 \quad (118)$$

$$i\gamma^\mu\nabla_\mu\psi_R^2 - \tfrac{3}{16}\left(\bar{\psi}_L^1\gamma_\mu\psi_L^1\gamma^\mu\psi_R^2 + \bar{\psi}_L^2\gamma_\mu\psi_L^2\gamma^\mu\psi_R^2 - \bar{\psi}_R^1\gamma_\mu\psi_R^1\gamma^\mu\psi_R^2 - \bar{\psi}_R^2\gamma_\mu\psi_R^2\gamma^\mu\psi_R^2\right) = 0 \quad (119)$$

which may be written in the form

$$i\gamma^\mu\nabla_\mu L - \tfrac{1}{2}g\vec{\sigma}\cdot\vec{A}_\mu\gamma^\mu L - \tfrac{1}{6}g'B_\mu\gamma^\mu L - G_u i\sigma^2\phi^* u - G_d\phi d + O_3 = 0 \quad (120)$$

$$i\gamma^\mu\nabla_\mu u - \tfrac{2}{3}g'B_\mu\gamma^\mu u + G_u i\phi^T\sigma^2 L + O_3 = 0 \quad (121)$$

$$i\gamma^\mu\nabla_\mu d + \tfrac{1}{3}g'B_\mu\gamma^\mu d - G_d\phi^\dagger L + O_3 = 0 \quad (122)$$

when we define

$$(\psi_R^1) = u \quad (\psi_R^2) = d \quad \begin{pmatrix}\psi_L^1\\ \psi_L^2\end{pmatrix} = L \quad (123)$$

with

$$\frac{15}{32G_d}\left(\bar{d}L\right)+\frac{3i}{16G_u}\left(\bar{u}\sigma^2 L\right)^*=\phi \tag{124}$$

and

$$\frac{9}{32}\left(\bar{L}\gamma_\mu\frac{\mathbb{I}}{2}L+2\bar{u}\gamma_\mu u-\bar{d}\gamma_\mu d\right)+q\Gamma_\mu=g'B_\mu \tag{125}$$

$$\frac{9}{32}\bar{L}\gamma_\mu\frac{\vec{\sigma}}{2}L=g\vec{A}_\mu \tag{126}$$

as new fields. In the field equations (120), (121) and (122), the terms O_3 represent residual interactions written in the form of three-field vertices. Under the reasonable assumption for which these terms may be negligible, this is formally identical to the system of field equations for the quark fields before the symmetry breaking in the Standard Model.

The Abelian field Γ_μ must be fundamental and its coupling is given in terms of the $-\frac{1}{6}q$, $-\frac{2}{3}q$ and $\frac{1}{3}q$ charges for the left-handed L, right-handed u and d quarks respectively; the Abelian field, given by the current $\frac{9}{32}(\bar{L}\gamma_\mu\frac{\mathbb{I}}{2}L+2\bar{u}\gamma_\mu u-\bar{d}\gamma_\mu d)$, is composite in terms of those quarks and what allows the present prescription to work is that this current is defined once but its appearance occurs in terms of the $-\frac{1}{6}$, $-\frac{2}{3}$ and $\frac{1}{3}$ factors in front of the L, u and d quarks respectively: this is the reason for which the current $\frac{9}{32}(\bar{L}\gamma_\mu\frac{\mathbb{I}}{2}L+2\bar{u}\gamma_\mu u-\bar{d}\gamma_\mu d)$ and the Abelian field $q\Gamma_\mu$ may always be reabsorbed into the definition of the Abelian field $g'B_\mu$ still fundamental and still occurring with the $-\frac{1}{6}$, $-\frac{2}{3}$ and $\frac{1}{3}$ factors in front of the L, u and d quarks again respectively. By following the same method presented above, it is possible to establish the form of the transformation laws for the quarks, the scalar and vector fields before the symmetry breaking in the Standard Model.

8.4. The Massive Spinor-Semispinor System

In this final example, we shall deal with the case in which one of the two spinors is massive while the other is massless and taken to be a semispinor having only the left-handed projection [98]. The aim is to compare this case with that of the electroweak interactions for leptons in a situation in which there is no symmetry breaking at all.

In the case is which one is a massive spinor and the other a massless semispinor having only the left-handed projection, the two different masses prevent them to mix forming doublets; this situation is such that the two fields are always asymmetric and they have to be identified with the electron and neutrino field right from the beginning.

In this case the field equations for the electron and neutrino fields e and ν are

$$i\gamma^\mu\nabla_\mu e-\frac{3}{16}\left(\bar{e}\gamma_\mu e\gamma^\mu e+\bar{\nu}\gamma_\mu\nu\gamma^\mu\gamma_5 e\right)-me=0 \tag{127}$$

$$i\gamma^\mu\nabla_\mu\nu-\frac{3}{16}\bar{e}\gamma_\mu\gamma_5 e\gamma^\mu\nu=0 \tag{128}$$

which can be transformed into the following

$$i\gamma^\mu\nabla_\mu e-\frac{3}{8}(\cos\theta)^2\bar{e}\gamma^\mu e\gamma_\mu e+q\tan\theta Z_\mu\gamma^\mu e-\frac{g}{2\cos\theta}Z_\mu\gamma^\mu e_L+\frac{g}{\sqrt{2}}W^*_\mu\gamma^\mu\nu-me=0 \tag{129}$$

$$i\gamma^\mu\nabla_\mu\nu+\frac{g}{2\cos\theta}Z_\mu\gamma^\mu\nu+\frac{g}{\sqrt{2}}W_\mu\gamma^\mu e_L=0 \tag{130}$$

once we define

$$Z^\mu=-\left[2(\sin\theta)^2\bar{e}\gamma^\mu e-\bar{e}_L\gamma^\mu e_L+\bar{\nu}\gamma^\mu\nu\right]\left(\frac{3\cot\theta}{16q}\right) \tag{131}$$

$$W^\mu=-\left(\bar{e}_L\gamma^\mu\nu\right)\left[\frac{12(\sin\theta)^2-3}{16q\sqrt{2}\sin\theta}\right] \tag{132}$$

as the new fields.

The field equations (129) and (130) are formally identical to the system of field equations for the lepton fields after the symmetry breaking and mass generation in the Standard Model; however, there has been no generation of any mass nor breaking of any symmetry because there has never been any symmetry in the model while there has always been mass for the fermion fields.

The lack of symmetry of the mediators comes from the fact that they are composite states of massive spinors due to the presence of torsion, undergoing to no transformation law. The lack of the standard Higgs field comes from the fact that, in this approach, it would be useless. These two features constitute important discrepancies with respect to the Standard Model: in fact, in this example, we must expect the weak bosons to display internal structure when the energy is high enough to probe their compositeness while in the Standard Model they must remain structureless at any energy scale; and because the present model predicts the absence of the Higgs field while the Standard Model needs its presence, then the detection of the Higgs or its persistent elusive character would settle the controversy definitively.

As an additional comment, we have to specify that, in this construction, there is a fundamental difference between the case of leptons and that of quarks, that is in the former, the weak interaction are reproduced exactly, whereas, in the latter, they have been reproduced only partially becoming exact when the three-field interactions are neglected: if, on the one hand, the corrections we have in the case of quarks may well be negligible, especially in situations in which also chromodynamics is present and where all quarks are confined as partons within nucleons, on the other hand there must be no correction for leptons since their interactions are relatively simple. In other words, we may say that in a comparison between the present approach and the Standard Model, we may have small deviations in situations in which the dynamics is dirty enough to keep them hidden, such as for quarks, but no deviation in situation the dynamics is clean enough to display them clearly, such as for leptons. Thus said, it is obvious how for our approach to work, the neutrino must be left-handed solely, and this means in particular massless. This situation may appear to be in contrast with what is becoming the accepted paradigm for which neutrinos appear instead to be massive; however the mass of neutrinos is seen as a consequence of the theory that links mass to oscillation and the observation of the oscillations themselves, the last being an irrefutable experimental evidence while the first being a theory that still has with many conceptual problems: these two problems may be solved at once by a model in which neutrino are massless and nevertheless still capable of oscillations. Incidentally, we do not have to look much far to find such a model, because in presence of torsion, neutrinos do oscillate precisely because they are massless, as discussed in [99].

Finally, we would like to spend a few words about the differences between the case in which fermions are initially massless yielding the dynamics of the Standard Model before the symmetry breaking, and that in which they are already massive yielding the dynamics of the Standard Model after the symmetry breaking; what we would like to specify is that, in this last instance, the field equations are those that correspond to that of the Standard Model after the symmetry breaking, not because we have witnessed the breakdown of an initial symmetry but rather because the system has never been symmetric. In this last example the problem of having a symmetry breaking will be circumvented.

These results show that torsion induces interactions among spinors whose form recalls that of the weak forces between fermions: what this implies is that for the Dirac fermionic field equations, the presence of torsion would be manifested disguised by weak forces: therefore the information about the weak forces is stored within torsion, in a similar way as the information about gravitation is stored within the metric; this suggests that the weak forces and gravity (although intrinsically different) may be accommodated into (different) parts of the same connection. This is not what we usually call unification, and yet it clearly is a stance of unification, inasmuch as a conceptual perspective is adopted.

9. Conclusion

We have shown that, starting from a theory of gravity formulated in 5-dimensions, the 4D-spacetime reduction procedure gives rise to extensions of GR in which the additional gravitational degrees of freedom can be considered as additional physical scalar fields. Such scalar fields are geometrical deformations undergoing the $GL(4)$-group of diffeomorphisms and are capable of generating the masses of particles. Furthermore, the $f(R)$ non-linearities encode energy-dependent running couplings able to fine-tune the scale. Finally, torsion gives rise to forces having the structure of weak force. In other words, for Dirac fermionic field equations, the presence of torsion would be manifested in the guise of weak forces: this means that, contrary to what is commonly believed, torsion might not be a quantity we never observed due to its weakness, instead, it might actually be a quantity whose effects are constantly observed in weak experiments, although we were not ready to think weak forces as a long-range manifestation of torsion. If this were true, it would give rise to some sort of unification of weak and gravitational forces, as they may be included respectively in torsion and metric degrees of freedom, that is in different parts of the same connection. This connection is naturally accompanied by electrodynamics, within the most general spinorial connection of the spinor field. The only thing that would be left out of this approach would be chromodynamics. However, this does not constitute a bad point of the present scheme compared to the commonly accepted Standard Model: in fact, neither here nor in the Standard Model chromodynamics is unified with all other interactions, and actually in the Standard Model, the weak and electrodynamic interactions are not unified as well, since the $U(1) \times SU(2)$ group, and for that matter the full $U(1) \times SU(2) \times SU(3)$ group are *no unification* of sub-groups into a larger group, but only a mere albeit quite elegant *patching* of sub-groups.

The philosophical implications of this approach are profound, as this would force us to rethink the issue of unification in physics under a conceptual perspective, rather than according to the usual mathematical symmetry-based way, a way that, despite the *plethora* of possible symmetry groups, we have been investigating in the last forty years, has given no real advance since the $SU(5)$-model predicted the decay of the proton, yet to be discovered. In this prescription there is no need to postulate any new ingredient, because torsion is naturally part of the most general connection, the non-linearity of the $f(R)$ function comes out from the requirement that the action of gravitational field be not restricted to be the simplest Hilbert-Einstein action. If the action has to be written in terms of the Ricci scalar curvature in the most general way, then $f(R)$ where R contains torsion does constitute the most general dynamics we may have for the gravitational background.

However, some crucial points have to be considered. On the one hand, torsion manufactures weak interactions, which, for fermions, are structurally identical to those arising from the Standard Model, and therefore any effect in the dynamics of fermions would be reproduced as in the Standard Model itself. Nevertheless, the gauge fields emerge as composite fields and, although at energies that are low enough not to probe their internal compositeness, they should look like those of the Standard Model. However, for higher energies discrepancies, the Standard Model must pop out. It is worth stressing that, in our approach, there is no standard Higgs field but a sort of gravitational Higgs mechanism is present. The validity of the presented scheme could be reasonably checked at LHC in short time, due to the increasing luminosities of the set-up. At the present stage, experiments are indicating, very preliminary, the presence of resonances and condensate states that cannot be considered final evidences for Higgs boson (at Fermi Lab, by CDF collaboration, and at CERN, in LHC, by CMS and ATLAS [92]).

Incoming years will offer new data carrying information about the structure of the elementary constituents of our universe, and although the full confirmation of the Standard Model and its most accredited extensions are expected, nevertheless they may fail to come; if so, new aspects of the problem of unification in physics will have to be faced, and they may turn out to be even more exciting. And despite the direction taken since the 1960s, the dream of unification in physics might be much more similar to what Einstein himself hoped it to be.

References

[1] G. Basini and S. Capozziello, "Open Quantum Relativity", in *Classical and Quantum Gravity Research* (Ed. Christiansen, Nova Science Publishers, New York, 2008).

[2] G. Basini and S. Capozziello, *Gen. Rel. Grav.*, 35, 2217 (2003).

[3] C. Rovelli, *Quantum Gravity*, Cambridge University Press, Cambridge (2004).

[4] G. Basini and S. Capozziello, *Mod. Phys. Lett.* A 20, 251 (2005).

[5] G. Basini and S. Capozziello, *Int. Jou. Mod. Phys.* D 15, 583 (2006).

[6] R. Utiyama, *Phys. Rev.* 101, 1597 (1956).

[7] C. N. Yang and R. L. Mills, *Phys. Rev.* 96, 191 (1954).

[8] T.W. Kibble, *J. Math. Phys.* 2, 212 (1960).

[9] E. Cartan, *Ann. Ec. Norm.* 42, 17 (1925).

[10] D.W. Sciama, On the analog between charge and spin in GR, in *Recent Developments in GR*, Festschrift for Leopold Infeld, (1962) 415, Pergamon Press, New York.

[11] R. Finkelstein, *Ann. Phys.* 12, 200 (1961).

[12] F. W. Hehl et al., *J. Math. Phys.* 12,1334 (1970).

[13] F.W. Hehl et al., *Rev. Mod. Phys.* 48, 393 (1976).

[14] F. Mansouri et. al., *Phys. Rev.* D 13, 3192 (1976) .

[15] F. Mansouri, *Phys. Rev. Lett.* 42, 1021 (1979).

[16] L. N. Chang et al., *Phys. Rev.* D 13, 235 (1976).

[17] G. Grignani et. al., *Phys. Rev.* D 45, 2719 (1992).

[18] F. W. Hehl and J. D. McCrea, *Found. Phys.* 16, 267 (1986).

[19] H. I. Arcos and J. G. Pereira, *Int. J. Mod.Phys.* D 13, 2193 (2004).

[20] A. Inomata et. al., *Phys. Rev.* D 19 , 1665 (1978).

[21] C. G. Callan, S. Coleman, J. Wess and B. Zumino, *Phys. Rev.* 117, 2247 (1969).

[22] S. Coleman, J. Wess and B. Zumino, *Phys. Rev.* 117, 2239 (1969).

[23] C. J. Isham, A. Salam and J. Strathdee, *Ann. of Phys.* 62, 98 (1971).

[24] A.B. Borisov and V.I. Ogievetskii, *Theor. Mat. Fiz.* 21, 329 (1974).

[25] E. A. Ivanov and V. I. Ogievetskii, *Gauge theories as theories of spontaneous breakdown*, Preprint of the Joint Institute of Nuclear Research, E2-9822 3-10 (1976).

[26] L. N. Chang et al., *Phys. Rev.* D 17, 3168 (1978).

[27] K. S. Stelle et al., *Phys. Rev.* D 21, 1466 (1980).

[28] E. A. Ivanov and J. Niederle, *Phys. Rev.* D 25, 976 (1982).

[29] E. A. Ivanov and J. Niederle, *Phys. Rev.* D 25, 988 (1982).

[30] D. Ivanenko and G. A. Sardanashvily, *Phys. Rep.* 94 , 1 (1983).

[31] A. Lord and P. Goswami, *J. Math. Phys.* 27, 3051 (1986).

[32] E. A. Lord and P. Goswami, *J. Math. Phys.* 29, 258 (1987).

[33] Y. Ne'eman and T. Regge, *Riv. Nuovo Cimento* 1 1, (1978).

[34] A. Lopez-Pinto, A. Tiemblo, R. Tresguerres, *Class. Quant. Grav.* 12, 1503 (1995).

[35] J. Julve et. al., *Class. Quantum Grav.* 12, 1327 (1995).

[36] R. Tresguerres and E. W. Mielke, *Phys.Rev.* D 62, 044004 (2000).

[37] R. Tresguerres, *Phys. Rev.* D 66, 064025 (2002).

[38] A. Tiemblo and R. Tresguerres, *Eur.Phys.J.* C 42, 437 (2005).

[39] A. Tiemblo and R. Tresguerres, *Recent Res.Devel.Phys.* 5, 1255 (2004).

[40] T. Appelquist, A. Chodos, P.G.O. Freund, Modern Kaluza-Klein Thories, *Frontiers in Physics*, Addison-Wesley Publishing Company, Inc. Redwood, CA (1987).

[41] G.G. Ross, *Grand Unified Theories*, Benjamin, Menlo Park, CA (1985).

[42] G. Basini, A. Morselli, M. Ricci, *La Riv. del N. Cimento* 12 (1989).

[43] S. Capozziello, G. Lambiase, C. Stornaiolo, *Ann. Phys. (Leipzig)* 10, 713 (2001).

[44] E. Kaehler, *Rendiconti di Matematica* 21, 425 (1962).

[45] V. De Sabbata and C. Sivaram, *"Spin and Torsion in Gravitation"* (World Scientific, Singapore, 1994).

[46] Y. Nambu and G. Jona–Lasinio, *Phys. Rev.* 124, 246 (1961).

[47] L. Fabbri and S. Vignolo, *Class. Quantum Grav.* 28, 125002 (2011).

[48] S. Capozziello, G. Basini, M. De Laurentis, *Eur. Phys. J.* C 71, 1679 (2011).

[49] A. Skopenkov, in: *Surveys in Contemporary Mathematics*, Ed. N. Young and Y. Choi, *London Math. Soc. Lect. Notes.* 347, London (2008).

[50] S. Capozziello and V. Faraoni, Beyond Einstein Gravity, *Fundamental Theories of Physics* Vol. 170, Springer Ed., Dordrecht (2011).

[51] S. Capozziello, M. De Laurentis, *Phys. Repts.* 509, 167 (2011).

[52] S. Nojiri, S.D. Odintsov, *Phys. Reps.* 505, 59 (2011).

[53] B. Coll, J. Llosa and D. Soler, *Gen. Rel. Grav.* 34, 269 (2002).

[54] S. Capozziello and C. Stornaiolo, *Int. Jou. Geom. Meth. Mod. Phys.* 5, 185 (2008).

[55] R. M. Wald, *GR*, The University of Chicago Press, (1984).

[56] J.E. Campbell *A Course of Differential Geometry*, Clarendon, Oxford (1926).

[57] S. Capozziello, J. Matsumoto, S. Nojiri, S. D. Odintsov, *Phys. Lett.* B 693, 198 (2010).

[58] N.Birrell and P.C.Davies, *Quantum Fields in Curved Space*, Cambridge Univ. Cambridge (1984).

[59] A. Pich, arXiv: 0705.4264[hep-ph] (2007).

[60] I. Antoniadis, S. Dimopoulos and A. Giveon, *JHEP* 0105, 055 (2001)

[61] X. Calmet, S. D. H. Hsu, D. Reeb *Phys. Rev.* D 77, 125015 (2008).

[62] P. Minkowski, *Phys. Lett.* B 71, 419 (1977).

[63] A. Zee, *Phys. Rev. Lett.* 42, 417 (1979).

[64] X. Calmet, S. D. H. Hsu, *Phys. Lett.* B 663, 95 (2008).

[65] X. Calmet, *Mod. Phys. Lett.* A 26, 1571 (2011).

[66] I. Antoniadis, N. Arkani-Hamed, S. Dimopoulos and G. R. Dvali, *Phys. Lett. B* 436, 257 (1998).

[67] S. Capozziello, R. Cianci, C. Stornaiolo, S. Vignolo, *Class. Quant. Grav.* 24, 6417 (2007).

[68] P. Meade and L. Randall, *Jou. High. Energy Phys.* 0805, 003, 2008)

[69] J. L. Feng and A. D. Shapere, *Phys. Rev. Lett.* 88, 021303 (2002)

[70] L. A. Anchordoqui, J. L. Feng, H. Goldberg and A. D. Shapere, *Phys. Rev.* D 65, 124027 (2002)

[71] L. A. Anchordoqui, J. L. Feng, H. Goldberg and A. D. Shapere, *Phys. Rev.* D 68, 104025 (2003)

[72] A. Ringwald and H. Tu, *Phys. Lett.* B 525, 135 (2002).

[73] M. Kowalski, A. Ringwald and H. Tu, *Phys. Lett.* B 529, 1 (2002)

[74] C. H. Llewellyn Smith, *Phys. Lett.* B 46, 233 (1973) .

[75] B. W. Lee, C. Quigg and H. B. Thacker, *Phys. Rev. Lett.* 38, 883 (1977).

[76] B. W. Lee, C. Quigg and H. B. Thacker, *Phys. Rev.* D 16, 1519 (1977).

[77] C. E. Vayonakis, *Lett. Nuovo Cim.* 17, 383 (1976).

[78] G. Dvali, G. F. Giudice, C. Gomez and A. Kehagias, CERN-PH-TH-226 (2010).

[79] V. Visnjic, *Nuovo Cim.* A 101, 385 (1989).

[80] G. 't Hooft, *"Topological aspects of quantum chromodynamics"*, Lectures given at International School of Nuclear Physics, Erice, Italy, 17-25 Sep. 1998; published in "Erice 1998, From the Planck length to the Hubble radius".

[81] G. 't Hooft, in "Recent Developments In Gauge Theories," Cargèse 1979, ed. G. 't Hooft et al. Plenum Press, New York, 1980, Lecture II, p.117; *proceedings, Nato Advanced Study Institute*, Cargèse, France, August 26 - September 8, 1979."

[82] S. Weinberg, in *Understanding the Fundamental Constituents of Matter*, ed. A. Zichichi (Plenum Press, New York, 1977).

[83] J. Hewett and T. Rizzo, *JHEP* 0712, 009 (2007).

[84] H. Arason, D. J. Castano, B. Keszthelyi, S. Mikaelian, E. J. Piard, P. Ramond and B. D. Wright, *Phys. Rev.* D 46, 3945 (1992).

[85] M. Fabbrichesi, R. Percacci, A. Tonero, O. Zanusso, *Phys.Rev.* D 83, 025016 (2011).

[86] R. Percacci and O. Zanusso, *Phys. Rev.* D 81, 065012 (2010)

[87] T. Appelquist and M. S. Chanowitz, *Phys. Rev. Lett.* 59, 2405 (1987).

[88] F. Maltoni, J. M. Niczyporuk and S. Willenbrock, *Phys. Rev. Lett.* 86, 212 (2001).

[89] F. Maltoni, J. M. Niczyporuk and S. Willenbrock, *Phys. Rev.* D 65, 033004 (2002).

[90] D. A. Dicus and H. J. He, *Phys. Rev.* D 71, 093009 (2005).

[91] D. A. Dicus and H. J. He, *Phys. Rev. Lett.* 94, 221802 (2005).

[92] G. Aad et al. (The ATLAS collaboration) *Eur. Phys. J.* C 71, 1512 (2011).

[93] V. De Sabbata and M. Gasperini, *Gen. Rel. Grav.* 10, 731 (1979).

[94] C. Sivaram and K. P. Sinha, *Lett. Nuovo Cim.* 13, 357 (1975).

[95] C. Sivaram and K. P. Sinha, *Phys. Rept.* 51, 111 (1979).

[96] L. Fabbri, *Mod. Phys. Lett.* A 26, 1697 (2011).

[97] L. Fabbri, arXiv:1009.4423 [hep-th].

[98] L. Fabbri, *Int. J. Theor. Phys.* 50, 3616 (2011).

[99] L. Fabbri, arXiv:1012.3862 [hep-th].

In: Einstein and Others: Unification
Editor: Valeriy V. Dvoeglazov ISBN: 978-1-63463-276-8

Chapter 2

GENERALIZED AFFINE GEOMETRIES, STRUCTURE OF SPACETIME AND UNIFICATION

Diego Julio Cirilo-Lombardo*
International Institute of Physics
Universidade Federal de Rio Grande do Norte, Natal - RN, Brazil
Bogoliubov Laboratory of Theoretical Physics
Joint Institute for Nuclear Research, Dubna (Moscow Region), Russian Federation

Abstract

Geometrical analysis of a new type of the Unified Field Theoretical models following the guidelines of previous works of the author is presented. These unified models are characterized by an underlying hypercomplex structure, zero non-metricity and their geometrical action which is determined essentially by the curvature provenient from the symmetry breaking of a group manifold in higher dimensions. The mechanism of Cartan-MacDowell-Mansouri type, allows us to construct geometrical actions of the determinantal type, leading to a non topological physical Lagrangian due to the splitting of a reductive geometry. Our goal is to take advantage of the geometrical and topological properties of this theory in order to determine the minimal group structure of the resultant spacetime Manifold requiered to support a fermionic structure. From this fact, the relation between the antisymmetric torsion and the Dirac structure of the spacetime is determined, and the existence of an important contribution of the torsion to the gyromagnetic factor of the fermions, shown. Also, we resume and analyze previous cosmological solutions in this new UFT where, as in our work [Class. Quantum Grav.**22** (2005) 4987–5004] for the non abelian Born-Infeld model, the Hosoya-Ogura ansatz is introduced for the important cases of tratorial, totally antisymmetric and general torsion fields. In the case of spacetimes with torsion, the real meaning of the spin-frame alignment is found and the question of the minimal coupling is discussed.

PACS: 04.20.Cv,04.20.Jb,04.20.Gz

Keywords: Unified field theories, affine geometries

*E-mail address: diego777jcl@gmail.com

1. Introduction

For a long time in the history of the modern theoretical physics the possibility of the unification of all fundamental forces has been considered from the mathematical and theoretical point of view. Several models, formulations and sophisticated mathematical tools have been developed and used in order to solve the intricate puzzle of conciliating gravity with the other fundamental forces of the nature: electromagnetic, weak and strong. Despite the fact that many attempts have been made, this issue remains still without concrete solution as is the string theory which is the typical case. In string theoretical frameworks it is common the claim that there are a consistent solution of the unification problem but, beside particular formulations, the theoretical and conceptual environment altogether with an obscure mathematical basis puts certainly some difficulties to the acceptance of such a claim.

As we pointed out in [1, 2], the cornerstone of the problem resides in to find the start-point from to reformulate conceptually the theoretical arena where the fundamental unified theory (with the geometry as the unifying essence) should be placed. According to Mach, there are not spacetime without matter. Then, two basic opposite ideas immediately arise:: the concept of dualistic or non-dualistic theories. In the first case, the simplest and most economical description can be formulated in terms of the gravitational field *without torsion* plus an energy-momentum tensor added 'by hand', in order to cover the lack of knowledge of the matter- energy spacetime distribution. In the second approach there are not prescriptions for the interaction between gravity and 'matter' fields: they arise from the same fundamental geometrical structure.

In some previous papers we presented a new model of non-dualistic Unified Field Theory. The preliminary model introduced firstly in [1], fully consistent from the mathematical and geometrical points of view, contains two main features: First, it is based on a manifold equipped with an *underlying hypercomplex structure* and *zero non-metricity*, which leads to the important fact that the torsion of the spacetime structure turns out to be totally antisymmetric. As is well known, in the particular case of a totally antisymmetric torsion tensor, this type of affine geometrical frameworks present geodesics that are equivalent to the minimal length equations and that is the most important, it is the only case in which the equivalence principle is fulfilled, as shown in [3, 4], and we will demonstrate also in the present chapter.

The second feature introduced in [1, 2] as a main ingredient is that the specific form of the action is determined by the curvature from the *symmetry breaking of a group manifold in higher dimensions*, via Cartan-MacDowell-Mansouri mechanism. This mechanism allows one to construct geometrical actions of determinantal type that, due to the splitting of a reductive geometry via the breaking of the higher dimensional group, leads to a non topological physical Lagrangian (*e.g.*, as in the typical case $SO(1,4) \to SO(1,3) \otimes \mathbb{M}_{1,3}$). Thus, following the guidelines of our papers [1, 2, 5], in the present chapter we review and complete the previous analysis of our fundamental model for an Unified Field Theory.

The presentation of the results of the chapter is organized as follows: in Section 2 the geometrical framework is introduced and the theoretical basis of the model, based on a geometrical action that acquires physical meaning through a symmetry breaking, is described. In Section 3, the dynamical equations are analyzed and the geometrical and physical meaning, elucidated. In Section 4 we resume and analyze previous cosmological solutions in the new UFT: as in our work [5], dealing with the non abelian Born-Infeld model: the Hosoya-Ogura ansatz is introduced in the important cases of tratorial and totally antisymmetric torsion. The meaning of the spin-frame alignment in the case with torsion is discussed. Also, we explicitly show that, contrarily to the case of the Poincare theory of gravitation (see [6]), the possibility of co-existence of both types of torsion in cosmological spacetimes is certainly allowed in our model.

Section 5 is the most important in the sense that the fermionic structure of the spacetime is described, and the possibility of geometrical unification realized: a unified theory of QED and GR can be derived from $P(G, M)$, the Principal Fiber Bundle of frames over the $4D$ spacetime manifold, with G as its structure group. In the subsections, the action of the UFT is analyzed from the group-theoretical point of view, considering the G-symmetry of the model. In Section 6 the derivation of the Dirac equation from the G-manifold, the relation between the electromagnetic field/fermionic structure of the spacetime, and the contribution of the torsion to the gyromagnetic factor are explicitly derived and the physical consequences are analyzed. Section 7 is devoted to a discussion of the cohomological interplay between the fields involved in the spacetime structure and, finally, in Section 8 we give our conclusion.

2. Spacetime Manifold and Geometrical Action

The starting point is a hypercomplex construction of the (metric compatible) spacetime manifold [1]

$$M, \; g_{\mu\nu} \equiv e_\mu \cdot e_\nu \tag{1}$$

where for each point $p \in M$ there exists a local affine space A. The connection over A, $\widetilde{\Gamma}$, define a generalized affine connection Γ on M, specified by (∇, K), where K is an invertible $(1,1)$ tensor over M. We will demand for the connection to be compatible and rectilinear, that is

$$\nabla K = KT, \qquad \nabla g = 0 \tag{2}$$

where T is the torsion, and g the spacetime metric (used to raise and lower the indices and determining the geodesics), that is preserved under parallel transport. This generalized compatibility condition ensures that the generalized affine connection Γ maps autoparallels of Γ on M into straight lines over the affine space A (locally). The first equation above is equal to the condition determining the connection in terms of the fundamental field in the UFT non-symmetric. Hence, K can be identified with the fundamental tensor in the non-symmetric fundamental theory. This fact gives us the possibility to restrict the connection to a (anti-)Hermitian theory.

The covariant derivative of a vector with respect to the generalized affine connection is given by

$$A^\mu{}_{;\nu} \equiv A^\mu{}_{,\nu} + \Gamma^\mu{}_{\alpha\nu} A^\alpha \tag{3}$$

$$A_{\mu;\nu} \equiv A_{\mu}{}_{,\nu} - \Gamma^\alpha{}_{\mu\nu} A_\alpha \tag{4}$$

The generalized compatibility condition (2) determines the 64 components of the connection by the 64 equations

$$K_{\mu\nu;\alpha} = K_{\mu\rho}T^{\rho}{}_{\nu\alpha} \quad \text{where} \quad T^{\rho}{}_{\nu\alpha} \equiv 2\Gamma^{\rho}{}_{[\alpha\nu]} \tag{5}$$

Notice that by contracting indices ν and α in the first equation above, an additional condition over this hypothetic fundamental (nonsymmetric) tensor K is obtained

$$K_{\mu\alpha;}{}^{\alpha} = 0$$

that, geometrically speaking, reads

$$d^{*}K = 0.$$

This is a current-free condition over the tensor K that can be nicely exemplified with the prototype of non-symmetric fundamental tensor: $K_{\mu\nu} = g_{\mu\nu} + f_{\mu\nu}$

$$d^{*}K = d^{*}g + d^{*}f \Rightarrow d^{*}f = 0 \quad \text{(current free e.o.m.)}$$

where usually $g_{\mu\nu}$ plays the role of the spacetime metric and $f_{\mu\nu}$ the role of electromagnetic field.

The metric is uniquely determined by the metricity condition, which puts 40 restrictions on the derivatives of the metric

$$g_{\mu\nu,\rho} = 2\Gamma_{(\mu\nu)\rho} \tag{6}$$

The spacetime curvature tensor, that is defined in the usual way, has two possible contractions: the Ricci tensor $R^{\lambda}_{\mu\lambda\nu} = R_{\mu\nu}$, and the second contraction $R^{\lambda}_{\lambda\mu\nu} = 2\Gamma^{\lambda}{}_{\lambda[\nu,\mu]}$, which is identically zero due to the metricity condition (2).

In order to find a symmetry of the torsion tensor, let us denote the inverse of K by $\widehat{K}$. Therefore, $\widehat{K}$ is uniquely specified by condition $\widehat{K}^{\alpha\rho}\, K_{\alpha\sigma} = K^{\alpha\rho}\widehat{K}_{\alpha\sigma} = \delta^{\rho}_{\sigma}$.

As it was pointed out in [1], inserting explicitly the torsion tensor as the antisymmetric part of the connection in (5), and multiplying by $\frac{1}{2}\widehat{K}^{\alpha\nu}$, results, after straighforward computations, in

$$\left(Ln\sqrt{-K}\right)_{,\mu} - \Gamma^{\nu}_{(\mu\nu)} = 0 \tag{7}$$

where $K = \det\left(K_{\mu\rho}\right)$. Notice that from expression (7) we arrive to the relation between the determinants K and g: $\frac{K}{g}$ =constant. Now we can write

$$\Gamma^{\nu}{}_{\alpha\nu,\beta} - \Gamma^{\nu}{}_{\beta\nu,\alpha} = \Gamma^{\nu}{}_{\nu\beta,\alpha} - \Gamma^{\nu}{}_{\nu\alpha,\beta}, \tag{8}$$

as the first term of (7) is the derivative of a scalar. Then, the torsion tensor has the symmetry

$$T^{\nu}{}_{\nu[\beta,\alpha]} = T^{\nu}{}_{\nu[\alpha,\beta]} = 0 \tag{9}$$

This implies that the trace of the torsion tensor, defined as $T^{\nu}{}_{\nu\alpha}$, is the gradient of a scalar field

$$T_{\alpha} = \nabla_{\alpha}\phi$$

In reference [7] an interesting geometrical analysis is presented of non-symmetric field structures. There, expressions precisely as (1) and (2) ensure that the basic non-symmetric field structures (*i.e.* K) take on a definite geometrical meaning when interpreted in terms of affine geometry.

The second important point to note is the following: let us consider the extended curvature [8]

$$\mathcal{R}^{ab}_{\mu\nu} = R^{ab}_{\mu\nu} + \Sigma^{ab}_{\mu\nu} \tag{10}$$

where

$$R^{ab}_{\mu\nu} = \partial_\mu \omega^{ab}_\nu - \partial_\nu \omega^{ab}_\mu + \omega^{ac}_\mu \omega_{\nu c}{}^b - \omega^{ac}_\nu \omega_{\mu c}{}^b \tag{11}$$

and

$$\Sigma^{ab}_{\mu\nu} = -\left(e^a_\mu e^b_\nu - e^a_\nu e^b_\mu\right)$$

We assume here that ω^{ab}_ν is a $SO(d-1,1)$ connection and e^a_μ is a vierbein field. The eqs. (10) and (11) can be obtained, for example, by using the formulation introduced in the seminal works by E. Cartan long time ago [1]. It is well known that in that formalism the gravitational field is represented as a connection 1-form, associated to some group containing the Lorentz group as subgroup. The typical example is provided by the $SO(d,1)$ de Sitter gauge theory of gravity. In this specific case, the $SO(d,1)$ the gravitational gauge field $\omega^{AB}_\mu = -\omega^{BA}_\mu$ is broken to the $SO(d-1,1)$ connection ω^{ab}_μ and the $\omega^{da}_\mu = e^a_\mu$ vierbein field, with the dimension d fixed. Then, the de Sitter (anti-de Sitter) curvature

$$\mathcal{R}^{AB}_{\mu\nu} = \partial_\mu \omega^{AB}_\nu - \partial_\nu \omega^{AB}_\mu + \omega^{AC}_\mu \omega_{\nu C}{}^B - \omega^{AC}_\nu \omega_{\mu C}{}^B \tag{12}$$

splits as in (10). At this point, our goal is to enlarge the group structure of the spacetime manifold in such a way that the curvature (12), (of course, after breaking the symmetry) allows us to define the geometrical Lagrangian of the theory by

$$L_g = \sqrt{det \mathcal{R}^a{}_\mu \mathcal{R}_{a\nu}} = \sqrt{det G_{\mu\nu}} \tag{13}$$

where we have defined the geometrical object

$$\mathcal{R}^a{}_\mu = \lambda\left(e^a{}_\mu + f^a{}_\mu\right) + R^a{}_\mu \qquad \left(M^a{}_\mu \equiv e^{a\nu} M_{\nu\mu}\right) \tag{14}$$

where $f^a{}_\mu$ (in sharp contrast to $e^a{}_\mu$) carries the following symmetry:

$$e_{a\mu} f^a{}_\nu = f_{\mu\nu} = -f_{\nu\mu}.$$

The action will contain, as usual, $\mathcal{R} = det\left(\mathcal{R}^a{}_\mu\right)$ as the geometrical object that defines the dynamics of the theory. The particularly convenient definition of $\mathcal{R}^a{}_\mu$ makes easy to establish the equivalent expression in the spirit of the unified theories developed some time ago by Eddington, Einstein and Born and Infeld, as for example:

$$\sqrt{det \mathcal{R}^a{}_\mu \mathcal{R}_{a\nu}} = \sqrt{det\left[\lambda^2\left(g_{\mu\nu} + f^a{}_\mu f_{a\nu}\right) + 2\lambda R_{(\mu\nu)} + 2\lambda f^a{}_\mu R_{[a\nu]} + R^a{}_\mu R_{a\nu}\right]} \tag{15}$$

where $R_{\mu\nu} = R_{(\mu\nu)} + R_{[\mu\nu]}$.

The important point to consider in this simple Cartan inspired model is that, although a cosmological constant λ is required, the expansion of the action in four dimensions leads automatically to the Einstein-Hilbert part when $f^a_{\ \mu} = 0$. Explicitly, we have

$$S = \int d^4x(e+f)\left\{\lambda^4 + \lambda^3\left(R + f^a_{\ \mu}R^\mu_{\ a}\right) + \frac{\lambda^2}{2!}\left[R^2 - R^{\mu\nu}R_{\mu\nu} + \left(f^a_{\ \mu}R^\mu_{\ a}\right)^2 - \right.\right.$$
$$\left. f^{\mu\nu}f^{\rho\sigma}R_{\mu\rho}R_{\nu\sigma}\right] + \frac{\lambda}{3!}\left[R^3 - 3RR^{\mu\nu}R_{\mu\nu} + 2R^{\mu\alpha}R_{\alpha\beta}R^\beta_\mu + \left(f^a_{\ \mu}R^\mu_{\ a}\right)^3 - \right. \tag{16}$$
$$\left.\left. 3\left(f^a_\mu R^\mu_a\right)f^{\mu\nu}f^{\rho\sigma}R_{\mu\rho}R_{\nu\sigma} + 2f^{\mu\nu}R_\mu^{\ \alpha}R_{\alpha\beta}R^\beta_{\ \nu}\right] + det\left(R_{\mu\nu}\right)\right\}$$

Notice that the tetrad property was used here and $R = g^{\alpha\beta}R_{\alpha\beta}$.

In the remaining part of the chapter, this property will be used or not, depending on the particular case.

3. The Dynamical Equations

In this case, the variation with respect to the metric remains the same as in previous papers – see [1], eq.(9) –, that is,

$$\delta_g\sqrt{G} = \frac{\sqrt{G}}{2}\left(G^{-1}\right)^{\mu\nu}\delta_g G = 0.$$

Then, the variation respect to the connection gives immediately

$$\frac{\delta\sqrt{G}}{\delta\Gamma^\omega_{\mu\nu}} = \left\{-\nabla_\sigma\left[\sqrt{G}\left(G^{-1}\right)^{\alpha\nu}\mathcal{R}^\sigma_{\ \alpha}\right]\delta^\mu_{\ \omega} + \right.$$
$$\left. \nabla_\omega\left[\sqrt{G}\left(G^{-1}\right)^{\alpha\nu}\mathcal{R}^\mu_{\ \alpha}\right] + \sqrt{G}\left(G^{-1}\right)^{\alpha\nu}\mathcal{R}^\sigma_{\ \alpha}\Gamma^\mu_{\ [\sigma\omega]}\right\} \tag{17}$$

where the general form of the Palatini's identity has been used and

$$G_{\mu\nu} \equiv \mathcal{R}^a_{\ \mu}\mathcal{R}_{a\nu}$$

with the $\mathcal{R}^a_{\ \mu}$ from (14). Defining $\Sigma^{\nu\sigma} \equiv \sqrt{G}\left(G^{-1}\right)^{\alpha\nu}\mathcal{R}^\sigma_{\ \alpha}$ the above equation can be written in a more suggestive form. However, the symmetric part of the variation with respect to the metric is identically zero (due to the lack of energy-momentum tensor). From the antisymmetric part of the variation with respect to the metric (see eq. (12) of Ref. [1]) we have

$$R_{\mu\nu} = -\lambda\left(g_{\mu\nu} + f_{\mu\nu}\right)$$
$$\Rightarrow R_{[\mu\nu]} = \left(\nabla_\alpha + 2T_\alpha\right)\left(T^\alpha_{\mu\nu} + T_\nu\delta^\alpha_\mu - T_\mu\delta^\alpha_\nu\right) = -\lambda f_{\mu\nu} \tag{18}$$

with T_α being the trace of the torsion tensor.

Let us now explore the role played by $f_{\mu\nu}$:

i) If $f_{\mu\nu}$ plays the role of the electromagnetic field we have a 1-form vector potential from which $f_{\mu\nu}$ is derived. Such an existence not necessarily follows *a priori* from the definition of $f_{\mu\nu}$,of course This fact leads to the usual Euler-Lagrange equations, where the variation is made with respect to the electromagnetic potential a_τ, namely

$$\frac{\delta\sqrt{G}}{\delta a_\tau} = \nabla_\rho \left(\frac{\partial\sqrt{G}}{\partial f_{\rho\tau}} \right) \equiv \nabla_\rho \mathbb{F}^{\rho\tau} = 0 \tag{19}$$

Explicitly

$$\nabla_\rho \left[\frac{\lambda^2 N^{\mu\nu} \left(\delta^\sigma_\mu f^\rho_{\ \nu} + \delta^\sigma_\nu f^\rho_{\ \mu} \right)}{2\mathbb{R}} \right] = 0\,, \tag{20}$$

where $N^{\mu\nu}$ is given by expression (120) in this chapter and (34) of ref.[1]. The set of equations to solve for this particular case is

$$\begin{aligned} R_{(\mu\nu)} = \overset{\circ}{R}_{\mu\nu} - T^{\alpha}_{\mu\rho} T^{\rho}_{\alpha\nu} &= -\lambda g_{\mu\nu} && (21a) \\ R_{[\mu\nu]} = \left(\nabla_\alpha + 2T_\alpha \right) \left(T^\alpha_{\mu\nu} + T_\nu \delta^\alpha_\mu - T_\mu \delta^\alpha_\nu \right) &= -\lambda f_{\mu\nu} && (21b) \\ \nabla_\rho \left[\frac{\lambda^2 N^{\mu\nu} \left(\delta^\sigma_\mu f^\rho_{\ \nu} + \delta^\sigma_\nu f^\rho_{\ \mu} \right)}{2\mathbb{R}} \right] &= 0 && (21c) \end{aligned}$$

where the quantities marked with the little circle ($^\circ$) are defined from the Christoffel connection (as in General Relativity). From this set (21), the link between T and f can be determined.

ii) If $f_{\mu\nu}$ has only the role to be the antisymmetric part of a fundamental (non-symmetric) tensor K: *i.e.*, $f_{\mu\nu}$ closed but not necessarily exact. Then, the variation of the geometrical Lagrangian $\delta_f\sqrt{G}$ gives the same information that $\delta_g\sqrt{G}$, that means that the remaining equations are

$$\begin{aligned} R_{(\mu\nu)} &= \overset{\circ}{R}_{\mu\nu} - T_{\mu\rho}^{\ \ \alpha} T_{\alpha\nu}^{\ \ \rho} = -\lambda g_{\mu\nu} && (22) \\ R_{[\mu\nu]} &= \left(\nabla_\alpha + 2T_\alpha \right) \left(T^\alpha_{\mu\nu} + T_\nu \delta^\alpha_\mu - T_\mu \delta^\alpha_\nu \right) = -\lambda f_{\mu\nu} && (23) \end{aligned}$$

3.1. Analysis and Reduction of the Dynamical Equations

An important equation appears in the two sets recently described, in (21b) and (23). This equation brings us a lot of information about the link between T and f .: the equation, $R_{[\mu\nu]} = -\lambda f_{\mu\nu}$, plus the condition $\nabla_\alpha T^\alpha_{\ \mu\nu} = 0$ lead immediately to

$$\nabla_\mu T_\nu - \nabla_\nu T_\mu = -\left(\lambda f_{\mu\nu} + 2T_\alpha T^\alpha_{\mu\nu} \right) \tag{24}$$

Then, the quantity that *naturally* appears in the RHS is the 'definition', in the current literature, of the *minimal coupling* electromagnetic tensor $\mathcal{F}_{\mu\nu}$, in a spacetime with torsion. Notice that $\nabla_\alpha T^\alpha_{\ \mu\nu} = 0$ is equivalent to

$$d^*T = 0.$$

in consequence the torsion is current free. Two cases naturally arise:

i) if we assume the existence of the potential vector we have

$$\nabla_\mu T_\nu - \nabla_\nu T_\mu \equiv \mathcal{F}_{\mu\nu} = -\lambda \left(\overbrace{\partial_\mu a_\nu - \partial_\nu a_\mu}^{f_{\mu\nu}} \right) - 2T_\alpha T^\alpha_{\mu\nu} \tag{25}$$

a link between a_ν and T_ν clearly appears: $T_\nu = -\lambda a_\nu$. The point to remark here is that, although in Ref. [12] the link between the trace of the torsion and the vector potential of the electromagnetic field was proposed, in our derivation this relation is derived automatically from a geometrical basis. Besides this, the suggestive form of $\mathcal{F}_{\mu\nu}$ as $F_{\mu\nu} + B_{\mu\nu}$ is remarkable, with $B_{\mu\nu}$ being a kind of 'background field' generated by the spacetime torsion.

ii) if $f_{\mu\nu}$ has the only role of being the antisymmetric part of a fundamental (non-symmetric) tensor K, it acquires a potential automatically, being therefore *an exact form*, where T_ν takes the role of vector potential. Clearly in this case f cannot be a potential for the torsion from this point of view (in a non-trivial topology it could be, of course).

From the above statements involving the 'trace' of the torsion, two ansatz appear as candidates for the torsion tensor structure: the 'tratorial' structure, $T^\alpha_{\mu\nu} \sim \left(\delta^\alpha_\mu a_\nu - \delta^\alpha_\nu a_\mu\right)$; and the 'product' structure, $T^\alpha_{\mu\nu} = k^\alpha f_{\mu\nu}$, where the vector k^α is, in general, eigenvector of the antisymmetric tensor $f_{\mu\nu}$ ..

The other possibility is to take $\nabla_\alpha T^\alpha{}_{\mu\nu} = -\lambda f_{\mu\nu}$ then $\nabla_\mu T_\nu - \nabla_\nu T_\mu = -2T_\alpha T^\alpha_{\mu\nu}$, but the interpretation in this case is not so clear as in the former. Even more, it probably leads us to a 'product structure' with the torsion tensor not fully antisymmetric.

3.2. A Potential for the Torsion

As was shown in [1], if we impose the restriction $T_{\alpha\beta\gamma} = T_{[\alpha\beta\gamma]}$ (*i.e.* totally antisymmetric torsion tensor), from eq. (2) for example, we note that only the antisymmetric part of the fundamental tensor $K_{\alpha\beta}$ determines completely the torsion tensor. Then, due to the assumption of a completely antisymmetric torsion tensor, the potential torsion $f_{\mu\nu}$ exists and arises in a natural way (the ∇ for the covariant derivative with respect to the full connection Γ). This potential torsion has the following properties:

$$f_{\mu\nu} = \overline{f}_{\mu\nu} = -f_{\nu\mu} \in \mathbb{HC}$$

$$\nabla_{[\rho} f_{\mu\nu]} = T_{\mu\nu\rho} = \varepsilon_{\mu\nu\rho\sigma} h^\sigma,$$

the last equality coming from the full antisymmetry of the Torsion field. Immediately we can see the following consequences:

i) the torsion is the dual of a vector h^σ

ii) from i), the existence in the spacetime of a completely antisymmetric tensor covariantly constant $\varepsilon_{\mu\nu\rho\sigma}$ $(\nabla\varepsilon = 0)$.

Notice that the choice for the real nature of the metric and the pure hypercomplex potential tensor comes from the Hermitian nature of the theory, as was clearly explained in [1].

For the variational equations (in the Palatini's sense [4, 13], see eqs. (12) and (13) of ref.[1]), despite their simple and compact form, it is necessary to show what is their deep physical and geometrical meaning.

In expression (13) of ref. [1] we have a highly nonlinear dynamical (propagating) equation for the torsion field, where the variation was performed with respect to the potential $f_{\mu\nu}$, and having a nonlinear term proportional to $f_{\mu\nu}$ which plays the role of current for the $\mathbb{T}^{\rho\sigma\tau}$. Thus, the potential 2-form is associated nonlinearly to the torsion field as its source, similarly to the relation between the electromagnetic field and the spin in particle physics.

For the expression (12) of ref. [1], it is useful to split the equation into the symmetric and antisymmetric parts using $R_{\mu\nu}$ explicitly as before

$$R_{(\mu\nu)} = \overset{\circ}{R}_{\mu\nu} - T_{\mu\rho}{}^{\alpha} T_{\alpha\nu}{}^{\rho} = -\lambda g_{\mu\nu} \tag{26}$$

$$\begin{aligned} R_{[\mu\nu]} &= \overset{\circ}{\nabla}_{\alpha} T^{\alpha}{}_{\mu\nu} = -\lambda f_{\mu\nu} \\ &= \nabla_{\alpha} T^{\alpha}{}_{\mu\nu}, \end{aligned} \tag{27}$$

the last equality coming from the full antisymmetry of the torsion.

Notice the important fact that $-\lambda f_{\mu\nu}$ is the 'current' for the torsion field as the 1-form potential vector a_{μ} acts as the current of the electromagnetic field $f_{\mu\nu}$ in the standard theory: $\nabla_{\alpha} f^{\alpha}{}_{\mu} = J_{\mu}$ (constants absorbed into J_{μ})

The symmetric part (26) can be written in a 'GR suggestive' fashion

$$\overset{\circ}{R}_{\mu\nu} = -\lambda g_{\mu\nu} + T_{\mu\rho}{}^{\alpha} T_{\alpha\nu}{}^{\rho}. \tag{28}$$

Notice that the above equation has the aspect of the Einstein equations with the cosmological term modified by the torsion symmetric term $T_{\mu\rho}{}^{\alpha} T_{\alpha\nu}{}^{\rho}$. This can be interpreted as the energy of the gravitational field itself – see [1].

The second antisymmetric part (27) is more involved. We can rewrite it using the language of differential forms that, beside their symbolic and conceptual simplicity, allows one to check consistency and covariance step by step. Hence, we write

$$\begin{aligned} \nabla_{\alpha} T^{\alpha}{}_{\mu\nu} &= -\lambda f_{\mu\nu} \\ d^{*}T &= -\lambda^{*} f. \end{aligned} \tag{29}$$

Now, using $T = {}^{*}h$ we have

$$dh = -\lambda^{*} f \Rightarrow {}^{*}f = -\frac{1}{\lambda}\, dh. \tag{30}$$

In a more familiar form

$$\nabla_{\mu} h_{\nu} - \nabla_{\nu} h_{\mu} = -\lambda\, {}^{*}f_{\mu\nu}, \tag{31}$$

and then using again $T = df =^{*} h$ and (29) follows

$$d^{*}f = 0 \tag{32}$$

and, fundamentally,

$$df = -\frac{1}{\lambda} d^{*}dh = T = {}^{*}h \tag{33}$$

$$d^{*}dh = -\lambda\, {}^{*}h. \tag{34}$$

We can recognize here the Laplace-de Rham operator that helps us to write the wave covariant equation

$$[(d\delta + \delta d) + \lambda]^{*}\, h = 0 \tag{35}$$
$$(\Delta + \lambda)^{*}\, h = 0$$

If we start with the potential it is not difficult to see that an equivalent equation can be found

$$(\Delta + \lambda)^{*}\, f = 0 \tag{36}$$

Notice that equation (35), coming from (30), is a consequence of the Tfh-relation, $(T = df = {}^{*}h)$. Now expression (36) comes directly from (29).

Thus, the geometric interplay between the involved geometrical entities (fields) can be represented with the following diagram

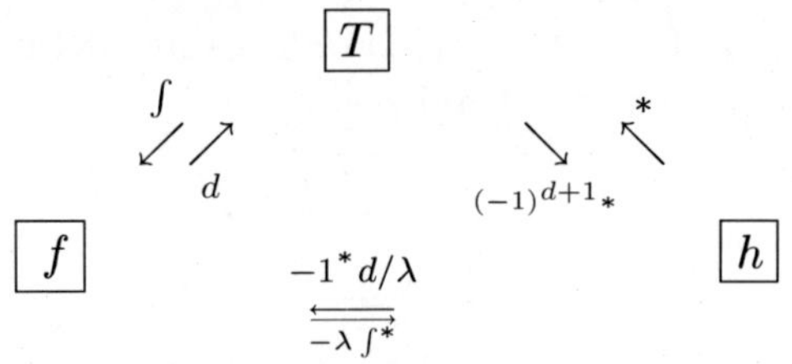

(37)

4. Exact Solutions in the New UFT Theory

The main motivation of this Section is clear: by studying wormhole solutions beyond Einstein equations we must to identify the fields in our 'theoretical arena' with the possible matter fields that in the standard General Relativistic framework are added "by hand" through the energy-momentum tensor. Then, let us now construct wormhole solutions in the viewpoint of the UFT model introduced here. The action in four dimensions is given by

$$S = -\frac{1}{16\pi G} \int d^4x \sqrt{det\,|G_{\mu\nu}|} \tag{38}$$

$$\mathbb{R} \equiv \sqrt{\gamma^4 - \frac{\gamma^2}{2}\overline{G}^2 - \frac{\gamma}{3}\overline{G}^3 + \frac{1}{8}\left(\overline{G}^2\right)^2 - \frac{1}{4}\overline{G}^4} \tag{39}$$

[2*] In order to be consistent with the action of the Hodge operator $(*)$, in this paragraph, we assume an even number of dimensions

4.1. Totally Antisymmetric Torsion

Scalar curvature R and the torsion 2-form field $T^{a}_{\mu\nu}$ with a $SU(2)$ −Yang-Mills structure are defined in terms of the affine connection $\Gamma^{\lambda}_{\mu\nu}$ and the $SU(2)$ potential torsion $f^{a}_{\ \mu}$ by

$$R = g^{\mu\nu} R_{\mu\nu} \qquad R_{\mu\nu} = R^{\lambda}_{\mu\lambda\nu} \tag{40}$$

$$R^{\lambda}_{\mu\lambda\nu} = \partial_{\nu}\Gamma^{\lambda}_{\ \mu\rho} - \partial_{\rho}\Gamma^{\lambda}_{\ \mu\nu} + ...$$

$$T^{a}_{\ \mu\nu} = \partial_{\mu} f^{a}_{\ \nu} - \partial_{\nu} f^{a}_{\ \mu} + \varepsilon^{a}_{bc} f^{b}_{\ \mu} f^{c}_{\ \nu}$$

G and Λ are the Newton gravitational constant and the cosmological constant, respectively. From the last equation for the Torsion 2-form, the potential $f^{a}_{\ \mu}$ must be proportional to the antisymmetric part of the affine connection $\Gamma^{\lambda}_{\mu\nu}$, as in the Strauss-Einstein UFT. Similarly to the case of Einstein-Yang-Mills systems, for our new UFT model it can be interpreted as a prototype of gauge theories interacting with gravity (*e.g.* QCD, GUTs, etc.). After varying the action, we obtain the gravitational 'Einstein-Eddington-like' equation

$$R_{\mu\nu} = -2\lambda \left(g_{\mu\nu} + f_{\mu\nu}\right) \tag{41}$$

and the field equation for the torsion 2-form in differential form

$$d^{*}\mathbb{T}^{a} + \frac{1}{2}\varepsilon^{abc}\left(f_{b} \wedge^{*} \mathbb{T}_{c} -^{*} \mathbb{T}_{b} \wedge f_{c}\right) = \mathbb{F}^{a} \tag{42}$$

where we define, as usual,

$$\mathbb{T}^{a}_{\ bc} \equiv \frac{\partial L_{G}}{\partial T_{a}^{\ bc}}, \mathbb{F}^{a}_{\ bc} \equiv \frac{\partial L_{G}}{\partial F_{a}}.$$

We are going to seek for a classical solution of (41) and (42) with the following spherically symmetric ansatz for the metric and gauge connection

$$ds^{2} = d\tau^{2} + a^{2}\left(\tau\right)\sigma^{i} \otimes \sigma^{i} \equiv d\tau^{2} + e^{i} \otimes e^{i}. \tag{43}$$

Here τ is the euclidean time and the dreibein is defined by $e^{i} \equiv a\left(\tau\right)\sigma^{i}$. The gauge connection is

$$f^{a} \equiv f^{a}_{\mu} dx^{\mu} = h\sigma^{a}, \tag{44}$$

for $a = 1, 2, 3$, and for $a = 0$ we have

$$f^{0} \equiv f^{0}_{\mu} dx^{\mu} = s\sigma^{0}. \tag{45}$$

This choice for the potential torsion is, as we will show soon, the most general and consistent from the physical and mathematical point of view due to the symmetries involved in the problem.

The σ^{i} 1-form satisfies the $SU(2)$ Maurer-Cartan structure equation

$$d\sigma^{a} + \varepsilon^{a}_{\ bc}\sigma^{b} \wedge \sigma^{c} = 0 \tag{46}$$

Notice that in the ansatz the frame and isospin indices are identified as for the case with the non-abelian-Born-Infeld (NBI) Lagrangian of ref. [5]. The torsion 2-form

$$T^{\gamma} = \frac{1}{2} T^{\gamma}_{\ \mu\nu} dx^{\mu} \wedge dx^{\nu} \tag{47}$$

becomes

$$T^a = df^a + \frac{1}{2}\varepsilon^a_{\ bc} f^b \wedge f^c \qquad (48)$$
$$= \left(-h + \frac{1}{2}h^2\right)\varepsilon^a_{\ bc}\sigma^b \wedge \sigma^c$$

Note that f^0 plays no role here because we take simply $ds = 0$ (the $U(1)$ component of $SU(2)$, in principle, does not belong to the space spherical symmetry), and the expression for the torsion is analogous to the non-abelian 2-form strength field of [5].

It is important to note that, when we go from the Lorentzian to the Euclidean gravitational regime, $it \rightarrow \tau$, the torsion passes from the field of the *Hypercomplex* to the *Complex* numbers[1].

Inserting T^a from (48) into the dynamical equation (42) we obtain

$$\begin{array}{l} d^*\mathbb{T}^a + \frac{1}{2}\varepsilon^{abc}\left(f_b \wedge^* \mathbb{T}_c -^* \mathbb{T}_b \wedge f_c\right) = {}^*\mathbb{F}^a \\ (-2h + h^2)(1-h)d\tau \wedge e^b \wedge e^c = -2\lambda d\tau \wedge e^b \wedge e^c, \end{array} \qquad (49)$$

where

$$^*\mathbb{T}^a \equiv \frac{\lambda\sqrt{|g|}}{\sqrt{3}} hA(-2h + h^2)d\tau \wedge \frac{e^a}{a^2} \qquad (50)$$

$$^*\mathbb{F}^a = -\frac{2\lambda^2\sqrt{|g|}}{\sqrt{3}} hA \frac{d\tau \wedge e^b \wedge e^c}{a^3} \qquad (51)$$

$$A \equiv \lambda^4\left[(1+\alpha)^2 + \alpha/2\right], \qquad (52)$$

and

$$\alpha = \frac{1}{2}\left(s^2 + 3h^2\right), \qquad (53)$$

from expression (49) we have an algebraic cubic equation for h

$$(-2h + h^2)(1-h) + 2\lambda = 0 \qquad (54)$$

We can see that, in contrast with our previous work with a dualistic theory [5]where the NBI energy-momentum tensor of Born-Infeld was considered, there exist three non trivial solutions for h, depending on the cosmological constant λ. In this preliminary analysis of the problem, only the values of h that make the quantity $\left(-h + \frac{1}{2}h^2\right) \in \mathbb{R}$ are relevant for our proposals. This fact is due to the pure imaginary character of T in the euclidean framework and to compare with the NABI wormhole solution of our previous work. (The case $h \in \mathbb{C}$ will be the focus of a further article [9]).

As the value of $h \in \mathbb{R}$ is -1 and in four spacetime dimensions we have $\lambda = |1-d| = 3$, then

$$T^a_{bc}\big|_{h_1} = \frac{3}{2}\frac{\varepsilon^a_{bc}}{a^2}; \qquad T^a_{0c} = 0 \qquad (55)$$

That is, only the magnetic field is non vanishing while the electric field vanishes. An analogous feature can be seen in the solution of Giddings and Strominger and in our previous

[1]Geometrically, multiplication of hypercomplex numbers preserves the (square) Minkowski norm $(x^2 - y^2)$ in the same way that multiplication of complex numbers preserves the (square) Euclidean norm $(x^2 + y^2)$

work [5]. Substituting the expression for the torsion 2-form (55) [1] into the symmetric part of the variational equation

$$R_{(\mu\nu)} = \overset{\circ}{R}_{\mu\nu} - T_{\mu\rho}{}^{\alpha}T_{\alpha\nu}{}^{\rho} = -2\lambda g_{\mu\nu}, \tag{56}$$

we reduce the equation (26) to an ordinary differential equation for the scale factor a,

$$\left[\left(\frac{\dot{a}}{a}\right)^2 - \frac{1}{a^2}\right] = \frac{2\lambda}{3} - \frac{9}{2a^4}, \tag{57}$$

$$\frac{Ln\left[1 + 4a^2 + 2\sqrt{-9 + 2a^2 + 4a^4}\right]}{2\sqrt{2}} = \tau - \tau_0 \tag{58}$$

$$T_{\mu\rho}{}^{\alpha}T_{\alpha\nu}{}^{\rho} = \frac{\left(-h + \frac{1}{2}h^2\right)^2}{a^4}2\delta_{\mu\nu} = \frac{9}{2a^4}\delta_{\mu\nu}$$

There are 2 values for the scale factor a: max. and min. respectively, namely

$$a = \mp\frac{e^{-\sqrt{2}(\tau-\tau_0)}\sqrt{37 - 2e^{2\sqrt{2}(\tau-\tau_0)} + e^{4\sqrt{2}(\tau-\tau_0)}}}{2\sqrt{2}} \tag{59}$$

Expression (59) for the scale factor a is depicted in the Figure 1 for the real value of h.

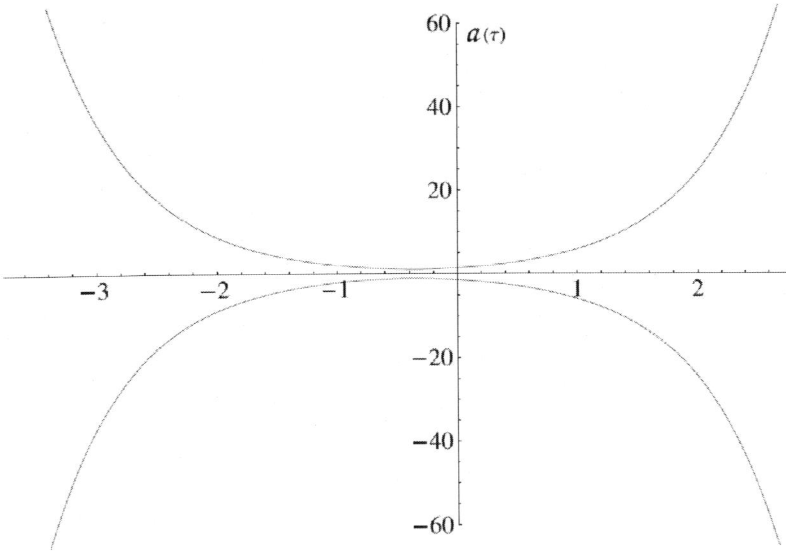

Figure 1. Wormhole solution profile for values of the Euclidean time and torsion $\tau_0 = 1$ and $T^a_{bc} = (3/2)\epsilon^a_{bc}$.

As it is easily seen from (59), the scale factor has an exponentially growing behavior, in sharp contrast to the wormhole solution from our previous work with the ‘dualistic’ non-abelian BI theory (figure 2). Also, for this particular value of the torsion, the wormhole tunneling interpretation (in the sense of the Coleman’s mechanism) is fulfilled.

[1] in the tetrad: $\overset{\circ}{R}_{00} = -3\frac{\ddot{a}}{a}$, $\overset{\circ}{R}_{ab} = -\left[\frac{\ddot{a}}{a} + 2\left(\frac{\dot{a}}{a}\right)^2 - \frac{2}{a^2}\right]$

Now we need to see what happens with the equation (29) in the particular case under consideration. Equation (29) takes the form

$$\begin{aligned} d^{*}T^{a}+\tfrac{1}{2}\varepsilon^{abc}\left(f_{b}\wedge^{*}T_{c}-^{*}T_{b}\wedge f_{c}\right) &= -2\lambda\,{}^{*}f^{a} \\ (-2h+h^{2})(1-h)d\tau\wedge e^{b}\wedge e^{c} &= -2\lambda d\tau\wedge e^{b}\wedge e^{c} \end{aligned} \tag{60}$$

$$^{*}T^{a}\equiv h(-2h+h^{2})d\tau\wedge\frac{e^{a}}{a^{2}} \tag{61}$$

$$^{*}f^{a}=-h\frac{d\tau\wedge e^{b}\wedge e^{c}}{a^{3}} \tag{62}$$

Thus, we arrived to the same equation for λ as in (54), corroborating the consistency of the procedure.

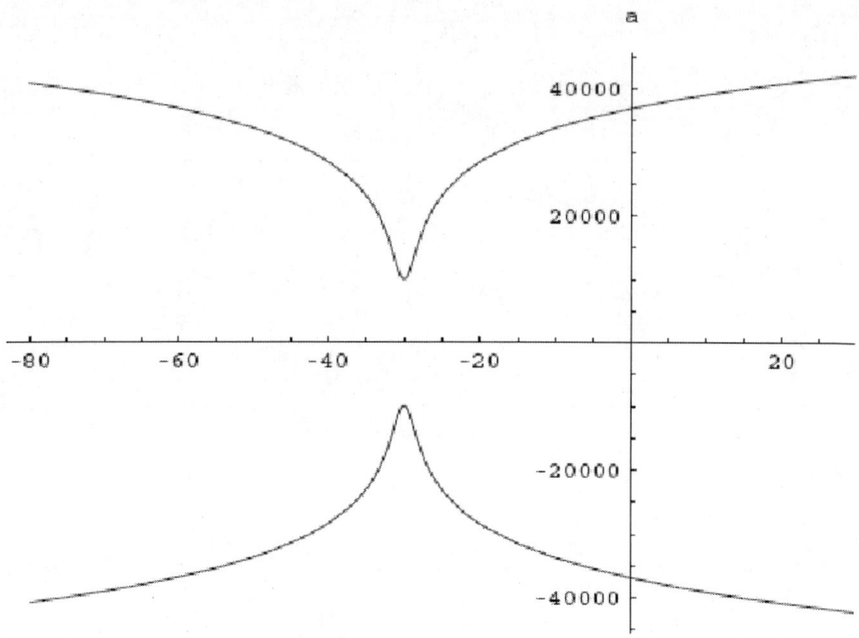

Figure 2. Profile of the instanton – wormhole solution for $r_0 = 70$ and $a = 40$.

4.2. 'Tratorial' Torsion

To begin with, let us consider the problem involving the set of equations (21), with the usual definition for the $SU(2)$ electromagnetic field strength

$$f^{\gamma}=\frac{1}{2}f^{\gamma}_{\ \mu\nu}dx^{\mu}\wedge dx^{\nu} \tag{63}$$

and, as before, we are going to seek for a classical solution of (21), with the following spherically symmetric ansatz for the metric and gauge connection

$$ds^{2}=d\tau^{2}+a^{2}\left(\tau\right)\sigma^{i}\otimes\sigma^{i}\equiv d\tau^{2}+e^{i}\otimes e^{i}. \tag{64}$$

Here τ is the euclidean time and the dreibein is defined by $e^i \equiv a(\tau)\sigma^i$. However, in the case of the set (21) we assumed that the 2-form f^γ comes from a 1-form potential A where, similarly to the non abelian Born-Infeld model of Ref. [5], it is defined as $A^a \equiv A^a_\mu dx^\mu = h\sigma^a$.

The important fact in this case is that we know that the σ^i 1-form satisfies the $SU(2)$ Maurer-Cartan structure equation, as fundamental geometrical structure of the non-abelian electromagnetic field

$$d_{su(2)}\sigma^a + \varepsilon^a_{\ bc}\sigma^b \wedge \sigma^c = 0, \tag{65}$$

but now, due to the identification assumed in (64), $e^i \equiv a(\tau)\sigma^i$,

$$de^a = T^a - e^a_{\ b} \wedge \sigma^b. \tag{66}$$

Here, we distinguish between the exterior derivatives in the spacetime with torsion and in the $SU(2)$ group manifold. It is clear that a question of compatibility involving the identification of the gauge group with the geometrical structure of the spacetime with torsion certainly exists. From (65-66) we see that

$$\partial_\tau a d\tau \wedge \sigma^a - a\varepsilon^a_{\ bc}\sigma^b \wedge \sigma^c = T^a - e^a_{\ b} \wedge \sigma^b \tag{67}$$

If

$$e^a_{\ b} = -\varepsilon^a_{\ bc}\sigma^c \tag{68}$$

and

$$T^a = \delta^a_b (\partial_\tau a)\, d\tau \wedge \sigma^b \tag{69}$$

then the spacetime and gauge group are fully compatible, and the relation

$$d\sigma^a + \varepsilon^a_{\ bc}\sigma^b \wedge \sigma^c = 0 \tag{70}$$

is restored. Hence, the general form assumed for the torsion field, due the symmetry conditions prescribed above, is

$$T^\alpha_{\beta\gamma} = \xi\left(\delta^\alpha_\beta u_\gamma - \delta^\alpha_\gamma u_\beta\right) + \varsigma h_\delta \varepsilon^{\delta\alpha}{}_{\beta\gamma} \quad (\xi, \varsigma : const.) \tag{71}$$

Notice that the condition of compatibility that imposes such type of the torsion tensor appears in a natural manner, without the introduction of any extra scalar field (dilaton) or the need to pass to another frame (i.e.: Jordan, Einstein, etc.). This fact happens in order to restore the behaviour of the volume form of the spacetime with respect to the covariant derivative. Moreover, if we have to continue without making the correspondences (68-69), the equations of motion for the electromagnetic field itself would bring automatically these conditions (see below).

In the Hosoya and Ogura (HO) ansatz, the frame and isospin indexes are identified as for the case with the NBI Lagrangian of ref.[5]. The electromagnetic field 2-form reads

$$\begin{aligned} f^a &= dA^a + \frac{1}{2}\varepsilon^a_{\ bc}A^b \wedge A^c \\ &= h\delta^a_b (\partial_\tau \ln a)\, d\tau \wedge \sigma^b + h\frac{T^a}{a} - \left(-h + \frac{1}{2}h^2\right)\varepsilon^a_{\ bc}\sigma^b \wedge \sigma^c \\ &= \left(-h + \frac{1}{2}h^2\right)\varepsilon^a_{\ bc}\sigma^b \wedge \sigma^c, \end{aligned} \tag{72}$$

where in the last equality conditions (68-69) have been assumed.

The dynamical eqs. are

$$\mathbb{F}^{a}_{\ bc} \equiv \frac{\partial L_G}{\partial F_a}$$

$$^{*}\mathbb{F}^{a} \equiv \frac{\lambda\sqrt{|g|}}{\sqrt{3}} h\mathbb{A}(-2h+h^{2})d\tau \wedge \frac{e^{a}}{a^{2}} \equiv Mh(-2h+h^{2})d\tau \wedge \frac{e^{a}}{a^{2}} \tag{73}$$

Inserting it in the Yang-Mills type field equation (21c) we obtain

$$\begin{aligned} d^{*}\mathbb{F}^{a} + \tfrac{1}{2}\varepsilon^{abc}\left(A_b \wedge^{*} \mathbb{F}_c -^{*} \mathbb{F}_b \wedge A_c\right) = 0 \\ = Mh\, d\tau \wedge \sigma^{b} \wedge \sigma^{c}\left(-2h+h^{2}\right)(h-1) \end{aligned} \tag{74}$$

$$\mathbb{A} \equiv \lambda^{4}\left[(1+\alpha)^{2} + \alpha/2\right]$$

Then, there exists a non trivial solution: $h = 1$, (with $s = 0 \in \mathbb{A}$ as in [1]). The electromagnetic field is immediately determined, and is as in the non abelian Born-Infeld model of our previous reference and in the result of Giddings and Strominger, namely

$$f^{a}_{\ bc} = -\frac{\varepsilon^{a}_{\ bc}}{a^{2}} \qquad f^{a}_{\ 0c} = 0, \tag{75}$$

we have only magnetic field.

Now considering only the torsion as (3b) it is identically null due the magnetic character of f^{a} and the particular form of the symmetric coefficients of the connection. Inserting the torsion (69) into the (21a), as in the previous section, we obtain

$$\left[\left(\frac{\dot{a}}{a}\right)^{2} - \frac{1}{a^{2}}\right] = \frac{\lambda}{3} \tag{76}$$

Integration of this last expression immediately leads to

$$a(\tau) = \left(\frac{\lambda}{3}\right)^{-1/2} \sinh\left[\left(\frac{\lambda}{3}\right)^{1/2}(\tau - \tau_0)\right] \tag{77}$$

It is quite evident that this particular case does not lead to wormhole configurations: we get only a 'eternal expansion', with $a(\tau_0) = 0$ (the origin of the euclidean time Fig. 3).

Now considering only the product form for the torsion, (21c) doesn't change but (21b) takes the form of a wave equation for the scale factor

$$\left[\Box a + (\partial_0 a)\left(\partial^{0} a\right)\right] = \lambda,$$

as $T^{\alpha}_{\beta\gamma} = \varsigma k^{\alpha}\varepsilon_{\beta\gamma} \rightarrow \varepsilon_{ab}\left(\partial^{0} a\right)$. Is not difficult to see that the $su(2)$ structure of the electromagnetic tensor is of some manner transferred to the structure of the torsion. But here we come into conflict because the system of (21) turns out to be *overdetermined*: we need probably more freedom in the ansatz for f^{a}_{bc} (s$\neq$ 0, or h=h(τ)). This issue will be studied in near future [9].

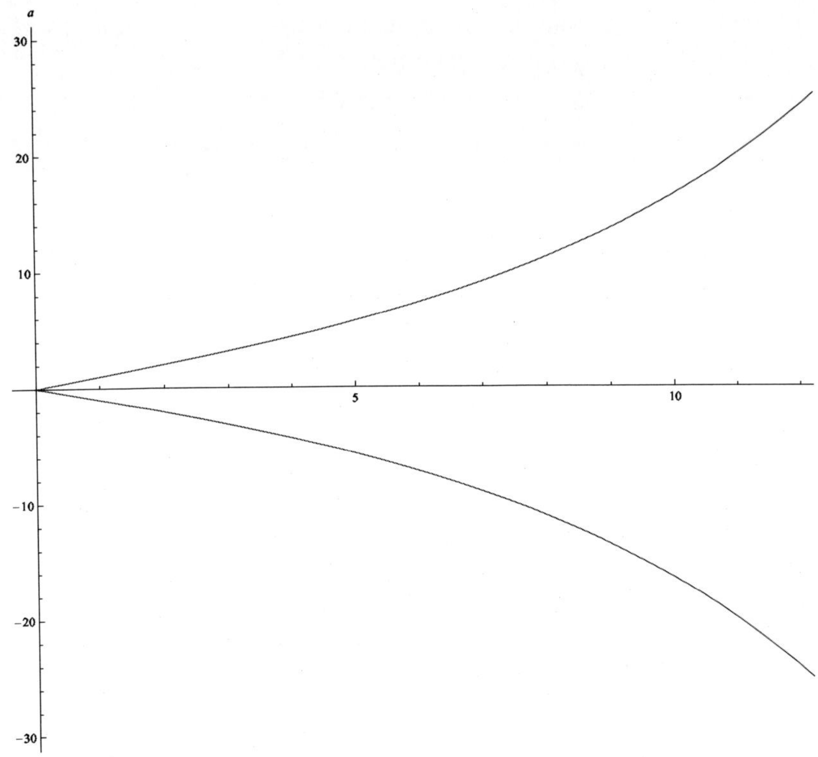

Figure 3. Cosmological solution shape for values of $|\lambda| = 3$, $d = 4$, and the Euclidean time $\tau > 0$.

4.3. General Case

Let us to assume the full form (71) for T^a

$$
{}^*\mathbb{F}^a \equiv Mh\left\{h\delta^a_d\left(\partial_\tau \ln a\right)\varepsilon^{d0}_{\ ed}\sigma^e\wedge\sigma^d+\right.
\tag{78}
$$

$$
\left.+\frac{h}{a}\left[\xi\left(\delta^a_i u_j-\delta^a_j u_i\right)+\varsigma h_\delta \varepsilon^{\delta a}{}_{ji}\right]\varepsilon^{ij}_{kl}\omega^k\wedge\omega^l+(-2h+h^2)\varepsilon^a_{bc}\varepsilon^{bc}_{0d}d\tau\wedge\sigma^a\right\}.
$$

In order to avoid the cumbersome expression in the second term due to the standard orthonormal splitting, $ij = 0, a, b, c$ and the ω^k are the corresponding 1-forms $(d\tau, \sigma^a..)$ whenever the case. The YM type equation can be written as

$$
d^*\mathbb{F}^a+\frac{1}{2}\varepsilon^{abc}\left(A_b\wedge^*\mathbb{F}_c-^*\mathbb{F}_b\wedge A_c\right)=
\tag{79}
$$

$$
Mh\left\{\left[h\delta^a_b\left(\partial_\tau\partial_\tau \ln a\right)+\partial_\tau\left(\frac{h}{a}\left(\xi\left(\delta^a_b u_0-\delta^a_0 u_b\right)+\varsigma h_c\varepsilon^{ca}{}_{b0}\right)\right)\right]\varepsilon^{b0}{}_{ed}d\tau\wedge\sigma^e\wedge\sigma^d+\right.
$$

$$
\left.\left[h\delta^a_b\left(\partial_\tau \ln a\right)+\frac{h}{a}\left(\xi\left(\delta^a_b u_0-\delta^a_0 u_b\right)+\varsigma h_c\varepsilon^{ca}{}_{b0}\right)\right]2d\left(\sigma^e\wedge\sigma^d\right)\right\}+
$$

$$
+M\left[\frac{h}{a}\left(\xi\left(\delta^a_b u_0-\delta^a_0 u_b\right)+\varsigma h_c\varepsilon^{ca}{}_{b0}\right)+(-2h+h^2)\right](h-1)d\tau\wedge\sigma^b\wedge\sigma^c=0
$$

from the above equation we obtain information about the determination of the f field and the torsion field, as in the previous cases. The first term

$$\left[h\delta_b^a\left(\partial_\tau\partial_\tau \ln a\right)+\partial_\tau\left(\frac{h}{a}\left(\xi\left(\delta_b^a u_0-\delta_0^a u_b\right)+\varsigma h_c\varepsilon^{ca}{}_{b0}\right)\right)\right]=0 \tag{80}$$

leads immediately to

$$\begin{aligned}\left[\eta_{ab}\partial_0 a+\left(\xi\left(\eta_{ab}u_0-\eta_{a0}u_b\right)+\varsigma h_c\varepsilon^{ca}{}_{b0}\right)\right]&=\Xi_{ab0}^A+\Xi_{ab0}^s \\ \Rightarrow \varsigma h_c\varepsilon^c{}_{ab0}&\equiv\Xi_{ab0}^A \\ \Rightarrow \eta_{ab}\partial_0 a+\xi\left(\eta_{ab}u_0-\eta_{a0}u_b\right)&=\Xi_{ab0}^S,\end{aligned} \tag{81}$$

where the tensor

$$\Xi_{ab0}=\Xi_{ab0}^A+\Xi_{ab0}^S$$

is independent of the time, and the superscripts A and S indicate the totally antisymmetric/symmetric parts. Then, the second and third equalities above follow.

It is not difficult to see that contracting indices, tracing and considering the symmetries involved, we obtain explicitly

$$T_{b0}^a=\delta_{[b}^a\partial_{0]}a-a\widetilde{\Xi}^{Sa}{}_{b0}+\Xi_{ab0}^A \tag{82}$$

$$T_{bc}^a=-a\widetilde{\Xi}^{Sa}{}_{bc}+\varsigma h_0\varepsilon^{0a}{}_{bc} \tag{83}$$

$$T_{bc}^0=-a\widetilde{\Xi}^{S0}{}_{bc}+\varsigma h_c\varepsilon^{c0}{}_{bc} \tag{84}$$

where the integration tensor (independent of time) are related with u_i and $\widetilde{\Xi}^{Sj}{}_{kl}$ $(ij..=0,a,b,c)$ as follows:

$$\begin{gathered} u_c=-\frac{a\Xi_c^S}{2\xi},\quad u_0=-\frac{1}{2\xi}\left(3\partial_0 a+a\Xi_c^S\right), \\ \Xi_c^S\equiv\Xi^{Sj}{}_{cj},\quad \Xi_0^S\equiv\Xi^{Sj}{}_{0j}\quad\text{and}\quad \widetilde{\Xi}_{kl}^{Sj}\equiv\frac{-1}{2}\left(\delta_k^j\Xi_l^S-\delta_l^j\Xi_k^S\right)\end{gathered} \tag{85}$$

The last term, however, indicates that there exists a simpler solution with $h=1$, as in the previous case for the non-abelian f. Then

$$f_{bc}^a=-\frac{\varepsilon^a{}_{bc}}{a^2},\quad f_{b0}^a=0$$

again, and the second is identically zero because of the symmetry of the torsion 2-form with respect to the tetrad defined by (64). Now the question is if the system of equations is overdetermined or not.(k^a and a are without determination). To this end, we put the information of expressions (82-84) into the second equation of the set (21b). Again, from the equations

$$\nabla_i T_{ab}^i+2T_iT_{ab}^i=-\lambda f_{ab}^c e_c \tag{86}$$

$$\nabla_i T_{ao}^i+2T_iT_{a0}^i=0 \tag{87}$$

the symmetry fix the torsion tensor components to

$$T^{a}_{b0} = \delta^{a}_{[b}\partial_{0]}a \tag{88}$$

$$T^{a}_{bc} = -a\widetilde{\Xi}^{Sa}{}_{bc} + \varsigma h_0 \varepsilon^{0a}{}_{bc} \tag{89}$$

$$T^{0}_{bc} = 0. \tag{90}$$

Expression (87) becomes a null identity, and (86) becomes

$$4T_i T^{i}_{ab} = 4a\widetilde{\Xi}^{S}_{c}\overbrace{\left(-a\widetilde{\Xi}^{Sa}{}_{bc} + \varsigma h_0 \varepsilon^{0a}{}_{bc}\right)}^{T^{a}_{bc}} = -\lambda f^{c}_{ab} e_c \tag{91}$$

$$\Rightarrow 4\left(a^2\widetilde{\Xi}^{S}_{c}\widetilde{\Xi}^{Sc}{}_{ab} - a\widetilde{\Xi}^{S}_{c}\varsigma h_0 \varepsilon^{0c}\varepsilon_{ab}\right) = -\lambda f^{c}_{ab} e_c$$

$$a\widetilde{\Xi}^{S}_{c}\varsigma h_0 \varepsilon^{0c}\varepsilon_{ab} = -\lambda f^{c}_{ab} e_c = \frac{\lambda \varepsilon^{c}_{ab}}{a^2} e_c$$

$$\widetilde{\Xi}^{S}_{c}\varsigma h_0 \varepsilon^{0c}\varepsilon_{ab} = \frac{\lambda \varepsilon^{c}_{ab}}{a^2}\sigma_c,$$

where in the last line we used the property $\widetilde{\Xi}^{S}_{c}\widetilde{\Xi}^{Sc}{}_{ab} = \Xi^{S}_{c}\left(\delta^{c}_{a}\Xi^{S}_{b} - \delta^{c}_{b}\Xi^{S}_{a}\right) \equiv 0$ (see definitions above).

Now it is easy to see that squaring both sides of (91), and from (90) we obtain

$$h_0 = \frac{\lambda \sigma_0}{a^2\left|\widetilde{\Xi}^{S}_{c}\right| 2\varsigma}, \qquad h_c = \frac{\lambda\left|\widetilde{\Xi}^{S}_{c}\right| a\sigma_c}{2\varsigma},$$

and analogously to the previous cases, from the (21a), the equation to be integrated takes the form

$$\frac{da}{d\tau} = \pm\frac{4}{5}\left[1 + \frac{\lambda}{3}a^2 + \frac{2}{3}a^4\left|\widetilde{\Xi}^{S}_{c}\right|^2 + \frac{3}{8}\left(\frac{\lambda}{\left|\widetilde{\Xi}^{S}_{c}\right| a}\right)^2\right]^{1/2}.$$

One interesting case when the above equation can be integrated exactly is precisely when $d = 4$. This condition, besides improving the integrability condition of the equation, fixes $\left|\widetilde{\Xi}^{S}_{c}\right|^2 > 3/2$. The scale factor $a(\tau)$ takes then the following form

$$a(\tau) = \sqrt{B + (A - B)Tanh^2\left[\frac{(\tau - \tau_0)\sqrt{(A - B)}}{2}\right]},$$

where A and B are nonlinear functions of the norm square $\left|\widetilde{\Xi}^{S}_{c}\right|^2$. The explicit form of these functions is not crucial: only the bound for $\left|\widetilde{\Xi}^{S}_{c}\right|^2 > 3/2$ needs to be preserved (Also through the normalization of A and B into the graphic representation in Fig. 4). Notice that the spacetime is asymptotically Minkowskian with a throat $a(\tau_0) = \sqrt{B}$ (the values of the constants have been selected according the previous remarks). Other possibilities, not enumerated here, lead to spacetimes with cyclic singularities due to transcendental functions appearing in the denominator of the expression for the scale factor $a(\tau)$. This issue will also be the focus of a future discussion elsewhere [9].

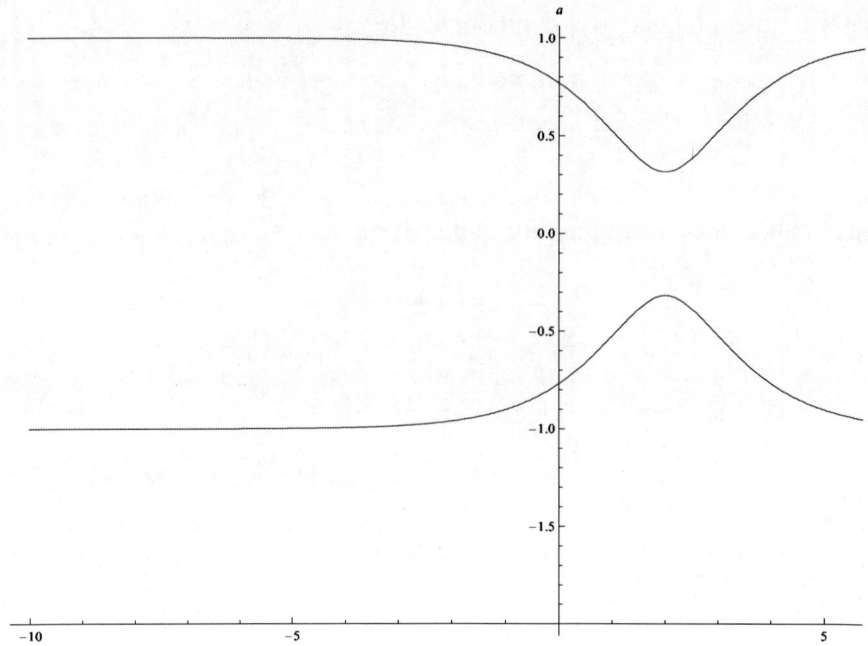

Figure 4. Cosmological solution profile for values of $|\lambda| = 3$, $d = 4$.

4.4. Coexistence of Both Types of Torsion in Cosmological Spacetimes

It is interesting to note that in reference [6] the field equations of vacuum quadratic Poincaré gauge field theory (QPGFT) were solved for purely null tratorial torsion. The author there describes the contortion tensor for such a case as

$$K_{\lambda\mu\nu} = -2(g_{\lambda\mu}a_\nu - g_{\lambda\nu}a_\mu)$$

However, the important thing is that the author discusses the relationship between this class (tratorial) and a similar class of solutions with null axial vector torsion, arriving to the conclusion that cosmological solutions with different type of torsion are forbidden. The main reason for this situation can have two different origins: the specific theory and action (QPGFT), or the Newman-Penrose method used in the computations that work, as is well know, with null geometric quantities. Here we show that this problem does not arise in our theory.

5. The Underlying Dirac Structure of the Spacetime Manifold

The real structure of the Dirac equation in the form

$$(\gamma_0 p_0 - i\gamma \cdot \mathbf{p})\,\mathbf{u} = m\mathbf{v} \tag{92}$$

$$(\gamma_0 p_0 + i\gamma \cdot \mathbf{p})\,\mathbf{v} = m\mathbf{u} \tag{93}$$

with

$$\gamma_0 = \begin{pmatrix} \sigma_0 & 0 \\ 0 & \sigma_0 \end{pmatrix}, \quad \gamma = \begin{pmatrix} 0 & -\sigma \\ \sigma & 0 \end{pmatrix}, \tag{94}$$

where σ are the Pauli matrices and $\mathbf{p} = (\widehat{p}_1, \widehat{p}_2, \widehat{p}_3)$, determines a $4D$ real vector space with G as its automorphism, such as $G \subset L(4)$. This real vector space can be made to coincide with the tangent space to the spacetime manifold M, being the idea of this. The principal fiber bundle (PFB) $P(G, M)$ with the structural group G, determines the (Dirac) geometry of the spacetime. We assume now G with the general form

$$G = \begin{pmatrix} A & B \\ -B & A \end{pmatrix}, \quad G^{+}G = I_4 \tag{95}$$

with A and B, 2×2 matrices. There also exists a fundamental tensor $J_{\mu}{}^{\lambda} J_{\lambda}{}^{\nu} = \delta^{\nu}_{\mu}$, invariant under G, with structure

$$J = \begin{pmatrix} 0 & \sigma_0 \\ -\sigma_0 & 0 \end{pmatrix} \tag{96}$$

where, however, the Lorentz metric $g_{\lambda\mu}$, is also invariant under G due its general form (95). Finally, there is a third fundamental tensor $\sigma_{\lambda\mu}$, also invariant under G. The following relations between the fundamental tensors are

$$J_{\lambda}{}^{\nu} = \sigma_{\lambda\mu} g^{\lambda\nu}, \qquad g_{\mu\nu} = \sigma_{\lambda\mu} J_{\nu}^{\lambda}, \qquad \sigma_{\lambda\mu} = J_{\lambda}^{\nu} g_{\mu\nu} \tag{97}$$

where

$$g^{\lambda\nu} = \frac{\partial g}{\partial g_{\lambda\nu}} \quad (g \equiv \det(g_{\mu\nu})) \,. \tag{98}$$

Then, the necessary fundamental structure is given by

$$G = L(4) \cap Sp(4) \cap K(4) \,, \tag{99}$$

which leaves concurrently invariant the three fundamental forms

$$ds^2 = g_{\mu\nu} dx^{\mu} dx^{\nu} \tag{100}$$

$$\sigma = \sigma_{\lambda\mu} dx^{\lambda} \wedge dx^{\mu} \tag{101}$$

$$\phi = J_{\nu}^{\lambda} w^{\nu} v_{\lambda}, \tag{102}$$

where w^{ν} are components of a vector $w^{\nu} \in V^*$ (the dual vector space). In expression (99) $L(4)$ is the Lorentz group in 4D, $Sp(4)$ is the Symplectic group in 4D real vector space and $K(4)$ denotes the almost complex group that leaves ϕ invariant [10].

For instance, G leaves the geometric product invariant [11]

$$\begin{aligned} \gamma_{\mu}\gamma_{\nu} &= \frac{1}{2}(\gamma_{\mu}\gamma_{\nu} - \gamma_{\nu}\gamma_{\mu}) + \frac{1}{2}(\gamma_{\mu}\gamma_{\nu} + \gamma_{\nu}\gamma_{\mu}) \\ &= \gamma_{\mu} \cdot \gamma_{\nu} - \gamma_{\mu} \wedge \gamma_{\nu} = g_{\mu\nu} + \sigma_{\mu\nu} \end{aligned} \tag{103}$$

where they are now regarded as a set of orthonormal basis vectors, in such a manner that any vector can be represented in the form $\mathbf{v} = v^{\lambda}\gamma_{\lambda}$, and

$$\varepsilon_{\alpha\beta\gamma\delta} \equiv \gamma_{\alpha} \wedge \gamma_{\beta} \wedge \gamma_{\gamma} \wedge \gamma_{\delta} \tag{104}$$

In resume, the fundamental structure of the spacetime is then represented by $\mathrm{P}(G, M)$, where G is given by (98), which leaves invariant the fundamental forms $(99 - 101)$, implying that

$$\nabla_{\lambda} g_{\mu\nu} = 0 \tag{105}$$

$$\nabla_{\nu} \sigma_{\lambda\mu} = 0 \tag{106}$$

$$\nabla_{\lambda} J_{\nu}^{\lambda} = 0 \tag{107}$$

where ∇_{λ} is the covariant derivative of the G connection. It is interesting to note that it is only necessary to consider two of above three equations: the third follows automatically. Then, we will consider (105) and (106) as, in some sense, they represent the boson and fermion symmetry respectively.

5.1. Field Equations and Group Structure

It is necessary to introduce now another antisymmetric tensor $\sigma'_{\mu\nu}$, which is not helical. That means that it is different of $\sigma_{\mu\nu}$ of (103), but also invariant with respect to the generalized connection G : $\nabla_{\nu} \sigma_{\lambda\mu} = 0$. For instance, we can construct also the antisymmetric tensor $\vartheta_{\mu\nu} \equiv \sigma'_{\mu\nu} - \sigma_{\lambda\mu} \neq 0$, that obeys $\nabla_{\nu} \vartheta_{\mu\nu} = 0$, and obviously $\frac{1}{6} \left(\partial_{\mu} \vartheta_{\nu\lambda} + \partial_{\nu} \vartheta_{\lambda\mu} + \partial_{\lambda} \vartheta_{\mu\nu} \right) = T^{\rho}_{\ \nu\mu} \vartheta_{\rho\lambda}$, given the completely antisymmetric nature of T.

5.2. Antisymmetric Torsion and Fermionic Structure of the Spacetime

We know that [8]

$$\Gamma^{\rho}_{\ \mu\lambda} = \left\{ {}^{\ \rho}_{\mu\lambda} \right\} + g^{\rho\nu} \left(T_{\mu\lambda\nu} + T_{\lambda\nu\mu} + T_{\nu\mu\lambda} \right) \tag{108}$$

where $\Gamma^{\rho}_{\ \mu\lambda}$ are the coefficients of the G-connection and $\left\{ {}^{\ \rho}_{\mu\lambda} \right\}$ denotes the coefficients of the Levi-Civita connection whose covariant derivative is denoted by $\overset{\circ}{\nabla}_{\lambda}$. From (106) we make the link between the fermionic structure of the fundamental geometry of the manifold and the torsion tensor

$$\nabla_{[\nu} \sigma_{\lambda\mu]} = 0 \quad \Rightarrow \quad \frac{1}{2} \partial_{[\nu} \sigma_{\lambda\mu]} = T^{\rho}_{[\nu\mu} \sigma_{\rho\lambda]} \tag{109}$$

A particularly simple solution for T arises when the torsion tensor is totally antisymmetric [3]

$$T_{\mu\lambda\nu} = T_{[\mu\lambda\nu]} \tag{110}$$

in order that the equivalence principle be obeyed [9, 3, 4]. In this case, as we have shown already in [1, 2, 3], we have

$$T_{\mu\lambda\nu} = \varepsilon_{\mu\lambda\nu\rho} h^{\rho}, \tag{111}$$

where the axial vector h^{ρ} is still to be determined. As it will be clear soon, it is useful to write it for d dimensions[9]

$$h^{\rho} = \frac{1}{\sqrt{w}} J^{\rho}_{\lambda} P^{\lambda} \tag{112}$$

where P^λ is the *generalized momentum vector.* If $d = 4$, $w = 6$.

Expression (109) can be simplified taking into account the symmetries of $T_{\mu\lambda\nu}$ and the contraction with the fundamental tensor J^λ_τ

$$T_{\lambda\mu\nu} = \frac{1}{w} J^\rho_\lambda \partial_{[\nu} \sigma_{\rho\mu]} \tag{113}$$

5.3. About the Equivalence Principle (EP) and the Antisymmetry of the Torsion Tensor: A Theorem

As it is well known, in order that experimental evidence constitutes the foundation of the theory, the EP has to be imposed, as well as the foregoing symmetry principles.

Because the G-connection contains a torsion tensor by specific requirements, it is currently suspected that the EP could be violated. Thus, a good question to ask is: what is the implication of EP as defined (or better described in this context) by the G-geometry?

Let us analyze specifically the question

i) The EP implies that the tangent space M_p must be a Minkowski space. Then, at M_p we have

$$(g_{\mu\nu})_p = \eta_{\mu\nu} \text{ and } \quad (\partial_\rho g_{\mu\nu})_p = 0, \tag{114}$$

where $\eta_{\mu\nu}$ is the Minkowski metric.

ii) The coefficients of the general affine connection are given by (eq. (17) in ref. [8], p.141)

$$\Gamma^\rho_{\ \mu\lambda} = \left\{ {\rho \atop \mu\lambda} \right\} + \overbrace{g^{\rho\nu} \left(T_{\mu\lambda\nu} + T_{\lambda\nu\mu} + T_{\nu\mu\lambda} \right)}^{\equiv S^\rho_{\ \mu\lambda}}, \tag{115}$$

where $T_{\nu\mu\lambda}$ is the torsion tensor and $S^\rho_{\ \mu\lambda}$ is the contortion.

iii) From

$$\nabla g = 0 \quad \text{we have}$$

$$\nabla_\lambda g_{\alpha\beta} = \overset{\circ}{\nabla}_\lambda g_{\alpha\beta} - T^\rho_{\ \lambda\alpha} g_{\rho\beta} - T^\rho_{\ \lambda\beta} g_{\alpha\rho} = 0 \tag{116}$$

which is valid also at p.

iv) from (114) and (116) we obtain

$$\left[T_{\beta\lambda\alpha} + T_{\alpha\lambda\beta} \right]_p = 0 \tag{117}$$

since (114) said

$$\left[\overset{\circ}{\nabla}_\lambda g_{\alpha\beta} \right]_p = 0 \tag{118}$$

v) The above relations have *tensorial character*, therefore they are valid in all coordinate systems (and in all points p), then

$$T_{\beta\lambda\alpha} = -T_{\alpha\lambda\beta} \tag{119}$$

and

$$\overset{\circ}{\nabla}_{\lambda} g_{\alpha\beta} = 0 \tag{120}$$

These equations show geometrically that the imposition of the PE implies the following equivalence

$$[\nabla_{\lambda} g_{\alpha\beta} = 0 \text{ and } PE] \Longleftrightarrow \text{(eqs. (119) and (120))} \tag{121}$$

vi) But, from (119) and (115) we have that the torsion tensor has the full antisymmetric property

$$T_{\alpha\lambda\beta} = T_{[\alpha\lambda\beta]} \tag{122}$$

With this proof we conclude that:

The full antisymmetry of the torsion tensor is the consequence of imposing the Equivalence Principle (EP) on the spacetime structure, and is not the result of a priori *assumptions concerning any hypothetical or possible physical meaning of the torsion tensor.*

5.4. The G-Invariance of the Action

As it is well known, the Palatini principle has the role of determining the connection required for the spacetime symmetry as the field equations.

By means of this principle, we were able to construct the action integral S. This action S necessarily needs to yield the G-invariant conditions (105-107) without prior assumptions; and then the Einstein, Dirac and Maxwell equations should arise from S as a causally connected closed system. These equations will be generalized unavoidably, so that causal connections between them can be established. Our action fulfill the above requirements, as the role of $f_{\mu\nu}$, which enters symmetrically with $g_{\mu\nu}$ in S is linked to the fundamental tensor $\vartheta_{\mu\nu}$ of the previous Section, denoting the dual of $\vartheta_{\mu\nu}$ by

$$f_{\mu\nu} \equiv \frac{1}{2}\varepsilon_{\mu\nu\rho\sigma}\vartheta^{\rho\sigma} = *\vartheta_{\mu\nu}$$

(where $\vartheta^{\mu\nu}$ is the inverse tensor of $\vartheta_{\mu\nu}$)

The usual Euler-Lagrange equations from the action will help us to compare the unitary model introduced here (in the sense of Eddington see [1, 2]) and the dualistic non abelian Born-Infeld model of [5]. With the explicit computation of the determinant in ($d = 4$) of expression (9), the equations take the familiar form [1, 2, 5]

$$S = \frac{b^2}{4\pi}\int \sqrt{-g}dx^4 \overbrace{\sqrt{\gamma^4 - \frac{\gamma^2}{2}\overline{G}^2 - \frac{\gamma}{3}\overline{G}^3 + \frac{1}{8}\left(\overline{G}^2\right)^2 - \frac{1}{4}\overline{G}^4}}^{\equiv\mathbb{R}} \tag{123}$$

$$G_{\mu\nu} \equiv \left[\lambda^2\left(g_{\mu\nu} + f^a{}_{\mu} f_{a\nu}\right) + 2\lambda R_{(\mu\nu)} + 2\lambda f^a{}_{\mu} R_{[a\nu]} + R^a{}_{\mu} R_{a\nu}\right] \tag{124}$$

$$G^{\nu}_{\nu} \equiv \left[\lambda^2\left(d + f_{\mu\nu}f^{\mu\nu}\right) + 2\lambda\left(R_S + R_A\right) + \left(R_S^2 + R_A^2\right)\right] \tag{125}$$

with (the bar on the tensorial quantities indicates traceless condition)

$$R_S \equiv g^{\mu\nu} R_{(\mu\nu)} \quad ; \quad R_A \equiv f^{\mu\nu} R_{[\mu\nu]} \quad ;$$
$$\gamma \equiv \frac{G^\nu_\nu}{d} \quad ; \quad \overline{G}_{\mu\nu} \equiv G_{\mu\nu} - \frac{g_{\mu\nu}}{4} G^\nu_\nu \quad ; \quad \overline{G}^\nu_\rho \overline{G}^\rho_\nu \equiv \overline{G}^2 \tag{126}$$
$$\overline{G}^\nu_\lambda \overline{G}^\lambda_\rho \overline{G}^\rho_\nu \equiv \overline{G}^3 \qquad \left(\overline{G}^\nu_\rho \overline{G}^\rho_\nu\right)^2 \equiv \left(\overline{G}^2\right)^2 \qquad \overline{G}^\nu_\mu \overline{G}^\mu_\lambda \overline{G}^\lambda_\rho \overline{G}^\rho_\nu \equiv \overline{G}^4.$$

The variation with respect to the electromagnetic potential a_τ was performed as follows

$$\frac{\delta\sqrt{G}}{\delta a_\tau} = \nabla_\rho \left(\frac{\partial \sqrt{G}}{\partial f_{\rho\tau}} \right) \equiv \nabla_\rho \mathbb{F}^{\rho\tau} = 0 \tag{127}$$

Explicitly,

$$\nabla_\rho \left[\frac{\lambda^2 N^{\mu\nu} \left(\delta^\sigma_\mu f^\rho_\nu + \delta^\sigma_\nu f^\rho_\mu \right)}{2\mathbb{R}} \right] = 0 \tag{128}$$

with $N^{\mu\nu}$ given by

$$N^{\mu\nu} = g \left[-\gamma^2 G^{\mu\nu} - \gamma \left(G^2\right)^{\mu\nu} + \frac{\left(G^2\right)^\mu_\mu G^{\mu\nu}}{2} - \left(G^3\right)^{\mu\nu} + \right. \tag{129}$$
$$\left. \frac{4\gamma^3 g^{\mu\nu}}{d} - \frac{\gamma \left(G^2\right)^\mu_\mu g^{\mu\nu}}{d} - \frac{\left(G^3\right)^\mu_\mu g^{\mu\nu}}{3d} \right] \tag{130}$$

Recall that the set of equations that solve for the action (15) in this particular case is (21)

$$R_{(\mu\nu)} = \overset{\circ}{R}_{\mu\nu} - T_{\mu\rho}{}^{\alpha} T_{\alpha\nu}{}^{\rho} = -\lambda g_{\mu\nu} \tag{20a}$$
$$R_{[\mu\nu]} = \nabla_\alpha T^\alpha_{\mu\nu} = -\lambda f_{\mu\nu} \tag{20b}$$
$$\nabla_\rho \left[\frac{\lambda^2 N^{\mu\nu} \left(\delta^\sigma_\mu f^\rho_\nu + \delta^\sigma_\nu f^\rho_\mu \right)}{2\mathbb{R}} \right] = 0 \tag{20c}$$

From this set, the link between T and f can be determined (f is not an *a priori* potential for the torsion T.)

The key point here is (112)

$$\overset{\circ}{R}_{\mu\nu} = -\lambda g_{\mu\nu} + T_{\mu\rho}{}^{\alpha} T_{\alpha\nu}{}^{\rho} \tag{131}$$
$$= -\lambda g_{\mu\nu} + w h_\mu h_\nu = -\lambda g_{\mu\nu} + P_\mu P_\nu \tag{132}$$

then we can obtain, as in mass shell condition

$$P^2 = m^2 \Rightarrow m = \pm\sqrt{\overset{\circ}{R} + \lambda d} \tag{133}$$

Notice that there exists a link between the dimension of the spacetime and the scalar Einstein's curvature $\overset{\circ}{R}$. Moreover, the curvature is constrained to take definite values $\in \mathbb{N}$, the natural number characteristic of the dimension. On the other hand, knowing that $|\lambda| = d-1$, and accepting that the parameter $m \in \mathbb{R}$, the limiting condition on the physical values for the mass is $\overset{\circ}{R} \geqslant (1-d)d$.

Introducing the geometric product in the above equation ($\gamma^\mu \gamma^\nu P_\mu P_\nu = m^2$), plus the quantum condition: $P_\mu \to \widehat{P}_\mu - e\widehat{A}_\mu$, we have

$$\left[\gamma^\mu \gamma^\nu \left(\widehat{P}_\mu - e\widehat{A}_\mu\right)\left(\widehat{P}_\nu - e\widehat{A}_\nu\right) - m^2\right]\Psi = 0 \tag{134}$$

where $\Psi = \mathbf{u} + i\mathbf{v}$ was given in (92, 93). That is

$$\left[\gamma^\mu \left(\widehat{P}_\mu - e\widehat{A}_\mu\right) + m\right]\left[\gamma^\nu \left(\widehat{P}_\nu - e\widehat{A}_\nu\right) - m\right] u^\lambda = 0 \tag{135}$$

which leads to the Dirac equation

$$\left[\gamma^\mu \left(\widehat{P}_\mu - e\widehat{A}_\mu\right) + m\right] u^\lambda = 0 \tag{136}$$

with m given by (133). This condition, in the Dirac case, is not only the way to pass from classical variables to quantum operators, but in the case that the action does not contains explicitly $\widehat{A}_\mu$, h_μ remains without specification due to the gauge freedom in the momentum. Applying the geometric product to (134) it is not difficult to see that

$$\left[\left(\widehat{P}_\mu - e\widehat{A}_\mu\right)^2 - m^2 - \frac{1}{2}e\sigma^{\mu\nu}F_{\mu\nu}\right]u^\lambda + \frac{1}{2}\sigma^{\mu\nu}R^\lambda_{\rho[\mu\nu]}u^\rho - \\ -\frac{1}{2}e\sigma^{\mu\nu}\left(\widehat{A}_\mu\widehat{P}_\nu - \widehat{A}_\nu\widehat{P}_\mu\right)u^\lambda = 0 \tag{137}$$

From which we can remark that:

i) The above formula is absolutely general for the type of geometrical Lagrangians considered, containing the generalized Ricci tensor.

ii) The variation of the action will carry the symmetric contraction of components of the torsion tensor (*i.e.* (131)), leading then to terms as $h_\mu\ h_\nu$.

iii) The only thing that changes is the mass (133) and the explicit form of the tensors considered, as $R^\lambda_{\rho[\mu\nu]}$, $F_{\mu\nu}$ etc., without variation of the Dirac general structure of the equation under consideration.

iv) Eq. (137) differs from the one obtained by Landau & Lifshitz by the occurrence of the last two terms: the term involving the curvature tensor is due to the spin interaction with the gravitational field (the torsion term in $R^\lambda_{\rho[\mu\nu]}$), and the last term is the spin interaction with the electromagnetic and mechanical momenta.

v) Expression (137) is valid for another vector v^λ, then it is valid for a bi-spinor of the form $\Psi = \mathbf{u} + i\mathbf{v}$.

vi) The meaning for a quantum measurement of the spacetime curvature is mainly due to the term in (137) involving explicitly the curvature tensor.

It is worth noting here that the spin-gravity interaction term is so easily derived because the spinors are represented as spacetime vectors, whose covariant derivatives are defined in terms of the G-(affine) connection. In their original form, the Dirac equations would have, in curved spacetime, their momentum operators replaced by covariant derivatives in terms of "spin-connection", whose relation is not immediately apparent.

6. Dirac Structure, Electromagnetic Field and Anomalous Gyromagnetic Factor

The following section is based on the observation that if we introduce expression (21b) in (137) then

$$\left[\left(\widehat{P}_\mu - e\widehat{A}_\mu\right)^2 - m^2 - \frac{1}{2}e\sigma^{\mu\nu}F_{\mu\nu}\right]u^\lambda - \frac{\lambda}{d}\frac{1}{2}\sigma^{\mu\nu}f_{[\mu\nu]}u^\lambda - \\ -\frac{1}{2}e\sigma^{\mu\nu}\left(\widehat{A}_\mu\widehat{P}_\nu - \widehat{A}_\nu\widehat{P}_\mu\right)u^\lambda = 0 \qquad (138)$$

$$\left[\left(\widehat{P}_\mu - e\widehat{A}_\mu\right)^2 - m^2 - \frac{1}{2}\sigma^{\mu\nu}\left(eF_{\mu\nu} + \frac{\lambda}{d}f_{\mu\nu}\right)\right]u^\lambda \\ -\frac{e}{2}\sigma^{\mu\nu}\left(\widehat{A}_\mu\widehat{P}_\nu - \widehat{A}_\nu\widehat{P}_\mu\right)u^\lambda = 0. \qquad (139)$$

We can see clearly that if $\widehat{A}_\mu = ja_\mu$ (with j arbitrary constant), then $F_{\mu\nu} = jf_{\mu\nu}$ and the last expression takes the suggestive form

$$\left[\left(\widehat{P}_\mu - e\widehat{A}_\mu\right)^2 - m^2 - \frac{1}{2}\left(ej + \frac{\lambda}{d}\right)\sigma^{\mu\nu}f_{\mu\nu}\right]u^\lambda \\ -\frac{e}{2}\sigma^{\mu\nu}\left(\widehat{A}_\mu\widehat{P}_\nu - \widehat{A}_\nu\widehat{P}_\mu\right)u^\lambda = 0 \qquad (140)$$

with the result that the gyromagnetic factor has been modified to $2/\left(j + \frac{\lambda}{ed}\right)$. Notice that in a Unified Theory with the characteristics introduced here, the identification $(F \leftrightarrows f)$ is reasonable in order that the fields arise from the same geometrical structure.

The concrete implications of this important contribution of the torsion to the gyromagnetic factor will be given elsewhere in great detail, mainly on the dynamical properties of the torsion field. Here we only remark the following:

i) there exists a substantial contribution of the torsion to the gyromagnetic factor, that can have implications on the problem of the anomalous momentum of the fermionic particles;

ii) this contribution appears (taking the second equality of expression (21b), as a modification of the vertex of interaction, almost from the effective point of view;

iii) it is quite evident that this contribution will justify probably the small contribution of the torsion at greater scales, because we can bound the torsion with the other well known contributions to the anomalous momentum of the elementary particles (QED, weak, hadronic contribution, etc.), which would therefore screen the effects;

iv) the form of the coupling spin-geometric structure comes from first principles, as the Dirac equation, not from prescriptions;

v) then, from iii) the way the covariant derivative works in the presence of torsion is totally determined by the G-structure of the spacetime;

vi) the Dirac equation (138) (where the second part of the equivalence (21b), coming from the equation of motion, was introduced), implies that the vertex was modified without a dynamical function of propagation. Then, another way to see the problem treated in this paragraph is by introducing the propagator for the torsion corresponding to the first part of the equivalence (21b). This important possibility will be studied elsewhere [9].

7. Spacetime and Structural Cohomologies

As it is well known from the physical and mathematical points of view, the cohomological interplay between the fields involved in any well posed geometrical and unified theory is crucial. This arises as a consequence of the logical (and causal) structure of the physical fields (sources, fields, conserved quantities), and not only as a mathematical play. In the theory presented here, there exist two cohomological structures: *spacetime cohomology* and *structural cohomology* The difference between them is that in the *Spacetime cohomology* the Dirac (fermionic) structure of the spacetime is not involved directly in the relations between the fields. The main equations necessary for the construction are

$$\begin{aligned} \nabla_\alpha T^\alpha{}_{\mu\nu} &= -\lambda f_{\mu\nu} \\ d^* T &= -\lambda^* f = dh. \end{aligned} \tag{141}$$

The interplay can be schematically represented as

$$\begin{array}{ccccc} & & \boxed{T} & & \\ & \overset{A_-}{\swarrow}\ \underset{A_+}{\nearrow} & & \overset{B_+}{\searrow}\ \underset{B_-}{\nwarrow} & \\ \boxed{f} & & \overset{C_+}{\underset{C_-}{\leftrightarrows}} & & \boxed{h} \end{array} \tag{142}$$

where the operators are

$$\begin{aligned} A_- &\equiv (-1)^{d+1}(-\lambda) * \textstyle\int * & \qquad A_+ &\equiv (-\lambda)^{-1} * d* \\ B_- &\equiv (-1)^{d+1} * & B_+ &\equiv * \\ C_- &\equiv -\lambda \textstyle\int^* & C_+ &\equiv \left[(-1)^{d+1}(-\lambda)\right]^{-1} * d \end{aligned} \tag{143}$$

$$
\begin{aligned}
D_- &\equiv (-1)^{d+1} * d & D_+ &\equiv (-1)^{d+1} * \int \\
E_- &\equiv d & E_+ &\equiv \int \\
G_- &\equiv \left[(-1)^{d+1}(-\lambda)\right]^{-1} * & G_+ &\equiv -\lambda *
\end{aligned}
$$

The Structural cohomology, in contrast, involves directly the fermionic structure of the spacetime, given that in the basic formulas $\vartheta_{\mu\nu}$ enters directly into the cohomological play, as can be seen below

$$
\begin{array}{ccccc}
 & & \boxed{a} & & \\
 & E_- \swarrow\nearrow E_+ & & D_+ \nwarrow\searrow D_- & \\
\boxed{f} & & \overset{B_-}{\underset{B_+}{\rightleftarrows}} & & \boxed{\vartheta} \\
 & C_+ \nwarrow\searrow C_- & & A_- \swarrow\nearrow A_+ & \\
 & & \boxed{h} & & \\
 & & G_- \uparrow\downarrow G_+ & & \\
 & & \boxed{a} & &
\end{array}
\qquad (144)
$$

Notice that in this case the degrees of the relations between the quantities involved are more fundamental than in the previous case (hierarchical sense).

8. Conclusion

In this chapter we have made an extensive analysis of the model based in the theory developed in early references of the authors. The simple structure of the spacetime described by this new theory makes, beside the connection between curvature and matter, the link between torsion and spin.

As it was explained, the mechanism of rupture of symmetry is responsible for the fact that the geometrical Lagrangian can be written in a suggestive Eddington-Born-Infeld like form. Three cases were treated from the point of view of the solutions, depending on the form of the torsion considered: totally antisymmetric (with torsion potential), not totally antisymmetric ('tratorial' type), and with a torsion tensor showing both characteristics. All the three cases were compared, from the point of view of the obtained solutions, with the non dualistic model of reference [5], the Non-Abelian Born-Infeld model.

In all these cases the (non-dualistic) unified model proposed here presents deep differences with the dualistic non-Abelian Born-Infeld model of Ref. [5].

The first obvious difference resides in the conceptual framework: the geometrical action will provide, besides the spacetime structure, the matter-energy spin distribution. This fact is the same basis of the unification: all the (apparently disconnected) theories and interactions of the natural world appear naturally as a consequence of the intrinsic spacetime geometry.

For the case of totally antisymmetric torsion tensor with torsion potential, several points were elucidated:

(i) about the Hosoya and Ogura ansatz, the natural question arising was: why does the identification of the isospin structure of the Yang-Mills field with the space frame, lead to a similar physical situation to that of a non-dualistic unified theory with torsion? The answer is: because once such identification is implemented, a potential torsion is introduced and the solution for the set of equations is the consistency between the definition of the torsion tensor from the potential and the Cartan structure equations [1, 2].

(ii) About the obtained solutions for the scale factor, the difference with our previous work is precisely the particular form of the energy-momentum tensor in the NABI case (in the UFT model presented here, there is no energy-momentum tensor, of course): both solutions describe a wormhole-instanton, but the final form of the differential equations for the scale factor are different. The scale factor in this theory has an exponentially growing behavior, in sharp contrast to the wormhole solution from our previous work with the "dualistic" non-Abelian BI theory. Also, for this particular value of the torsion, the wormhole tunneling interpretation (in the sense of the Coleman's mechanism) is fulfilled.

(iii) The contact point between the compared models, however, is the dynamical equations that are very similar although the existence of a "current term" in the UFT model (cf. (47)), that does not appear in the NABI case. This fact was pointed out in a slightly different context by N. Chernikov.

For the case of non-totally antisymmetric (tratorial type) the spacetime structure was analyzed from the point of view of the interacting fields arising from the same geometry of the space time, and relaxing now the condition of a totally antisymmetric torsion and then, the prior existence of an antisymmetric 2-form potential for it. The precise results can be enumerated as:

(i) the notion of minimal coupling has been elucidated and come naturally from the compatibility condition between the gauge field structure of the antisymmetric part of the fundamental tensor and the $SL(2, C)$ structure of the base manifold,

(ii) through exact cosmological solutions for this model, where the geometry is Euclidean $R \otimes O(3) \sim R \otimes SU(2)$, the relation between the spacetime geometry and the gauge group structure was explicitly shown,

(iii) this relation is directly connected to the relation between the spin and torsion fields.

From the point of view of the obtained solutions, a solution of this model was explicitly compared with our previous ones and we find that:

(i) the torsion is not identified directly with the Yang-Mills type strength field,

(ii) there exists a compatibility condition connected to the identification of the gauge group with the geometric structure of the spacetime: this fact leads to the identification between derivatives of the scale factor a, with the components of the torsion, in order to establish the Hosoya-Ogura ansatz (namely, the alignment of the isospin with the frame geometry of the spacetime),

(iii) this compatibility condition precisely mark the fact that local gauge covariance, coordinate independence and arbitrary spacetime geometries are harmonious concepts;

(iv) also, of two possible structures of the torsion, the "tratorial" form forbids wormhole configurations, leading only to cosmological instanton spacetimes in eternal expansion.

For the general case, *i.e.* a torsion with totally antisymmetric and tratorial parts, the full analysis was given in a clear manner in Section VI. Here we point out that the Hosoya and Ogura anzats can be implemented as in the previous cases, and, the most important, that wormhole solutions can be obtained for some particular cases. The solutions are asymptotically flat, where vector and tensor integration constants appear, that are constrained in norm, to bring physical consistency to the solution.

About the problem of the possibility of coexistence of the trace of the torsion due to the tratorial part, and the axial vector from the totally antisymmetric part of the torsion, we showed here that there are no problems within the new theory: there are tratorial and antisymmetric torsion fields without contradictions.

The fact that in reference [6] the field equations of vacuum quadratic Poincare gauge field theory (QPGFT) were solved for purely null tratorial torsion, if well permit to express the contortion tensor for such a case as (tratorial form, with notation of ref. [6])

$$K_{\lambda\mu\nu} = -2(g_{\lambda\mu}a_\nu - g_{\lambda\nu}a_\mu),$$

it does not allow the coexistence with an axial torsion vector, as was clearly shown by Singh in the beautiful paper [6]. The two points that lead to such discrepancy are:

i) the different theories described, not only in foundations but also because one is unitary and the other, of [6], dualistic;

ii) and the fact that the Newman-Penrose formulation was used in [6]and, as is well known, that method works in a null tetrad.

8.1. On the Geometrical Structure

From the point of view of the concrete structure able to explain the content of the bosonic and fermionic matter of the universe, the present work is left open-ended, as many physical consequences need to be explored. Some words concerning the realization and the choice of the correct group structure of the tangent space to M is that $G = L(4) \cap Sp(4) \cap K(4)$ preserves the boson and fermion symmetry simultaneously, without implying direct supersymmetry of the model. As we would like to show in a future paper, the super-gravitational extension of the model will be discussed jointly with the problem of its quantization, where the key point will be precisely the group structure of the tangent space of the spacetime manifold M. Here we conclude enumerating the main results concerning the basic structure of the Manifold supporting a Unified Field Theoretical model:

i) the simplest geometrical structure able to support the fermionic fields was constructed based on a tangent space with a group structure $G = L(4) \cap Sp(4) \cap K(4)$

ii) then, the explicit link between the fermionic structure and the torsion field was implemented, and the Dirac type equation was obtained from the same spacetime manifold.

iii) Matter was not included in the Geometrical Lagrangian of the Unified theory presented here: only symmetry arguments (that lead to the correct dynamical equations for the material fields, arising from the same manifold) are needed in order to allow the emergence of matter, and this fact is the essence of the unification (notice, in contrast, that several references are trying to include matter into the Eddington type theories 'by hand', without any physical or symmetry principles).

It is worth pointing here an interesting recent discussion between Steven Weinberg and Frederik Hehl that appeared in [14], regarding the question of the difference between the standard GR theory with the addition of 'some tensor' (quoting Steven Weinberg), and a theory of gravitation with torsion. As the debate was established in the context of a dualistic theory, the lack of a final conclusion arising from a such discussion is clear: the addition 'by hand' of a Lagrangian of matter fields in a "dualistic" theory, makes the inclusion of the torsion in the theory obscure, unclear, cumbersome and in some sense unnecessary. However, as we try to demonstrate in this chapter, if the theory is "non dualistic", torsion and curvature *must* appear: all the "matter" and the "energy" of the universe (natural world) coming from the geometry eliminates in the formulation, any place for the ambiguities arising by the freedom in the choice of the form of the energy momentum tensor.

8.2. On the Energy Concept

1) Looking at the equation

$$\overset{\circ}{R}_{\mu\nu} = -\lambda g_{\mu\nu} + T_{\mu\rho}{}^{\alpha} T_{\alpha\nu}{}^{\rho}$$

we can affirm that, conceptually, the terms that arise as 'energy-momentum' part coming from the symmetric contraction of the torsion components, are different, in essence, of the ones coming from the inclusion of an energy-momentum tensor in the Einstein theory. The conceptual framework that 'matter and energy curve the spacetime' implicitly carries the idea of some 'embedding', where the matter and energy are put into some Minkowskian flexible carpet, and one sees how it is curved under the 'weight' of the 'ball' (matter+energy). However, in the theory presented here, the situation is that the torsion terms (contributing as energy momentum in above equation) arise from the same geometry. Then, the picture contains a unique entity: the interplay fields-spacetime. The idea can be illustrated thinking in a solitonic vortex in the water.

This fact can be also interpreted as the concept of force which is introduced due to the torsion in the unified model, the thing that is lost in the Einstein theory [4], where there is no force but only curvature.

2) Some remarks on the general Hodge-de Rham decomposition of $h = h_\alpha dx^\alpha$.

Theorem 1. *if* $h = h_\alpha dx^\alpha \notin F'(M)$ *is a 1-form on* M*, then there exist a zero-form* Ω*, a 2-form* $\alpha = A_{[\mu\nu]} dx^\mu \wedge dx^\nu$ *and a harmonic 1-form* $q = q_\alpha dx^\alpha$ *on* M *that*

$$h = d\Omega + \delta\alpha + q \to h_\alpha = \nabla_\alpha \Omega + \varepsilon_\alpha^{\beta\gamma\delta} \nabla_\beta A_{\gamma\delta} + q_\alpha$$

Notice that even if is not harmonic, and assuming that q_α is a polar vector, an axial vector can be added so that the above expression takes the form

$$h_\alpha = \nabla_\alpha \Omega + \varepsilon_\alpha^{\beta\gamma\delta} \nabla_\beta A_{\gamma\delta} + \varepsilon_\alpha^{\beta\gamma\delta} M_{\beta\gamma\delta} + q_\alpha$$

where $M_{\beta\gamma\delta}$ is a completely antisymmetric tensor.

3) Notice the important fact that when the torsion is a totally antisymmetric tensor field, $-2\lambda f_{\mu\nu}$ assumes the role of a 'current' for the torsion field, similarly to the terms proportional to the 1-form potential vector a_μ that usually act as currents of the electromagnetic field $f_{\mu\nu}$, in the field equation of the standard theory: $\nabla_\alpha f^\alpha{}_\mu = J_\mu$ (constants absorbed into the J_μ). The interpretation and concrete consequences of this question will be presented in [9].

Acknowledgments

I am very grateful to Professor Yu Xin who introduced me into the subject of the Unified Theories based only in geometrical concepts and the Mach principle. Many thanks are given to Professors Salvatore Cappoziello, and Victor Gershun for their interest and discussions, and to Prof. V. I. Afonso for his help in the preparation of this text.

This work was partially supported by Brazilian Ministry of Science and Technology (MCT-CNPQ).

References

[1] D. J. Cirilo-Lombardo, *Int. J. Theor. Phys.* (2010) 49: 1288–1301. (and references therein)

[2] D. J. Cirilo-Lombardo, *Int. J. Theor. Phys.* DOI 10.1007/s10773-011-0678-1 (and references therein)

[3] Yu Xin, "General Relativity on Spinor-Tensor Manifold", in: *"Quantum Gravity - Int. School on Cosmology & Gravitation"*, XIV Course. Eds. P.G. Bergman, V. de. Sabbata & H.J. Treder, pp. 382-411, World Scientific (1996).

[4] Yu Xin, private communication.

[5] D.J. Cirilo-Lombardo, *Class. Quantum Grav.* 22 (2005) 4987–5004 (and references therein)

[6] P Singh , *Class. Quantum Grav.*7 (1990) 2125.

[7] B.T. McInnes, *Class. Quantum Grav.* 1 (1984) 105-113.

[8] Schouten J., *Ricci-Calculus* Heidelberg, Springer (1954).

[9] D.J. Cirilo-Lombardo, in preparation.

[10] Shoshichi Kobayashi and Katsumi Nomizu, *Foundations of Differential Geometry*, Volume 1 and 2, Wiley-Interscience (1996).

[11] David Hestenes, *Spacetime Algebra, Gordon & Breach* (1966).

[12] K. Borchsenius, Gen. Relativ. Gravit. 7, 527 (1976); R. T. Hammond, *Class. Quantum Gravity* 6, L195 (1989); N. J. Poplawski, *Int. J. Theor. Phys.* 49, 1481 (2010).

[13] T. W. B. Kibble, *J. Math. Phys.* 2, 212 (1961); D. W. Sciama, *Rev. Mod. Phys.* 36, 463 (1964); 36, 1103 (1964); F. W. Hehl, *Gen. Relativ. Gravit.* 4, 333 (1973); 5, 491 (1974); F. W. Hehl, P. von der Heyde, G. D. Kerlick, and J. M. Nester, *Rev. Mod. Phys.* 48, 393 (1976).

[14] A.S. Goldhaber, T. Cornsweet, H.C. Ohanian, R. Gomatam, R. Larson et al., *Phys. Today* 59(4), 10 (2006); F.W. Hehl and S. Weinberg, *Phys. Today* 60(3), 16 (2007).

In: Einstein and Others: Unification
Editor: Valeriy V. Dvoeglazov

ISBN: 978-1-63463-276-8

Chapter 3

MODIFIED GRAVITY: $F(R) = R \exp(\alpha R)$

S. I. Kruglov
Department of Chemical and Physical Sciences, University of Toronto,
Mississauga, Ontario, Canada

Abstract

A modified theory of gravity with the function $F(R) = R \exp(\alpha R)$ instead of Ricci scalar R in the Einstein-Hilbert action is analyzed. The action is converted into Einstein-Hilbert action at small value of the parameter α. From local tests we obtain a bound on the parameter $\alpha \leq 10^{-6}$ cm^2. The Jordan and Einstein frames are considered and the potential of the scalar field in Einstein's frame is found. The constant curvature solutions of the model are obtained corresponding to the Schwarzschild-de Sitter space. We show that the de Sitter space is unstable but a solution with zero curvature is stable. It was demonstrated that the model passes the matter stability test.

1. Introduction

According to astronomical data the Universe accelerates at the present time. The nature of the driving force that results the accelerated expansion is unknown yet. There are several approaches to explain cosmic acceleration. The interpretation of currently accelerating Universe, within the General Relativity (GR), requires the introduction of dark energy - exotic substance with large negative pressure P_{DE} so that $P_{DE} \simeq -\rho_{DE}$ (ρ_{DE} is the dark energy density). The similar scheme uses the cosmological constant Λ [1]. Such a model gives a good description of all observational data. However, with theoretical point of view, it is not clear how to explain the introduction of a new physical constant Λ which is very small compared with vacuum energy of elementary particles. Models with dynamical dark energy include a new scalar field [2]. Another way to describe the acceleration of the early and late Universe is the modification of GR. So-called $F(R)$-gravity theories replace the Ricci scalar in Einstein-Hilbert action by the function $F(R)$ [3], [4], [5]. Such purely gravitational models are an alternative to ΛCDM model and may solve the coincidence problem, and to describe the inflation and late-time acceleration. In $F(R)$-gravity models the cosmic acceleration is due to new gravitational physics.

In this paper a model of exponential-like $F(R)$-gravity is analyzed. The paper is organized as follows. In Sec.2, we consider a model of modified gravity with the exponential-like Lagrangian density. A bound on the parameter α with the dimension (length)^2 is obtained. We find constant curvature Schwarzschild-de Sitter solutions in Jordan's frame and describe FRW (Friedmann-Robertson-Walker) cosmology in Sec.3. The potential of the scalar field in the scalar-tensor form of the model (Einstein's frame) is obtained in Sec.4. It is shown that the de Sitter space is unstable and the Minkowski space corresponding to a solution with zero Ricci scalar is stable. In Sec.5 the matter stability of the model is investigated and we demonstrate that the model passes the matter stability test at $R_0 < 2/\alpha$. We discuss results obtained in Sec.6.

The Minkowski metric $\eta_{\mu\nu}$=diag(-1, 1, 1, 1) is used and we imply that $c = \hbar = 1$ throughout the paper.

2. The Modified Gravity Model

We study the modified gravitational theory with the function $F(R)$ instead of the Ricci curvature R ($R \to F(R)$) in the Einstein-Hilbert action (see also [6]):

$$F(R) = R\exp(\alpha R), \tag{1}$$

so that the action is

$$S = \int d^4x\sqrt{-g}\mathcal{L} = \int d^4x\sqrt{-g}\left[\frac{1}{2\kappa^2}F(R) + \mathcal{L}_m\right], \tag{2}$$

where $\kappa = \sqrt{8\pi} m_{Pl}^{-1}$, g=det$g_{\alpha\beta}$ ($g_{\alpha\beta}$ is a metric tensor), $m_{Pl} = G^{-1/2}$ is the Planck mass, G is the gravitation (Newton) constant, $\mathcal{L}_m$ is the matter Lagrangian density. The action (2) is written in the Jordan frame. It should be mentioned that other variants of exponential gravity were considered in [7], [8]. Thus, the constant α with the dimension of (length)^2 is introduced. To pass the Solar System tests the constant α should be small compared with R^{-1} ($\alpha R \ll 1$) because the deviation from GR based on the Einstein-Hilbert action has to be diminutive. As a result one can obtain from (1) the Taylor series

$$F(R) = R + \alpha R^2 + \frac{1}{2}\alpha^2 R^3 + ... \tag{3}$$

Because $\lim_{\alpha\to 0} F(R) = R$ action (2) of our model approaches to the Einstein-Hilbert action at $\alpha R \ll 1$. GR passes local tests and, therefore, one may obtain a restriction on the parameter α from observational data. From the Eöt-Wash experiment [9], [10] (see also [11], [12]), we obtain a laboratory bound on the parameter α:

$$\alpha \leq 10^{-6}\text{cm}^2. \tag{4}$$

It should be mentioned that F(R)-gravity with the function $F(R) = R + R^2/6M^2$ (M has a dimension of the mass) was considered in [13] which is the approximation to series (3) at small αR (the value αR is dimensionless). The modified R^2-gravity is insensitively investigated [14], [15] because quantum corrections to GR include R^2 counter term [16],

[17], [18] and such a model possesses attractive features: an absence singularity in the past and in the future and gives the self-consistent inflation, etc. Let us consider the matter Lagrangian density $\mathcal{L}_m$, entering (2), which represents the perfect fluid with the energy-momentum tensor

$$T_{\alpha\beta}^{mat} = \left(P^{mat} + \rho^{mat}\right) u_\alpha u_\beta + P^{mat} g_{\alpha\beta}, \quad (5)$$

where P^{mat} is a pressure, ρ^{mat} is the energy density, and the four-velocity of the fluid obeys the relation $u^\alpha u_\alpha = -1$. Then equations of motion following from Eq.(2) are given by

$$R_{\mu\nu}F'(R) - \frac{1}{2}g_{\mu\nu}F(R) + g_{\mu\nu}g^{\alpha\beta}\nabla_\alpha\nabla_\beta F'(R) - \nabla_\mu\nabla_\nu F'(R) = \kappa^2 T_{\mu\nu}^{mat}, \quad (6)$$

where a covariant derivative is ∇_μ, $F'(R) = dF(R)/dR$. For FRW metric, the conservation of the energy-momentum tensor $\nabla^\mu T_{\mu\nu}^{mat} = 0$ results

$$\dot{\rho}^{mat} + 3H\left(\rho^{mat} + P^{mat}\right) = 0. \quad (7)$$

Here the Hubble parameter is $H = \dot{a}(t)/a(t)$, where a(t) is a scale factor and over dot denotes the differentiation with respect to the time. It follows from Eq.(7) that for the fluid with the property of the dark energy, when the equation of state (EoS) is $P^{mat} = -\rho^{mat}$, the energy density ρ^{mat} is a constant.

3. Constant Curvature Solutions

Let us consider solutions to Eq. (6) for action (2) in a case with a constant Ricci scalar $R = R_0$. For EoS: $P^{mat} = -\rho^{mat}$ (ρ^{mat} is a constant) Eq. (6) reads [4]

$$2F(R_0) - R_0F'(R_0) = 4\kappa^2\rho^{mat}. \quad (8)$$

From Eq.(8), we obtain the equation as follows:

$$R_0\left(1 - \alpha R_0\right) = 4\kappa^2\rho^{mat}\exp\left(-\alpha R_0\right). \quad (9)$$

Because transcendent equation (9) can be solved only by the numerical computation, we restrict our consideration by the case of free space ($\rho^{mat} = 0$). Then there are two solutions to Eq.(9) at $\rho^{mat} = 0$:

$$R_0 = 0, \qquad R_0 = \frac{1}{\alpha}. \quad (10)$$

The conditions of classical and quantum stability [14] $F'(R) > 0$, $F''(R) > 0$ are realized in this model for $\alpha > 0$, $R > 0$. As a result, both solutions lead to the Schwarzschild-de Sitter space. Non-trivial solution (10) $R_0 = 1/\alpha$ corresponds to early-time inflation. The solution with vanishing curvature $R_0 = 0$ leads to Minkowski space. Of course the concrete scenario of the Universe evolution can be described after finding the exact solutions to Eq.(6) for Ricci scalar depending on time. The Schwarzschild spherically symmetric metric is given by

$$ds^2 = -B(r)dt^2 + \frac{dr^2}{B(r)} + r^2\left(d\theta^2 + \sin^2\theta d\phi^2\right). \quad (11)$$

For the constant Ricci scalar R_0, $F(R)$-gravity theories possess Schwarzschild-(anti-)de Sitter solutions [4], [5] with the function $B(r)$:

$$B(r) = 1 - \frac{2MG}{r} - \frac{R_0}{12}r^2, \tag{12}$$

with the mass of the black hole M. For $R_0 > 0$ the de Sitter space is realized, and the case $R_0 < 0$ corresponds to the anti-de Sitter space. As in our model the non-trivial solution (10) is $R_0 = 1/\alpha > 0$, it corresponds to the de Sitter space and the function (12) is

$$B(r) = 1 - \frac{2MG}{r} - \frac{1}{12\alpha}r^2. \tag{13}$$

Because $\alpha > 0$ we have the classical stability of Schwarzschild black holes. Solutions of Einstein's equation with cosmological constant Λ have also the function of the form (12) with $R_0 = 4\Lambda$. Therefore, the model under consideration leads to the dynamical cosmological constant $\Lambda = 1/(4\alpha)$ at the time when $R = R_0 = 1/\alpha$. Thus, even for the space without any matter, the model mimics the dark energy (the cosmological constant). The similar property of other $F(R)$ models was discussed in [19], [20], [4], [5].

In $F(R)$-gravity the entropy S is given as follows [21], [22], [23], [4], [5]:

$$S = \frac{F'(R)A}{4G}, \tag{14}$$

where A is the area of the horizon . Eq.(14) is the generalization of the Bekenstein-Hawking formula [24], [25] on the case of $F(R)$-gravity. One obtains from equation (1) $F'(R) = (1 + \alpha R)\exp(\alpha R)$, and entropy (14) becomes

$$S = \frac{(1 + \alpha R)\exp(\alpha R)\, A}{4G}. \tag{15}$$

As a result, one can introduce the effective gravitational coupling $G_{eff} = G/\,(1 + \alpha R)\exp(\alpha R)$. The nontrivial solution (10) $R_0 = 1/\alpha$ gives the effective gravitational constant $G_{eff} = G/(2e)$ at the time of inflation when $R = R_0 = 1/\alpha$.

3.1. FRW Cosmology

The homogeneous, isotropic and spatially flat FRW cosmology is described by the space-time metric

$$ds^2 = -dt^2 + a^2(t)\left(dx^2 + dy^2 + dz^2\right). \tag{16}$$

In this case the Ricci scalar R is given by $R = 12H^2 + 6\dot{H}$. For constant curvature solutions H_0=const, and $\dot{H}_0 = 0$. Then $H_0 = \sqrt{R_0/12}$ and we obtain from Eq.(10) for a de Sitter phase $H_0 = 1/\sqrt{12\alpha}$, and a scale factor becomes

$$a(t) = a_0 \exp\left(\frac{t}{2\sqrt{3\alpha}}\right), \tag{17}$$

where $a(0)$ is a scale factor at a cosmic time $t = 0$. Solution (17) describes the inflation phase.

4. The Scalar-Tensor Form

We have formulated modified $F(R)$-gravity in the Jordan frame with tensor variables $g_{\mu\nu}$. Another description, in the Einstein frame, corresponds to the scalar-tensor theory of gravity [4], [5] with conformally transformed metric [26]

$$\widetilde{g}_{\mu\nu} = F'(R)g_{\mu\nu} = (1+\alpha R)\exp(\alpha R)\, g_{\mu\nu}. \tag{18}$$

In new variables equation (2) for $\mathcal{L}_m = 0$, in the Einstein frame, is given by

$$S = \int d^4x\sqrt{-g}\left[\frac{1}{2\kappa^2}\widetilde{R} - \frac{1}{2}\widetilde{g}^{\mu\nu}\nabla_\mu\varphi\nabla_\nu\varphi - V(\varphi)\right], \tag{19}$$

where $\widetilde{R}$ is defined by new metric (18). The scalar field φ and the potential $V(\varphi)$ are given by equations

$$\varphi = \frac{\sqrt{3}\ln F'(R)}{\sqrt{2}\kappa} = \frac{\sqrt{3}}{\sqrt{2}\kappa}\left[\ln(1+\alpha R) + \alpha R\right], \tag{20}$$

$$V(\varphi) = \frac{RF'(R) - F(R)}{2\kappa^2 F'^2(R)}|_{R=R(\varphi)} = \frac{\alpha R^2\exp(-\alpha R)}{2\kappa^2(1+\alpha R)^2}|_{R=R(\varphi)}, \tag{21}$$

where the scalar curvature R in Eq.(21) being the solution of transcendental Eq.(20), $R(\varphi)$. It should be mentioned that in the Einstein frame free particles of matter do not move in space-time geodesics because of interactions with the scalar field φ. There is a correction in the right hand side of the geodesic equation representing a fifth force. Because the fifth force depends on space-time (and proportional to $\nabla^\mu\varphi$) the universality of free fall (Weak Equivalence Principle) is violated. The mass squared of a scalar state is defined by the equation [3]

$$m_\varphi^2 = \frac{d^2V}{d\varphi^2} = \frac{1}{3}\left(\frac{1}{F''(R)} + \frac{R}{F'(R)} - \frac{4F(R)}{F'^2(R)}\right). \tag{22}$$

We obtain from Eqs.(1), (22) the mass squared of a scalar field

$$m_\varphi^2 = \frac{\left[1 - 4\alpha R + (\alpha R)^3\right]\exp(-\alpha R)}{3\alpha(2+\alpha R)(1+\alpha R)^2}. \tag{23}$$

For $R_0 = 0$ the value $m_\varphi^2 = 1/(6\alpha)$ is positive and the solution with zero scalar curvature, corresponding to the Minkowski space, gives a stabile state. The de Sitter solution (10) $R_0 = 1/\alpha$ leads to the negative mass squared $m_\varphi^2 = -1/(18e\alpha)$. Thus, the de Sitter space with $R_0 = 1/\alpha$ is unstable. For small value of α corrections to the Newton law are negligible. One can say that after the Big Bang the Universe is in unstable de Sitter's phase and inflates (rapidly expands) having the positive curvature $R_0 = 1/\alpha$. Then the curvature decreases and the Universe approaches to the stable Minkowski space.

5. Matter Stability

From equation (6), after taking the trace, one obtains the equation of motion for a curvature scalar

$$3g^{\alpha\beta}\nabla_\alpha\nabla_\beta F'(R) + F'(R)R - 2F(R) = \kappa^2 T^{mat}, \tag{24}$$

where $T^{mat} = T^{mat}_{\mu\nu} g^{\mu\nu}$. Following [27], to investigate the matter stability, we consider Eq.(24) for weak gravity objects. For a flat Minkowski metric and spatially constant distribution, Eq.(24) reads

$$-3F^{(2)}(R)\ddot{R} - 3F^{(3)}(R)\dot{R}^2 + F^{(1)}(R)R - 2F(R) = \kappa^2 T^{mat}, \tag{25}$$

with the notation $F^{(n)}(R) = d^n F(R)/dR^n$. From Eq.(1), we find

$$F^{(n)}(R) = \alpha^{n-1}(n + \alpha R)\exp(\alpha R). \tag{26}$$

Let us consider a perturbation so that $R = R_0 + R_1$, $|R_1| \ll |R_0|$, and according to GR $R_0 = -\kappa^2 T^{mat}$. Then Eq.(25) leads to [5]

$$\ddot{R}_0 + \ddot{R}_1 + \frac{F^{(3)}(R_0)}{F^{(2)}(R_0)}\left(\dot{R}_0^2 + 2\dot{R}_0\dot{R}_1\right)$$
$$+\frac{2F(R_0) - R_0\left[1 + F^{(1)}(R_0)\right]}{3F^{(2)}(R_0)} = U(R_0)R_1, \tag{27}$$

with

$$U(R_0) = \frac{F^{(3)2} - F^{(2)}F^{(4)}}{F^{(2)2}}\dot{R_0}^2$$
$$+\frac{\left(R_0F^{(2)} - F^{(1)}\right)F^{(2)} + \left(2F - R_0F^{(1)} - R_0\right)F^{(3)}}{3F^{(2)2}}. \tag{28}$$

For $U(R_0) > 0$, R_1 exponentially increases in the time and the system is unstable. From Eqs.(26),(28) we obtain

$$U(R_0) = \frac{\left(\alpha\dot{R}_0\right)^2}{(2 + \alpha R_0)^2} + \frac{\alpha R_0 - 2}{3\alpha(2 + \alpha R_0)^2}. \tag{29}$$

Implying that the rate $(\alpha\dot{R}_0)^2$ is small compared to $1/\alpha$, $(\alpha\dot{R}_0)^2 \ll 1/\alpha$, for a matter stability $U(R_0) < 0$, one comes to the condition

$$R_0 < \frac{2}{\alpha}. \tag{30}$$

For constant curvature solutions (10) the inequality (30) is satisfied and the model passes the matter stability test. Eq.(30) gives the restriction on the biggest curvature of the Universe. One can introduce the fundamental length $L = \sqrt{\alpha}$ so that the smallest size of the Universe is L. It should be mentioned that the Born-Infeld-like modified gravity model [28] possesses the similar features.

6. Conclusion

A modified theory of gravity with exponential-like Lagrangian density and the fundamental length $L = \sqrt{\alpha}$ is analyzed. We have considered $F(R)$-model that admits de Sitter solutions without a cosmological constant. Therefore, possibly the cosmic acceleration arises from new theory of gravity and GR is only an approximation describing the Universe at the intermediate cosmic time. At $\alpha \to 0$ the action (2) approaches to Einstein-Hilbert action. From the bound (4) obtained, we find the restriction on the fundamental length $L \leq 10^{-3}$ cm. We have found the constant curvature Schwarzschild-de Sitter solutions of the model and the potential of the scalar field in the scalar-tensor form of the theory. It was demonstrated that, for constant curvature solutions obtained, the de Sitter space is unstable and the Minkowski space with zero Ricci scalar is stable. Vacuum solutions are important in investigations of early and late time Universe. The model under consideration passes the matter stability test at $R_0 < 2/\alpha$. One can interpret this as a restriction on the size of the Universe at the beginning (the size of the Universe was greater than L). To describe precisely the cosmological scenario, one needs to obtain exact solutions of equations of motion depending on time. This and other problems we leave for further investigations. We intend to verify the model introduced to be viable and an alternative to GR.

References

[1] J. A. Frieman, M. S. Turner and D. Huterer, *Ann. Rev. Astron. Astroph.* 46, 385 (2008).

[2] Andrei D. Linde, *Particle physics and inflationary cosmology* (Chur, Switzerland: Harwood Academic Publishers, 1992).

[3] T. P. Sotiriou and V. Faraoni, *Rev. Mod. Phys.* 82, 451 (2010).

[4] S. Capozziello and V. Faraoni, *Beyond Einstein Gravity: A Survey of Gravitational Theories for Cosmology and Astrophysics* (Springer Science+Business Media B.V. 2011).

[5] S. Nojiri and S. D. Odintsov, *Phys. Rept.* 505, 59 (2011).

[6] L. Amendola, R. Gannouji, D. Polarski and S. Tsujikawa, *Phys. Rev.* D 75, 083504 (2007).

[7] G. Cognola, E. Elizalde, S. Nojiri, S. D. Odintsov, L. Sebastiani and S. Zerbini, *Phys. Rev.* D 77, 046009 (2008).

[8] E. V. Linder, *Phys. Rev*. D 80 (2009) 123528.

[9] D. J. Kapner, T. S. Cook, E. G. Adelberger, J. H. Gundlach, B. R. Heckel, C. D. Hoyle and H. E. Swanson, *Phys. Rev. Lett.* 98, 021101 (2007).

[10] J. Näf and P. Jetzer, *Phys. Rev.* D 81, 104003 (2010).

[11] C. P. L. Berry and J. R. Gair, *Phys. Rev.* D 83, 104022 (2011).

[12] M. Eingorn and A. Zhuk, *Phys. Rev.* D 84, 024023 (2011).

[13] A. A. Starobinsky, *Phys. Lett.* B 91, 99 (1980).

[14] S. A. Appleby, R. A. Battye and A. A. Starobinsky, *JCAP* 1006, 005 (2010).

[15] E. V. Arbusova, A. D. Dolgov and L. Reverberi, *J. Cosmol. Astropat. Phys.* 02, 049 (2012).

[16] K. S. Stelle, *Phys. Rev.* D 16, 953 (1977).

[17] N. D. Birrel and P. C. W. Davies, *Quantum Fields in Curved Space* (Cambrige University Press, Cambrige, 1982).

[18] I. L. Buchbinder, S. D. Odintsov and I. L. Shapiro, *Effective action in quantum gravity* (IOP Publishing Ltd., Btistol, UK, 1992).

[19] J. D. Barrow and A. C. Ottewill, *J. Phys.* A 16, 2757 (1983).

[20] S. M. Carroll, V. Duvvuri, M. Trodden and M. S. Turner, *Phys. Rev.* D 70, 043528 (2004).

[21] M. Akbar and R.-G. Cai, *Phys. Lett.* B 635, 7 (2006).

[22] Y. Gong and A. Wang, *Phys. Rev. Lett.* 99, 211301 (2007).

[23] R. Brustein, D. Gorbonos and M. Hadad, *Phys. Rev.* D 79, 044025 (2009).

[24] J. D. Bekenstein, *Phys. Rev.* D 7, 2333 (1973).

[25] S. Hawking, *Phys. Rev.* D 191, 13 (1976).

[26] G. Magnano and L. M. Sokolowski, *Phys. Rev.* D 50, 5039 (1994).

[27] A. D. Dolgov and M. Kawasaki, *Phys. Lett.* B 573, 1 (2003).

[28] S. I. Kruglov, *Int. J. Theor. Phys.* 52, 2477 (2013).

In: Einstein and Others: Unification
Editor: Valeriy V. Dvoeglazov

ISBN: 978-1-63463-276-8

Chapter 4

TOWARDS A QUANTUM DESCRIPTION OF EXTENDED PARTICLES

V. V. Varlamov* and M. V. Beloborodova
Department of Mathematics, Siberian State Industrial University,
Novokuznetsk, Russia

Abstract

Extended particles are considered in terms of the fields on the Poincaré group. Dirac like wave equations for extended particles of any spin are defined on the various homogeneous spaces of the Poincaré group. Free fields of the spin 1/2 and 1 (Dirac and Maxwell fields) are considered in detail on the eight-dimensional homogeneous space, which is equivalent to a direct product of Minkowski spacetime and two-dimensional complex sphere. It is shown that a massless spin-1 field, corresponding to a photon field, should be defined within principal series representations of the Lorentz group. Interaction between spin-1/2 and spin-1 fields is studied in terms of a trilinear form. An analogue of the Dyson formula for S-matrix is introduced on the eight-dimensional homogeneous space. It is shown that in this case elements of the S-matrix are defined by convergent integrals.

PACS: 02.30.Gp, 02.60.Lj, 03.65.Pm, 12.20.-m

Keywords: Extended particles, fields on the Poincaré group, homogeneous spaces, wave equations, quantum electrodynamics

1. Introduction

It is well-known that the representation of a particle as an idealized point has been used in physics for a long time. Historically, such a description follows from classical (celestial) mechanics, where the distances between explored objects are much greater than their sizes. However, in quantum field theory a point-like particle description meets with some conceptual difficulties. For example, ultra-violet divergences follow directly from

*E-mail address: vadim.varlamov@mail.ru

the point-like description. Refusing from the point-like idealization, we come to *extended objects*.

Our consideration based on the concept of generalized wave functions introduced by Ginzburg and Tamm in 1947 [24], where the wave function depends both coordinates x_μ and additional internal variables u_μ which describe spin of the particle, $\mu = 0, 1, 2, 3$. In 1955, Finkelstein showed [18] that elementary particles models with internal degrees of freedom can be described on manifolds larger then Minkowski spacetime (homogeneous spaces of the Poincaré group). The quantum field theories on the Poincaré group were discussed in the papers [38, 35, 12, 5, 36, 61, 39, 16, 25, 29]. A consideration of the field models on the homogeneous spaces leads naturally to a generalization of the concept of wave function (fields on the Poincaré group). The general form of these fields relates closely with the structure of the Lorentz and Poincaré group representations [20, 43, 11, 25] and admits the following factorization $f(x, \mathbf{z}) = \phi^n(\mathbf{z})\psi_n(x)$, where $x \in T_4$ and $\phi^n(\mathbf{z})$ form a basis in the representation space of the Lorentz group. At this point, four parameters x^μ correspond to position of the point-like object, whereas remaining six parameters $\mathbf{z} \in \mathbf{Spin}_+(1, 3)$ define orientation in quantum description of orientable (extended) object [26, 27] (see also [33]). It is obvious that the point-like object has no orientation, therefore, orientation is an intrinsic property of the extended object. Taking it into account, we come to consideration of physical quantity as an extended object, the generalized wave function of which is described by the field

$$\boldsymbol{\psi}(\boldsymbol{\alpha}) \equiv \langle x, g\,|\boldsymbol{\psi}\rangle$$

on the homogeneous space of some orthogonal group $\mathrm{SO}(p, q)$, where $x \in T_n$ (position) and $g \in \mathbf{Spin}_+(p, q)$ (orientation), $n = p + q$. So, in [54, 55] Segal and Zhou proved convergence of quantum field theory, in particular, quantum electrodynamics, on the homogeneous space $R^1 \times S^3$ of the conformal group $\mathrm{SO}(2, 4)$, where S^3 is the three-dimensional real sphere.

On the other hand, measurements in quantum field theory lead to extended objects. As is known, loop divergences emerging in the Green functions in quantum field theory originate from correspondence of the Green functions to *unmeasurable* (and hence unphysical) point-like quantities. This is because no physical quantity can be measured in a point, but in a region, the size of which (or 'diameter'[1] of the extended object) is constrained by the resolution of measuring equipment [2]. Ordinary quantum field theory defines the field function $\phi(x)$ as a scalar product of the state vector of the system and the state vector corresponding to the localization at the point:

$$\phi(x) \equiv \langle x|\phi(x)\rangle.$$

This field describes the point-like object. On the other hand, *resolution-dependent fields*

$$\phi_a(x) \equiv \langle x, a; g|\phi(x)\rangle,$$

introduced in the work [2], describe in essence extended objects, here $\langle x, a; g|$ is the bra-vector corresponding to localization of the measuring device around the point with the

[1]For example, this diameter can be defined by a characteristic rest-frame kernel dimension $\sim \hbar/mc$ (Compton wavelength). The magnitude of the 'kernel' radius of extended particles was first proposed by de Broglie [15], see also [60, 52].

spatial resolution a, g labels the apparatus function of the equipment (an aperture). It is easy to see that the resolution-dependent fields have the same mathematical structure as the fields on the Poincaré group (more generally, *fields on the groups*).

The present paper is organized as follows. Basic facts concerning fields on the Poincaré group are considered in the section 2. Dirac-like wave equations for extended objects are presented in the section 3. Solutions for the fields $\psi(\boldsymbol{\alpha}) = \langle x, \mathfrak{g} \,|\psi\rangle$ of spin 1/2 and 1 (Dirac and Maxwell fields) are given in terms of associated hyperspherical functions defined on the two-dimensional complex sphere, where $\mathfrak{g} \in \mathbf{Spin}_+(1,3)$. Interaction between these fields is introduced in the section 4 in terms of a trilinear form.

2. Fields on the Poincaré Group

As it mentioned above, Ginzburg and Tamm proposed to consider new internal continuous variables u_μ. It allows one to generalize the Klein-Gordon wave equation. The Ginzburg-Tamm equation has the form

$$\left[\Box - m^2 + \frac{\beta}{2} M_{\mu\nu} M_{\mu\nu}\right] \Psi(x_\rho, u_\gamma) = 0, \tag{1}$$

where

$$M_{\mu\nu} = -i\left(u_\mu \frac{\partial}{\partial u_\nu} - u_\nu \frac{\partial}{\partial u_\mu}\right) \quad (\mu, \nu = 0, 1, 2, 3)$$

are rotation generators of the four-dimensional space, the constant β is analogous to the momentum J in the rotator hamiltonian $\mathcal{H} = (2J)^{-1} L_{ik} L_{ik}$, where L_{ik} are generators of the Lorentz group, m^2 is a constant. The wave function Ψ depends on coordinates x_μ of mass centre of the particle and internal variables u_μ. The 4-vector u_μ is a 4-vector of relative motion of the structural component of the particle around its mass centre. The wave function Ψ can be factored:

$$\Psi(x_\mu, u_\mu) = \Psi(x_\mu)\Phi(u_\mu).$$

Then the Ginzburg-Tamm equation is decomposed in the following two equations:

$$(\Box - m^2 + \lambda_1 \beta)\Psi(x_\mu) = 0, \quad L_1 \Phi(u_\mu) = \lambda_1 \Phi(u_\mu). \tag{2}$$

At this point, the function $\Phi(u_\mu)$ is an eigenfunction of the operator L_1,

$$L_1 = M_{\mu\nu} M_{\mu\nu}/2, \quad L_1 \Phi = \lambda_1 \Phi.$$

The first equation from (2) has the following solution:

$$\Psi(x_\mu) = C \exp(-i m_0 t),$$

where

$$m_0^2 = m^2 - \lambda_1 \beta.$$

Thus, a mass spectrum, described by the equation (1), is defined by an eigenvalue of the operator L_1. The mass operator L_1 in spherical coordinates on the one-sheeted hyperboloid H^3,

$$\begin{aligned} u_0 &= r\sinh\chi; & u_2 &= r\cosh\chi\sin\theta\sin\phi; \\ u_1 &= r\cosh\chi\sin\theta\cos\phi; & u_3 &= r\cosh\chi\cos\theta, \end{aligned}$$

is defined by an expression

$$L_1 = \frac{1}{r}\frac{\partial}{\partial r}\left(r^3\frac{\partial}{\partial r}\right) - \frac{1}{\cosh^2\chi}\frac{\partial}{\partial\chi}\left(\cosh^2\chi\frac{\partial}{\partial\chi}\right) -$$
$$- \frac{1}{\cosh^2\chi}\left[\frac{1}{\sin\theta}\frac{\partial}{\partial\theta}\left(\sin\theta\frac{\partial}{\partial\theta}\right) + \frac{1}{\sin^2\theta}\frac{\partial^2}{\partial\phi^2}\right] - \left(r\frac{\partial}{\partial r}\right)^2 - 2r\frac{\partial}{\partial r}.$$

The hyperboloid H^3 is a homogeneous space of the Lorentz group. Solutions of the Ginzburg-Tamm equation have the form of the fields on the Poincaré group:

$$\Psi(x_\mu, u_\mu) = C\exp(-im_0t)Y_{l,m}(\theta,\phi)P_l^j(\tanh\chi),$$

where $Y_{l,m}(\theta,\phi)$ is a spherical function, and $P_l^j(\tanh\chi)$ is a Legendre polynomial.

Further, fields on the Poincaré group present itself a natural generalization of the concept of wave function. These fields (generalized wave functions) were introduced independently by several authors [24, 8, 70, 56] mainly in connection with constructing relativistic wave equations. In essence, this generalization consists in replacing the Minkowski space by a larger space on which the Poincaré group acts. If this action is to be transitive, one is lead to consider the homogeneous spaces of the Poincaré group. All the homogeneous spaces of this type were listed by Finkelstein [18] and by Bacry and Kihlberg [6] and the fields on these spaces were considered in the works [38, 7, 45, 34, 35, 61, 25].

A homogeneous space $\mathcal{M}$ of a group G has the following properties:
a) It is a topological space on which the group G acts continuously, that is, let y be a point in $\mathcal{M}$, then gy is defined and is again a point in $\mathcal{M}$ ($g \in G$).
b) This action is transitive, that is, for any two points y_1 and y_2 in $\mathcal{M}$ it is always possible to find a group element $g \in G$ such that $y_2 = gy_1$.
There is a one-to-one correspondence between the homogeneous spaces of G and the coset spaces of G. Let H_0 be a maximal subgroup of G which leaves the point y_0 invariant, $gy_0 = y_0$, $g \in H_0$, then H_0 is called the stabilizer of y_0. Representing now any group element of G in the form $g = g_cg_0$, where $g_0 \in H_0$ and $g_c \in G/H_0$, we see that, by virtue of the transitivity property, any point $y \in \mathcal{M}$ can be given by $y = g_cg_0y_0 = g_cy$. Hence it follows that the elements g_c of the coset space give a parametrization of $\mathcal{M}$. The mapping $\mathcal{M} \leftrightarrow G/H_0$ is continuous since the group multiplication is continuous and the action on $\mathcal{M}$ is continuous by definition. The stabilizers H and H_0 of two different points y and y_0 are conjugate, since from $H_0g_0 = g_0$, $y_0 = g^{-1}y$, it follows that $gH_0g^{-1}y = y$, that is, $H = gH_0g^{-1}$.

Returning to the Poincaré group $\mathcal{P}$, we see that the enumeration of the different homogeneous spaces $\mathcal{M}$ of $\mathcal{P}$ amounts to an enumeration of the subgroups of $\mathcal{P}$ up to a conjugation. Following to Finkelstein, we require that $\mathcal{M}$ always contains the Minkowski space $\mathbb{R}^{1,3}$ which means that four parameters of $\mathcal{M}$ can be denoted by x (x^μ). This means that the stabilizer H of a given point in $\mathcal{M}$ can never contain an element of the translation subgroup of $\mathcal{P}$. Thus, the stabilizer must be a subgroup of the homogeneous Lorentz group $\mathfrak{G}_+$.

In such a way, studying different subgroups of $\mathfrak{G}_+$, we obtain a full list of homogeneous spaces $\mathcal{M} = \mathcal{P}/H$ of the Poincaré group. In the present paper we restrict ourselves by a consideration of the following four homogeneous spaces:

$$\begin{aligned}
\mathcal{M}_{10} &= \mathbb{R}^{1,3} \times \mathfrak{L}_6, \quad H = 0; \\
\mathcal{M}_8 &= \mathbb{R}^{1,3} \times S^2_c, \quad H = \Omega^c_\psi; \\
\mathcal{M}_7 &= \mathbb{R}^{1,3} \times H^3, \quad H = \mathrm{SU}(2); \\
\mathcal{M}_6 &= \mathbb{R}^{1,3} \times S^2, \quad H = \{\Omega^c_\psi, \Omega_\tau, \Omega_\epsilon\}.
\end{aligned}$$

Hence it follows that a group manifold of the Poincaré group, $\mathcal{M}_{10} = \mathbb{R}^{1,3} \times \mathfrak{L}_6$, is a maximal homogeneous space of $\mathcal{P}$, $\mathfrak{L}_6$ is a group manifold of the Lorentz group. The fields on the manifold $\mathcal{M}_{10}$ were considered by Lurçat [38]. These fields depend on all the ten parameters of $\mathcal{P}$:

$$\boldsymbol{\psi}(\boldsymbol{\alpha}) = \langle x, \mathfrak{g}\,|\boldsymbol{\psi}\rangle = \psi(x_0, x_1, x_2, x_3)\psi(\mathfrak{g}_1, \mathfrak{g}_2, \mathfrak{g}_3, \mathfrak{g}_4, \mathfrak{g}_5, \mathfrak{g}_6),$$

where an explicit form of $\psi(x)$ is given by the exponentials, and the functions $\psi(\mathfrak{g})$ are expressed via the generalized hyperspherical functions $\mathfrak{M}^l_{mn}(\mathfrak{g})$. As is known, the universal covering of the proper Poincaré group is isomorphic to a semidirect product $\mathrm{SL}(2;\mathbb{C}) \odot T_4$ or $\mathbf{Spin}_+(1,3) \odot T_4$. Since the group T_4 is Abelian, then all its representations are one-dimensional. Thus, all the finite-dimensional representations of the proper Poincaré group in essence are equivalent to the representations $\mathfrak{C}$ of the group $\mathbf{Spin}_+(1,3)$. In the case of finite-dimensional representations of $\mathbf{Spin}_+(1,3)$,

$$T_{\mathfrak{g}} q(\xi, \overline{\xi}) = (\gamma\xi + \delta)^{l_0+l_1-1}\overline{(\gamma\xi + \delta)}^{l_0-l_1+1} q\left(\frac{\alpha\xi + \beta}{\gamma\xi + \delta}; \overline{\frac{\alpha\xi + \beta}{\gamma\xi + \delta}}\right), \tag{3}$$

basic functions (matrix elements) of the finite-dimensional representation of $\mathcal{P}$ have the form

$$t^l_{mn}(\boldsymbol{\alpha}) = e^{-imx}\mathfrak{M}^l_{mn}(\mathfrak{g}),$$

where

$$\mathfrak{M}^l_{mn}(\mathfrak{g}) = e^{-im\varphi^c}\mathfrak{Z}^l_{mn}(\cos\theta^c)e^{-in\psi^c}.$$

Hyperspherical functions $\mathfrak{Z}^l_{mn}(\cos\theta^c)$ can be expressed via the hypergeometric functions [65]. So, at $m \geq n$ and $m \geq k$, $k \geq n$ we have

$$\mathfrak{Z}^l_{mn}(\cos\theta^c) = i^{m-n}\sqrt{\frac{\Gamma(l+m+1)\Gamma(l-n+1)}{\Gamma(l-m+1)\Gamma(l+n+1)}}\cos^{2l}\frac{\theta}{2}\cosh^{2l}\frac{\tau}{2}\times$$

$$\times \sum_{k=-l}^{l} \tan^{m-k}\frac{\theta}{2}\tanh^{k-n}\frac{\tau}{2}\times$$

$$\times\, {}_2F_1\left(\begin{matrix} m-l, -k-l \\ m-k+1 \end{matrix}\,\middle|\, -\tan^2\frac{\theta}{2}\right){}_2F_1\left(\begin{matrix} k-l, -n-l \\ k-n+1 \end{matrix}\,\middle|\, \tanh^2\frac{\tau}{2}\right). \tag{4}$$

There also exist three expressions of the hypergeometric type for the functions $\mathfrak{Z}^{l}_{mn}(\cos\theta^c)$ with the index values $m \geq n,\ m \geq k,\ n \geq k,\ n \geq m,\ k \geq m,\ n \geq k$ and $n \geq m,\ k \geq m,\ k \geq n$. For the unitary representations, that is, in the case of principal series representations of the group $\mathrm{SO}_0(1,3)$, there exists an analogue of the formula (3),

$$V_a f(z) = (a_{12}z + a_{22})^{\frac{\lambda}{2}+i\frac{\rho}{2}-1}\overline{(a_{12}z + a_{22})}^{-\frac{\lambda}{2}+i\frac{\rho}{2}-1} f\left(\frac{a_{11}z + a_{21}}{a_{12}z + a_{22}}\right), \tag{5}$$

where $f(z)$ is a measurable function of the Hilbert space $L_2(Z)$, satisfying the condition $\int |f(z)|^2 dz < \infty$, $z = x + iy$. A totality of all representations $a \to T^{\alpha}$, corresponding to all possible pairs λ, ρ, is called a principal series of representations of the group $\mathrm{SO}_0(1,3)$ and denoted as $\mathfrak{S}_{\lambda,\rho}$. At this point, a comparison of (5) with the formula (3) for the spinor representation $\mathfrak{S}_l$ shows that the both formulas have the same structure; only the exponents at the factors $(a_{12}z + a_{22})$, $\overline{(a_{12}z + a_{22})}$ and the functions $f(z)$ are different. In the case of spinor representations the functions $f(z)$ are polynomials $p(z,\bar{z})$ in the spaces $\mathrm{Sym}_{(k,r)}$, and in the case of a representation $\mathfrak{S}_{\lambda,\rho}$ of the principal series $f(z)$ are functions from the Hilbert space $L_2(Z)$. Further, we know that a representation S_l of the group $\mathrm{SU}(2)$ is realized in terms of the functions $t^l_{mn}(u) = e^{-im\varphi}P^l_{mn}(\cos\theta)e^{-in\psi}$ [68]. It is well known that the representation S_k of $\mathrm{SU}(2)$ is contained in $\mathfrak{S}_{\lambda,\rho}$ not more then one time [43]. At this point, S_k is contained in $\mathfrak{S}_{\lambda,\rho}$, when $\frac{\lambda}{2}$ is one from the numbers $-k, -k+1, \ldots, k$. Matrix elements of the principal series representations of the group $\mathrm{SO}_0(1,3)$ have the form [65]

$$\mathfrak{M}^{-\frac{1}{2}+i\rho,l_0}_{mn}(\mathfrak{g}) = e^{-m(\epsilon+i\varphi)-n(\varepsilon+i\psi)}\mathfrak{Z}^{-\frac{1}{2}+i\rho,l_0}_{mn} = e^{-m(\epsilon+i\varphi)-n(\varepsilon+i\psi)}\times$$

$$\sum^{l_0}_{t=-l_0} i^{m-t}\sqrt{\Gamma(l_0-m+1)\Gamma(l_0+m+1)\Gamma(l_0-t+1)\Gamma(l_0+t+1)}\times$$

$$\cos^{2l_0}\frac{\theta}{2}\tan^{m-t}\frac{\theta}{2}\times$$

$$\sum^{\min(l_0-m,l_0+t)}_{j=\max(0,t-m)}\frac{i^{2j}\tan^{2j}\frac{\theta}{2}}{\Gamma(j+1)\Gamma(l_0-m-j+1)\Gamma(l_0+t-j+1)\Gamma(m-t+j+1)}\times$$

$$\sqrt{\Gamma(\tfrac{1}{2}+i\rho-n)\Gamma(\tfrac{1}{2}+i\rho+n)\Gamma(\tfrac{1}{2}+i\rho-t)\Gamma(\tfrac{1}{2}+i\rho+t)}\cosh^{-1+2i\rho}\frac{\tau}{2}\tanh^{n-t}\frac{\tau}{2}\times$$

$$\sum^{\infty}_{s=\max(0,t-n)}\frac{\tanh^{2s}\frac{\tau}{2}}{\Gamma(s+1)\Gamma(\frac{1}{2}+i\rho-n-s)\Gamma(\frac{1}{2}+i\rho+t-s)\Gamma(n-t+s+1)}, \tag{6}$$

where $l_0 = \left|\frac{\lambda}{2}\right|$ and $\frac{\lambda}{2}$ is one from the numbers $-k, -k+1, \ldots, k$. It is obvious that $\mathfrak{M}^{-\frac{1}{2}+i\rho,l_0}_{mn}(\mathfrak{g})$ cannot be attributed as matrix elements to single irreducible representation. From the latter expression it follows that relativistic spherical functions $f(\mathfrak{g})$ of the principal series can be defined by means of the function

$$\mathfrak{M}^{-\frac{1}{2}+i\rho,l_0}_{mn}(\mathfrak{g}) = e^{-m(\epsilon+i\varphi)}\mathfrak{Z}^{-\frac{1}{2}+i\rho,l_0}_{mn}(\cos\theta^c)e^{-n(\varepsilon+i\psi)}, \tag{7}$$

where

$$\mathfrak{Z}_{mn}^{-\frac{1}{2}+i\rho,l_0}(\cos\theta^c) = \sum_{t=-l_0}^{l_0} P_{mt}^{l_0}(\cos\theta)\mathfrak{P}_{tn}^{-\frac{1}{2}+i\rho}(\cosh\tau).$$

In turn, the functions $\mathfrak{M}_{mn}^{-\frac{1}{2}+i\rho,l_0}(\mathfrak{g})$ can be expressed via the hypergeometric functions. So, at $m \geq t,\ t \geq n$ we have [65]

$$\mathfrak{M}_{mn}^{-\frac{1}{2}+i\rho,l_0}(\mathfrak{g}) = i^{m-n}e^{-m(\epsilon+i\varphi)-n(\varepsilon+i\psi)}\sqrt{\frac{\Gamma(l_0+m+1)\Gamma(i\rho-n+\frac{1}{2})}{\Gamma(l_0-m+1)\Gamma(i\rho+n+\frac{1}{2})}}\times$$
$$\cos^{2l_0}\frac{\theta}{2}\cosh^{-1+2i\rho}\frac{\tau}{2}\sum_{t=-l_0}^{l_0}\tan^{m-t}\frac{\theta}{2}\tanh^{t-n}\frac{\tau}{2}\times$$
$${}_2F_1\left(\begin{matrix} m-l_0, -t-l_0 \\ m-t+1 \end{matrix}\middle| -\tan^2\frac{\theta}{2}\right){}_2F_1\left(\begin{matrix} t-i\rho+\frac{1}{2}, -n-i\rho+\frac{1}{2} \\ t-n+1 \end{matrix}\middle| \tanh^2\frac{\tau}{2}\right). \quad (8)$$

There also exist three expressions of the hypergeometric type for the functions $\mathfrak{M}_{mn}^{-\frac{1}{2}+i\rho,l_0}(\mathfrak{g})$ with the index values $m \geq t,\ n \geq t, t \geq m,\ n \geq t$ and $t \geq m,\ t \geq n$.

The following eight-dimensional homogeneous space $\mathcal{M}_8 = \mathbb{R}^{1,3} \times S_c^2$ is a direct product of the Minkowski space $\mathbb{R}^{1,3}$ and the complex two-sphere S_c^2. In this case the stabilizer H consists of the subgroup Ω_ψ^c of the diagonal matrices $\begin{pmatrix} e^{\frac{i\psi^c}{2}} & 0 \\ 0 & e^{-\frac{i\psi^c}{2}} \end{pmatrix}$. Bacry and Kihlberg [6] claimed that the space $\mathcal{M}_8$ is the most suitable for a description of both half-integer and integer spins. The fields, defined in $\mathcal{M}_8$, depend on the eight parameters of $\mathcal{P}$:

$$\langle x, \varphi^c, \theta^c|\psi\rangle = \psi(x)\psi(\varphi^c, \theta^c) = \psi(x_0, x_1, x_2, x_3)\psi(\varphi, \epsilon, \theta, \tau), \quad (9)$$

where the functions $\psi(\varphi^c, \theta^c)$ are expressed via the associated hyperspherical functions defined on the surface of the complex two-sphere S_c^2. The sphere S_2^c can be constructed from the quantities $z_k = x_k + iy_k$, $\overset{*}{z}_k = x_k - iy_k$ $(k = 1, 2, 3)$ as follows:

$$S_2^c: \ z_1^2 + z_2^2 + z_3^2 = \mathbf{x}^2 - \mathbf{y}^2 + 2i\mathbf{x}\mathbf{y} = r^2. \quad (10)$$

The complex conjugate (dual) sphere $\dot{S}_2^c$ is

$$\dot{S}_2^c: \ \overset{*}{z}_1{}^2 + \overset{*}{z}_2{}^2 + \overset{*}{z}_3{}^2 = \mathbf{x}^2 - \mathbf{y}^2 - 2i\mathbf{x}\mathbf{y} = \overset{*}{r}{}^2. \quad (11)$$

For more details about the two-dimensional complex sphere see [30, 31, 59]. The surface of S_2^c is homeomorphic to the space of the pairs (z_1, z_2), where z_1 and z_2 are the points of a *complex projective line*, $z_1 \neq z_2$. This space is a homogeneous space of the Lorentz group [21]. In Euler parametrization we have $(z_1, z_2) \sim (\varphi^c, \theta^c) = (\varphi - i\epsilon, \theta - i\tau)$. It is well-known that both quantities $\mathbf{x}^2 - \mathbf{y}^2$, $\mathbf{x}\mathbf{y}$ are invariant with respect to the Lorentz transformations, since a surface of the complex sphere is invariant (Casimir operators of the Lorentz group are constructed from such quantities). Moreover, since the real and imaginary parts of the complex two-sphere transform like the electric and magnetic fields, respectively, the invariance of $\mathbf{z}^2 \sim (\mathbf{E} + i\mathbf{B})^2$ under proper Lorentz transformations is evident. At this point, the quantities $\mathbf{x}^2 - \mathbf{y}^2$, $\mathbf{xy}$ are similar to the well known electromagnetic invariants $E^2 - B^2$, $\mathbf{EB}$. This intriguing relationship between the Laplace-Beltrami operators, Casimir operators of the Lorentz group and electromagnetic invariants $E^2 - B^2 \sim \mathbf{x}^2 - \mathbf{y}^2$, $\mathbf{EB} \sim \mathbf{xy}$ leads naturally to a Riemann-Silberstein representation of the electromagnetic field (see, for example, [69, 57, 10]). Further, associated hyperspherical functions

$$\mathfrak{M}_l^m(\varphi^c, \theta^c, 0) = e^{-im\varphi^c}\mathfrak{Z}_l^m(\cos\theta^c), \tag{12}$$

where

$$\mathfrak{Z}_l^m(\cos\theta^c) = \sum_{k=-l}^{l} P^l_{mk}(\cos\theta)\mathfrak{P}_l^k(\cosh\tau),$$

are defined on the surface of the two-dimensional complex sphere (10). In turn, the functions $\mathfrak{M}_{\dot{l}}^{\dot{m}}(\dot{\varphi}^c, \theta^c, 0) = e^{im\dot{\varphi}^c}\mathfrak{Z}_{\dot{l}}^{\dot{m}}(\cos\dot{\theta}^c)$ are defined on the surface of the dual sphere (11). Explicit expressions of the hypergeometric type for the functions $\mathfrak{Z}_l^m(\cos\theta^c)$ and $\mathfrak{Z}_{\dot{l}}^{\dot{m}}(\cos\dot{\theta}^c)$ follow directly from the previous expressions (4) and (8) at $n = 0$. In the case of the principal series representations of the group $\mathrm{SO}_0(1,3)$ we have

$$\mathfrak{Z}^m_{-\frac{1}{2}+i\rho,l_0}(\cos\theta^c) = \sum_{t=-l_0}^{l_0} P^{l_0}_{mt}(\cos\theta)\mathfrak{P}^t_{-\frac{1}{2}+i\rho}(\cosh\tau), \tag{13}$$

where $\mathfrak{P}^t_{-\frac{1}{2}+i\rho}(\cosh\tau)$ are *conical functions* (see [9]).

In turn, a seven-dimensional homogeneous space $\mathcal{M}_7 = \mathbb{R}^{1,3} \times H^3$ is a direct product of $\mathbb{R}^{1,3}$ and a three-dimensional timelike (two-sheeted) hyperboloid H^3. The stabilizer H consists of the subgroup of three-dimensional rotations, $\mathrm{SU}(2)$. Quantum field theory on the space $\mathcal{M}_7$ was studied by Boyer and Fleming [12]. They showed that the fields built over $\mathcal{M}_7$ are in general nonlocal and become local only when the finite dimensional representations of the Lorentz group are used. It is easy to see that the fields $\boldsymbol{\psi} \in \mathcal{M}_7$ depend on the seven parameters of the Poincaré group:

$$\langle x, \tau, \epsilon, \varepsilon | \boldsymbol{\psi}\rangle = \psi(x_0, x_1, x_2, x_3)\psi(\tau, \epsilon, \varepsilon),$$

where the functions $\psi(\tau, \epsilon, \varepsilon)$ are expressed via $e^{-(m\epsilon+n\varepsilon)}\mathfrak{P}^l_{mn}(\cosh\tau)$ in the case of finite dimensional representations, and also via $e^{-(m\epsilon+n\varepsilon)}\mathfrak{P}^{-\frac{1}{2}+i\rho}_{m\dot{n}}(\cosh\tau)$ in the case of principal series of unitary representations, and via $e^{-(m\epsilon+n\varepsilon)}\mathfrak{P}^{l,\pm}_{mn}(\cosh\tau)$ in the case of the discrete representation series of the subgroup $\mathrm{SU}(1,1)$. The spherical function $\mathfrak{P}^l_{mn}(\cosh\tau)$ on the group $\mathrm{SU}(1,1)$ has the form

$$\mathfrak{P}^l_{mn}(\cosh\tau) = \sqrt{\Gamma(l-n+1)\Gamma(l+n+1)\Gamma(l-m+1)\Gamma(l+m+1)}\times$$
$$\cosh^{2l}\frac{\tau}{2}\tanh^{n-m}\frac{\tau}{2}\times$$
$$\sum_{s=\max(0,m-n)}^{\min(l-n,l+m)} \frac{\tanh^{2s}\frac{\tau}{2}}{\Gamma(s+1)\Gamma(l-n-s+1)\Gamma(n-m+s+1)\Gamma(l+m-s+1)}. \tag{14}$$

Further, a six-dimensional space $\mathcal{M}_6 = \mathbb{R}^{1,3} \times S^2$ is a minimal homogeneous space of the Poincaré group, since the real two-sphere S^2 has a minimal possible dimension among the homogeneous spaces of the Lorentz group. In this case, the stabilizer H consists of the subgroup Ω^c_ψ and the subgroups Ω_τ and Ω_ϵ formed by the matrices $\begin{pmatrix} \cosh\frac{\tau}{2} & \sinh\frac{\tau}{2} \\ \sinh\frac{\tau}{2} & \cosh\frac{\tau}{2} \end{pmatrix}$ and $\begin{pmatrix} e^{\frac{\epsilon}{2}} & 0 \\ 0 & e^{-\frac{\epsilon}{2}} \end{pmatrix}$, respectively. Field models on the homogeneous space $\mathcal{M}_6$ have been considered in the works [36, 39, 16]. In the paper [16] Drechsler considered the real two-sphere as a 'spin shell' $S^2_{r=2s}$ of radius $r = 2s$, where $s = 0, \frac{1}{2}, 1, \frac{3}{2}, \ldots$. The fields, defined on $\mathcal{M}_6$, depend on the six parameters of $\mathcal{P}$:

$$\langle x, \varphi, \theta | \boldsymbol{\psi} \rangle = \psi(x_0, x_1, x_2, x_3)\psi(\varphi, \theta),$$

where the functions $\psi(\varphi, \theta)$ are expressed via the generalized spherical functions of the type $e^{-im\varphi} P^l_{m0}(\cos\theta)$ or via the Wigner D-functions, here

$$P^l_{mn}(\cos\theta) = e^{-i(m\varphi+n\psi)} i^{m-n}\sqrt{\Gamma(l-m+1)\Gamma(l+m+1)\Gamma(l-n+1)\Gamma(l+n+1)}\times$$
$$\cos^{2l}\frac{\theta}{2}\tan^{m-n}\frac{\theta}{2}\times$$
$$\sum_{j=\max(0,n-m)}^{\min(l-n,l+n)} \frac{i^{2j}\tan^{2j}\frac{\theta}{2}}{\Gamma(j+1)\Gamma(l-m-j+1)\Gamma(l+n-j+1)\Gamma(m-n+j+1)} \tag{15}$$

is a spherical function on the group $\mathrm{SU}(2)$.

Further, let $\mathcal{H}$ be a Hilbert space and let $U(g)$ be a square integrable representation of a locally compact Lie group G acting transitively on $\mathcal{H}$, $\forall\phi \in \mathcal{H}$, $g \in G : U(g)\phi \in \mathcal{H}$. Let there exist such a vector $|\psi\rangle \in \mathcal{H}$ that satisfies the admissibility condition:

$$C_\psi = ||\psi||^{-2}\int_{g\in G} |\langle| U(g)\psi\rangle|^2 d\mu(g) < \infty,$$

where $d\mu(g)$ is the left-invariant Haar measure on G. Then, any vector $|\phi\rangle \in \mathcal{H}$ can be represented in the form

$$|\phi\rangle = C_\psi^{-1}\int_G | U(g)\psi\rangle d\mu(g)\langle\psi U^*(g) | \phi\rangle.$$

At the restriction of G on the affine subgroup $x' = ax + b$ (the group of translations and dilatations of the real axis) a harmonic analysis on the group G is reduced to the wavelet-transformation:

$$\begin{aligned} \phi(x) &= \frac{1}{C_\psi}\int \frac{1}{a^d}\psi\left(\frac{x-b}{a}\right)\phi_a(b)\frac{dad^db}{a}, \\ \phi_a(b) &= \int \overline{\frac{1}{a^d}\psi\left(\frac{x-b}{a}\right)}\phi(x)d^dx, \\ C_\psi &= \int_0^\infty |\widetilde{\psi}(ak)|^2\frac{da}{a} < \infty. \end{aligned} \tag{16}$$

At this restriction the field $\boldsymbol{\psi}(\boldsymbol{\alpha}) = \langle x, g\,|\boldsymbol{\psi}\rangle$ on the group G is reduced to the field $\boldsymbol{\psi}(\boldsymbol{\alpha}) = \langle x, a\,|\boldsymbol{\psi}\rangle \sim \langle x, a; g|\phi(x)\rangle$ on the affine group, where a is a scale factor which can be associated with the resolution of measuring equipment [2]. It should be noted that further restriction of G on the translation subgroup leads to the restriction of $\psi(\boldsymbol{\alpha}) = \langle x, g\,|\psi\rangle$ to the field $\psi(x) = \langle x\,|\psi\rangle$ describing the point-like object, and the wavelet-transformation in this case is reduced to the Fourier-transformation. Thus, the affine group is a minimal group which can be used for description of the extended object. The restriction of G to the affine subgroup is equivalent formally to the Faddeev-Popov method considering translation and dilatation invariance in quantum field calculations related with the renormalization group, at this point, $a = 1/\Lambda$, where Λ is a cut-off parameter. Moreover, in the framework of wavelet-analysis, loop Green functions emerging in ϕ^4-model, are free from ultra-violet divergences [1, 2].

3. Free Fields on the Two-Dimensional Complex Sphere

We will start with a more general homogeneous space of the group $\mathcal{P}$, $\mathcal{M}_{10} = \mathbb{R}^{1,3}\times\mathfrak{L}_6$ (group manifold of the Poincaré group). Let $\mathcal{L}(\boldsymbol{\alpha})$ be a Lagrangian on the group manifold $\mathcal{M}_{10}$ (in other words, $\mathcal{L}(\boldsymbol{\alpha})$ is a 10-dimensional point function), where $\boldsymbol{\alpha}$ is the parameter set of this group. Then an integral for $\mathcal{L}(\boldsymbol{\alpha})$ on some 10-dimensional volume Ω of the group manifold we will call *an action on the Poincaré group*:

$$A = \int\limits_{\Omega} d\boldsymbol{\alpha}\mathcal{L}(\boldsymbol{\alpha}),$$

where $d\boldsymbol{\alpha}$ is a Haar measure on the group $\mathcal{P}$.

Let $\psi(\boldsymbol{\alpha}) = \langle x, \mathfrak{g}\,|\psi\rangle$ be a function on the group manifold $\mathcal{M}_{10}$ (now it is sufficient to assume that $\psi(\boldsymbol{\alpha})$ is a square integrable function on the Poincaré group) and let

$$\frac{\partial\mathcal{L}}{\partial\psi} - \frac{\partial}{\partial\boldsymbol{\alpha}}\frac{\partial\mathcal{L}}{\partial\frac{\partial\psi}{\partial\boldsymbol{\alpha}}} = 0 \qquad (17)$$

be Euler-Lagrange equations on $\mathcal{M}_{10}$ (more precisely speaking, the equations (17) act on the tangent bundle $T\mathcal{M}_{10} = \bigcup\limits_{\boldsymbol{\alpha}\in\mathcal{M}_{10}} T_{\boldsymbol{\alpha}}\mathcal{M}_{10}$ of the manifold $\mathcal{M}_{10}$, see [4]). Let us introduce a Lagrangian $\mathcal{L}(\boldsymbol{\alpha})$ depending on the field function $\psi(\boldsymbol{\alpha})$ as follows

$$\mathcal{L}(\boldsymbol{\alpha}) = -\frac{1}{2}\left(\psi^*(\boldsymbol{\alpha})B_\mu\frac{\partial\psi(\boldsymbol{\alpha})}{\partial\boldsymbol{\alpha}_\mu} - \frac{\partial\psi^*(\boldsymbol{\alpha})}{\partial\boldsymbol{\alpha}_\mu}B_\mu\psi(\boldsymbol{\alpha})\right) - \kappa\psi^*(\boldsymbol{\alpha})B_{11}\psi(\boldsymbol{\alpha}),$$

where B_ν ($\nu = 1, 2, \ldots, 10$) are square matrices. The number of rows and columns in these matrices is equal to the number of components of $\psi(\boldsymbol{\alpha})$, κ is a non-null real constant.

Further, if B_{11} is non-singular, then we can introduce the matrices

$$\Gamma_\mu = B_{11}^{-1}B_\mu, \quad \mu = 1, 2, \ldots, 10,$$

and represent the Lagrangian $\mathcal{L}(\boldsymbol{\alpha})$ in the form

$$\mathcal{L}(\boldsymbol{\alpha}) = -\frac{1}{2}\left(\overline{\psi}(\boldsymbol{\alpha})\Gamma_\mu\frac{\partial\psi(\boldsymbol{\alpha})}{\partial\boldsymbol{\alpha}_\mu} - \frac{\overline{\psi}(\boldsymbol{\alpha})}{\partial\boldsymbol{\alpha}_\mu}\Gamma_\mu\psi(\boldsymbol{\alpha})\right) - \kappa\overline{\psi}(\boldsymbol{\alpha})\psi(\boldsymbol{\alpha}), \qquad (18)$$

where

$$\overline{\psi}(\boldsymbol{\alpha}) = \psi^*(\boldsymbol{\alpha})B_{11}.$$

Varying independently $\psi(x)$ and $\overline{\psi}(x)$, we obtain from (18) in accordance with (17) the following equations:

$$\begin{aligned} \Gamma_i \frac{\partial \psi(x)}{\partial x_i} + \kappa\psi(x) &= 0, \\ \Gamma_i^T \frac{\partial \overline{\psi}(x)}{\partial x_i} - \kappa\overline{\psi}(x) &= 0. \end{aligned} \qquad (i = 0, \ldots, 3) \tag{19}$$

The matrix Γ_0 in equations (19) can be written in the form (see [20, 3, 48])

$$\Gamma_0 = \text{diag}\left(C^0 \otimes I_1, C^1 \otimes I_3, \ldots, C^s \otimes I_{2s+1}, \ldots\right) \tag{20}$$

for integer spin and

$$\Gamma_0 = \text{diag}\left(C^{\frac{1}{2}} \otimes I_2, C^{\frac{3}{2}} \otimes I_4, \ldots, C^s \otimes I_{2s+1}, \ldots\right) \tag{21}$$

for half-integer spin, where C^s is a spin block. If the spin block C^s has non-null roots, then the particle possesses the spin s. The spin block C^s in (20)–(21) consists of the elements $c^s_{\boldsymbol{\tau}\boldsymbol{\tau}'}$, where $\boldsymbol{\tau}_{l_1,l_2}$ and $\boldsymbol{\tau}_{l_1',l_2'}$ are interlocking irreducible representations of the Lorentz group, that is, such representations, for which $l_1' = l_1 \pm \frac{1}{2}$, $l_2' = l_2 \pm \frac{1}{2}$. At this point, the block C^s contains only the elements $c^s_{\boldsymbol{\tau}\boldsymbol{\tau}'}$ corresponding to such interlocking representations $\boldsymbol{\tau}_{l_1,l_2}$, $\boldsymbol{\tau}_{l_1',l_2'}$ which satisfy the conditions

$$|l_1 - l_2| \leq s \leq l_1 + l_2, \quad |l_1' - l_2'| \leq s \leq l_1' + l_2'.$$

The two most full schemes of the interlocking irreducible representations of the Lorentz group for integer and half-integer spins are shown on the Fig. 1 and Fig. 2.

As follows from Fig. 1 the simplest field is the scalar field

$$(0, 0).$$

This field is described by the Fock-Klein-Gordon equation. In its turn, the simplest field from the Fermi-scheme (Fig. 2) is the electron-positron (spinor) field corresponding to the following interlocking scheme:

$$\left(\frac{1}{2}, 0\right) \longleftrightarrow \left(0, \frac{1}{2}\right).$$

This field is described by the Dirac equation. Further, the next field from the Bose-scheme (Fig. 1) is a photon field (Maxwell field) defined within the interlocking scheme

$$(1, 0) \longleftrightarrow \left(\frac{1}{2}, \frac{1}{2}\right) \longleftrightarrow (0, 1)$$

This interlocking scheme leads to the Maxwell equations. The fields $(1/2, 0) \oplus (0, 1/2)$ and $(1, 0) \oplus (0, 1)$ (Dirac and Maxwell fields) are particular cases of fields $\psi(\boldsymbol{\alpha}) = \langle x, \mathfrak{g} \,|\psi\rangle$

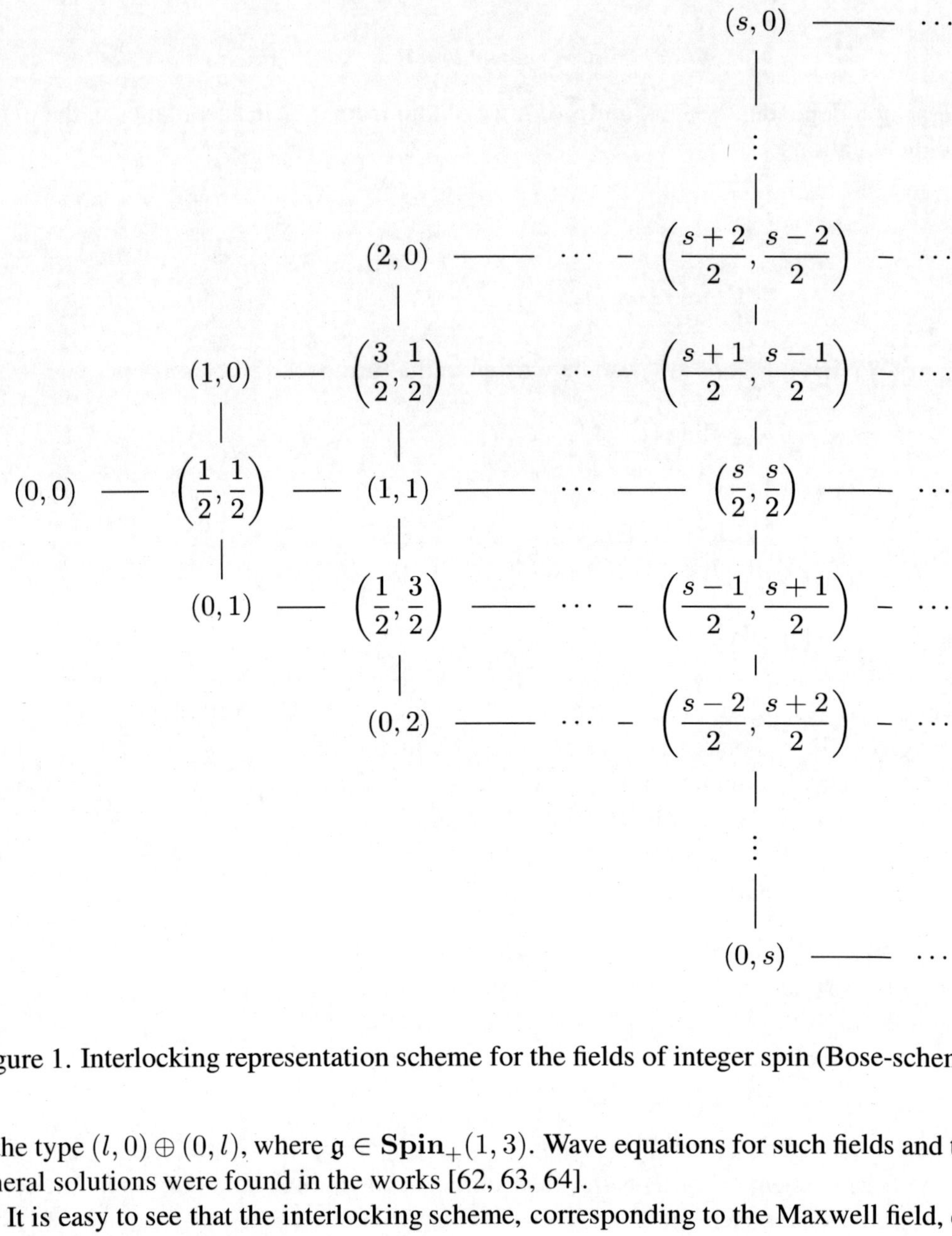

Figure 1. Interlocking representation scheme for the fields of integer spin (Bose-scheme).

of the type $(l,0)\oplus(0,l)$, where $\mathfrak{g}\in\mathbf{Spin}_+(1,3)$. Wave equations for such fields and their general solutions were found in the works [62, 63, 64].

It is easy to see that the interlocking scheme, corresponding to the Maxwell field, contains the field of tensor type:

$$\left(\frac{1}{2},\frac{1}{2}\right).$$

Further, the next interlocking scheme (see Fig. 2)

$$\left(\frac{3}{2},0\right)\longleftrightarrow\left(1,\frac{1}{2}\right)\longleftrightarrow\left(\frac{1}{2},1\right)\longleftrightarrow\left(0,\frac{3}{2}\right)$$

corresponding to the Pauli-Fierz equations [19], contains a chain of the type

$$\left(1,\frac{1}{2}\right)\longleftrightarrow\left(\frac{1}{2},1\right)$$

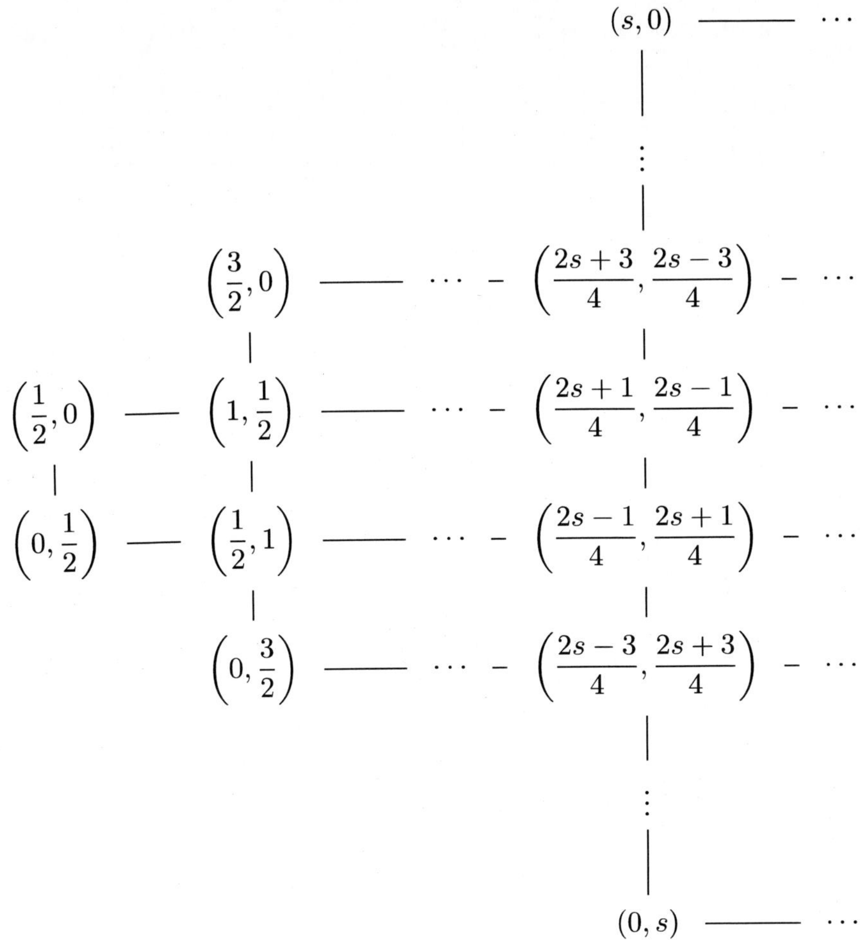

Figure 2. Interlocking representation scheme for the fields of half-integer spin (Fermi-scheme).

In such a way we come to wave equations for the fields $\boldsymbol{\psi}(\boldsymbol{\alpha}) = \langle x, \mathfrak{g}\,|\boldsymbol{\psi}\rangle$ of tensor type $(l_1, l_2) \oplus (l_2, l_1)$. Wave equations for such fields and their general solutions were found in the work [67].

Further, varying independently $\psi(\mathfrak{g})$ and $\overline{\psi}(\mathfrak{g})$ one gets from (18) the following equations:

$$\begin{aligned} \Gamma_k \frac{\partial \psi(\mathfrak{g})}{\partial \mathfrak{g}_k} + \kappa\psi(\mathfrak{g}) &= 0, \\ \Gamma_k^T \frac{\overline{\psi}(\mathfrak{g})}{\partial \mathfrak{g}_k} - \kappa\overline{\psi}(\mathfrak{g}) &= 0, \end{aligned} \qquad (k = 1, \ldots, 6) \tag{22}$$

where

$$\psi(\mathfrak{q}) = \begin{pmatrix} \psi(\mathfrak{q}) \\ \dot{\psi}(\mathfrak{q}) \end{pmatrix}, \quad \Gamma_k = \begin{pmatrix} 0 & \Lambda_k^* \\ \Lambda_k & 0 \end{pmatrix}.$$

Non-zero elements of the matrices Λ_k and Λ_k^* have the form (for more details see [62])

$$\Lambda_1: \left\{ \begin{array}{rcl} a^{k'k}_{l-1,l,m-1,m} & = & -\frac{c_{l-1,l}}{2}\sqrt{(l+m)(l+m-1)}, \\ a^{k'k}_{l,l,m-1,m} & = & \frac{c_{ll}}{2}\sqrt{(l+m)(l-m+1)}, \\ a^{k'k}_{l+1,l,m-1,m} & = & \frac{c_{l+1,l}}{2}\sqrt{(l-m+1)(l-m+2)}, \\ a^{k'k}_{l-1,l,m+1,m} & = & \frac{c_{l-1,l}}{2}\sqrt{(l-m)(l-m-1)}, \\ a^{k'k}_{l,l,m+1,m} & = & \frac{c_{ll}}{2}\sqrt{(l+m+1)(l-m)}, \\ a^{k'k}_{l+1,l,m+1,m} & = & -\frac{c_{l+1,l}}{2}\sqrt{(l+m+1)(l+m+2)}. \end{array} \right. \tag{23}$$

$$\Lambda_2: \left\{ \begin{array}{rcl} b^{k'k}_{l-1,l,m-1,m} & = & -\frac{ic_{l-1,l}}{2}\sqrt{(l+m)(l+m-1)}, \\ b^{k'k}_{l,l,m-1,m} & = & \frac{ic_{ll}}{2}\sqrt{(l+m)(l-m+1)}, \\ b^{k'k}_{l+1,l,m-1,m} & = & \frac{ic_{l+1,l}}{2}\sqrt{(l-m+1)(l-m+2)}, \\ b^{k'k}_{l-1,l,m+1,m} & = & -\frac{ic_{l-1,l}}{2}\sqrt{(l-m)(l-m-1)}, \\ b^{k'k}_{l,l,m+1,m} & = & -\frac{ic_{ll}}{2}\sqrt{(l+m+1)(l-m)}, \\ b^{k'k}_{l+1,l,m+1,m} & = & \frac{ic_{l+1,l}}{2}\sqrt{(l+m+1)(l+m+2)}. \end{array} \right. \tag{24}$$

$$\Lambda_3: \left\{ \begin{array}{rcl} c^{k'k}_{l-1,l,m} & = & c^{k'k}_{l-1,l}\sqrt{l^2-m^2}, \\ c^{k'k}_{l,l,m} & = & c^{k'k}_{ll}m, \\ c^{k'k}_{l+1,l,m} & = & c^{k'k}_{l+1,l}\sqrt{(l+1)^2-m^2}. \end{array} \right. \tag{25}$$

$$\Lambda_1^*: \left\{ \begin{array}{rcl} d^{\dot{k}'\dot{k}}_{\dot{l}-1,\dot{l},\dot{m}-1,\dot{m}} & = & -\frac{c_{\dot{l}-1,\dot{l}}}{2}\sqrt{(\dot{l}+\dot{m})(\dot{l}-\dot{m}-1)}, \\ d^{\dot{k}'\dot{k}}_{\dot{l},\dot{l},\dot{m}-1,\dot{m}} & = & \frac{c_{\dot{l}\dot{l}}}{2}\sqrt{(\dot{l}+\dot{m})(\dot{l}-\dot{m}+1)}, \\ d^{\dot{k}'\dot{k}}_{\dot{l}+1,\dot{l},\dot{m}-1,\dot{m}} & = & \frac{c_{\dot{l}+1,\dot{l}}}{2}\sqrt{(\dot{l}-\dot{m}+1)(\dot{l}-\dot{m}+2)}, \\ d^{\dot{k}'\dot{k}}_{\dot{l}-1,\dot{l},\dot{m}+1,\dot{m}} & = & \frac{c_{\dot{l}-1,\dot{l}}}{2}\sqrt{(\dot{l}-\dot{m})(\dot{l}-\dot{m}-1)}, \\ d^{\dot{k}'\dot{k}}_{\dot{l},\dot{l},\dot{m}+1,\dot{m}} & = & \frac{c_{\dot{l}\dot{l}}}{2}\sqrt{(\dot{l}+\dot{m}+1)(\dot{l}-\dot{m})}, \\ d^{\dot{k}'\dot{k}}_{\dot{l}+1,\dot{l},\dot{m}+1,\dot{m}} & = & -\frac{c_{\dot{l}+1,\dot{l}}}{2}\sqrt{(\dot{l}+\dot{m}+1)(\dot{l}+\dot{m}+2)}. \end{array} \right. \tag{26}$$

$$\Lambda_2^*: \left\{ \begin{array}{rcl} e^{\dot{k}'\dot{k}}_{\dot{l}-1,\dot{l},\dot{m}-1,\dot{m}} & = & -\frac{ic_{\dot{l}-1,\dot{l}}}{2}\sqrt{(\dot{l}+\dot{m})(\dot{l}-\dot{m}-1)}, \\ e^{\dot{k}'\dot{k}}_{\dot{l},\dot{l},\dot{m}-1,\dot{m}} & = & \frac{ic_{\dot{l}\dot{l}}}{2}\sqrt{(\dot{l}+\dot{m})(\dot{l}-\dot{m}+1)}, \\ e^{\dot{k}'\dot{k}}_{\dot{l}+1,\dot{l},\dot{m}-1,\dot{m}} & = & \frac{ic_{\dot{l}+1,\dot{l}}}{2}\sqrt{(\dot{l}-\dot{m}+1)(\dot{l}-\dot{m}+2)}, \\ e^{\dot{k}'\dot{k}}_{\dot{l}-1,\dot{l},\dot{m}+1,\dot{m}} & = & \frac{-ic_{\dot{l}-1,\dot{l}}}{2}\sqrt{(\dot{l}-\dot{m})(\dot{l}-\dot{m}-1)}, \\ e^{\dot{k}'\dot{k}}_{\dot{l},\dot{l},\dot{m}+1,\dot{m}} & = & \frac{-ic_{\dot{l}\dot{l}}}{2}\sqrt{(\dot{l}+\dot{m}+1)(\dot{l}-\dot{m})}, \\ e^{\dot{k}'\dot{k}}_{\dot{l}+1,\dot{l},\dot{m}+1,\dot{m}} & = & -\frac{ic_{\dot{l}+1,\dot{l}}}{2}\sqrt{(\dot{l}+\dot{m}+1)(\dot{l}+\dot{m}+2)}. \end{array} \right. \tag{27}$$

$$\Lambda_3^*: \left\{ \begin{array}{rcl} f^{\dot{k}'\dot{k}}_{\dot{l}-1,\dot{l},\dot{m}} & = & c^{\dot{k}'\dot{k}}_{\dot{l}-1,\dot{l}}\sqrt{\dot{l}^2-\dot{m}^2}, \\ f^{\dot{k}'\dot{k}}_{\dot{l}\dot{l},\dot{m}} & = & c^{\dot{k}'\dot{k}}_{\dot{l}\dot{l}}\dot{m}, \\ f^{\dot{k}'\dot{k}}_{\dot{l}+1,\dot{l},\dot{m}} & = & c^{\dot{k}'\dot{k}}_{\dot{l}+1,\dot{l}}\sqrt{(\dot{l}+1)^2-\dot{m}^2}. \end{array} \right. \tag{28}$$

In general, the matrix Λ_3 must be a reducible representation of the proper Lorentz group $\mathfrak{G}_+$, and can always be written in the form

$$\Lambda_3 = \begin{bmatrix} \Lambda_3^{l_1} & & & & \\ & \Lambda_3^{l_2} & & 0 & \\ & & \Lambda_3^{l_3} & & \\ & 0 & & \ddots & \\ & & & & \Lambda_3^{l_n} \end{bmatrix}, \tag{29}$$

where $\Lambda_3^{l_i}$ is a spin block (the matrix $\overset{*}{\Lambda}_3$ has the same decompositions). It is obvious that the matrices Λ_1, Λ_2 and $\overset{*}{\Lambda}_1$, $\overset{*}{\Lambda}_2$ admit also the decompositions of the type (29) by definition. If the spin block $\Lambda_3^{l_i}$ has non-null roots, then the particle possesses the spin l_i. Corresponding to the decomposition (29), the wave function also decomposes into a direct sum of component wave functions which we write

$$\psi = \psi_{l_1 m_1} + \psi_{l_2 m_2} + \psi_{l_3 m_3} + \ldots .$$

According to a de Broglie theory of fusion [14], interlocking representations give rise to indecomposable relativistic wave equations. Otherwise, we have decomposable equations. As is known, the indecomposable wave equations correspond to composite particles. A relation between indecomposable wave equations and composite particles will be studied in a separate work.

Returning to the equations (22), we see that the first equation can be written in the form

$$\begin{aligned} \sum_{j=1}^{3} \Lambda_j^* \frac{\partial \dot{\psi}}{\partial \widetilde{a}_j} + i \sum_{j=1}^{3} \Lambda_j^* \frac{\partial \dot{\psi}}{\partial \widetilde{a}_j^*} + \kappa^c \dot{\psi} &= 0, \\ \sum_{j=1}^{3} \Lambda_j \frac{\partial \psi}{\partial a_j} - i \sum_{j=1}^{3} \Lambda_j \frac{\partial \psi}{\partial a_j^*} + \dot{\kappa}^c \dot{\psi} &= 0, \end{aligned} \tag{30}$$

where $\mathfrak{g}_1 = a_1$, $\mathfrak{g}_2 = a_2$, $\mathfrak{g}_3 = a_3$, $\mathfrak{g}_4 = ia_1$, $\mathfrak{g}_5 = ia_2$, $\mathfrak{g}_6 = ia_3$, $a_1^* = -i\mathfrak{g}_4$, $a_2^* = -i\mathfrak{g}_5$, $a_3^* = -i\mathfrak{g}_6$, and $\widetilde{a}_j$, $\widetilde{a}_j^*$ are the parameters corresponding to the dual basis. In essence, the equations (30) are defined in a three-dimensional complex space $\mathbb{C}^3$. In turn, the space $\mathbb{C}^3$ is isometric to a 6-dimensional bivector space $\mathbb{R}^6$ (a parameter space of the Lorentz group [47]). The bivector space $\mathbb{R}^6$ is a tangent space of the group manifold $\mathfrak{L}_6$ of the Lorentz group, that is, the manifold $\mathfrak{L}_6$ in the each its point is equivalent locally to the space $\mathbb{R}^6$. Thus, for all $\mathfrak{g} \in \mathfrak{L}_6$ we have $T_{\mathfrak{g}}\mathfrak{L}_6 \simeq \mathbb{R}^6$. There exists a close relationship between the metrics of the Minkowski spacetime $\mathbb{R}^{1,3}$ and the metrics of $\mathbb{R}^6$ (see Appendix). In the case of $\mathbb{R}^{1,3}$ with the metric tensor

$$g_{\alpha\beta} = \begin{pmatrix} -1 & 0 & 0 & 0 \\ 0 & -1 & 0 & 0 \\ 0 & 0 & -1 & 0 \\ 0 & 0 & 0 & 1 \end{pmatrix}$$

in virtue of (63) for the metric tensor of $\mathbb{R}^6$ we obtain

$$g_{ab} = \begin{bmatrix} -1 & 0 & 0 & 0 & 0 & 0 \\ 0 & -1 & 0 & 0 & 0 & 0 \\ 0 & 0 & -1 & 0 & 0 & 0 \\ 0 & 0 & 0 & 1 & 0 & 0 \\ 0 & 0 & 0 & 0 & 1 & 0 \\ 0 & 0 & 0 & 0 & 0 & 1 \end{bmatrix}, \tag{31}$$

where the order of collective indices in $\mathbb{R}^6$ is $23 \rightarrow 0$, $10 \rightarrow 1$, $20 \rightarrow 2$, $30 \rightarrow 3$, $31 \rightarrow 4$, $12 \rightarrow 5$. The system (30) in the bivector space is written as follows:

$$\sum_i \Big[g_{i\overset{-}{1}}\Lambda_1 T_{\mathfrak{g}}^{-1}\frac{\partial\psi'}{\partial a'_i} + g_{i\overset{-}{2}}\Lambda_2 T_{\mathfrak{g}}^{-1}\frac{\partial\psi'}{\partial a'_i} + g_{i\overset{-}{3}}\Lambda_3 T_{\mathfrak{g}}^{-1}\frac{\partial\psi'}{\partial a'_i} -$$

$$-ig_{i\overset{-}{1}}\Lambda_1 T_{\mathfrak{g}}^{-1}\frac{\partial\psi'}{\partial a^{*\prime}_i} - ig_{i\overset{-}{2}}\Lambda_2 T_{\mathfrak{g}}^{-1}\frac{\partial\psi'}{\partial a^{*\prime}_i} - ig_{i\overset{-}{3}}\Lambda_3 T_{\mathfrak{g}}^{-1}\frac{\partial\psi'}{\partial a^{*\prime}_i}\Big] + \kappa^c T_{\mathfrak{g}}^{-1}\psi' = 0,$$

$$\sum_i \Big[g_{i\overset{+}{1}}\Lambda^*_1 \overset{*}{T}{}_{\mathfrak{g}}^{-1}\frac{\partial\dot{\psi}'}{\partial\widetilde{a}'_i} + g_{i\overset{+}{2}}\Lambda^*_2 \overset{*}{T}{}_{\mathfrak{g}}^{-1}\frac{\partial\dot{\psi}'}{\partial\widetilde{a}'_i} + g_{i\overset{+}{3}}\Lambda^*_3 \overset{*}{T}{}_{\mathfrak{g}}^{-1}\frac{\partial\dot{\psi}'}{\partial\widetilde{a}'_i} +$$

$$+ig_{i\overset{+}{1}}\Lambda^*_1 \overset{*}{T}{}_{\mathfrak{g}}^{-1}\frac{\partial\dot{\psi}'}{\partial\widetilde{a^*}_i{}'} + ig_{i\overset{+}{2}}\Lambda^*_2 \overset{*}{T}{}_{\mathfrak{g}}^{-1}\frac{\partial\dot{\psi}'}{\partial\widetilde{a^*}_i{}'} + ig_{i\overset{+}{3}}\Lambda^*_3 \overset{*}{T}{}_{\mathfrak{g}}^{-1}\frac{\partial\dot{\psi}'}{\partial\widetilde{a^*}_i{}'}\Big] + \dot{\kappa}^c \overset{*}{T}{}_{\mathfrak{g}}^{-1}\dot{\psi}' = 0,$$

where

$$\frac{\partial}{\partial a_1} = -\frac{\sin\varphi^c}{r\sin\theta^c}\frac{\partial}{\partial\varphi} + \frac{\cos\varphi^c\cos\theta^c}{r}\frac{\partial}{\partial\theta} + \cos\varphi^c\sin\theta^c\frac{\partial}{\partial r}, \tag{32}$$

$$\frac{\partial}{\partial a_2} = \frac{\cos\varphi^c}{r\sin\theta^c}\frac{\partial}{\partial\varphi} + \frac{\sin\varphi^c\cos\theta^c}{r}\frac{\partial}{\partial\theta} + \sin\varphi^c\sin\theta^c\frac{\partial}{\partial r}, \tag{33}$$

$$\frac{\partial}{\partial a_3} = -\frac{\sin\theta^c}{r}\frac{\partial}{\partial\theta} + \cos\theta^c\frac{\partial}{\partial r}. \tag{34}$$

$$\frac{\partial}{\partial a^*_1} = i\frac{\partial}{\partial a_1} = -\frac{\sin\varphi^c}{r\sin\theta^c}\frac{\partial}{\partial\epsilon} + \frac{\cos\varphi^c\sin\theta^c}{r}\frac{\partial}{\partial\tau} + i\cos\varphi^c\sin\theta^c\frac{\partial}{\partial r}, \tag{35}$$

$$\frac{\partial}{\partial a^*_2} = i\frac{\partial}{\partial a_2} = \frac{\cos\varphi^c}{r\sin\theta^c}\frac{\partial}{\partial\epsilon} + \frac{\sin\varphi^c\cos\theta^c}{r}\frac{\partial}{\partial\tau} + i\sin\varphi^c\sin\theta^c\frac{\partial}{\partial r}, \tag{36}$$

$$\frac{\partial}{\partial a^*_3} = i\frac{\partial}{\partial a_3} = -\frac{\sin\theta^c}{r}\frac{\partial}{\partial\tau} + i\cos\theta^c\frac{\partial}{\partial r}. \tag{37}$$

$$\frac{\partial}{\partial\widetilde{a}_1} = -\frac{\sin\dot{\varphi}^c}{r^*\sin\dot{\theta}^c}\frac{\partial}{\partial\varphi} + \frac{\cos\dot{\varphi}^c\cos\dot{\theta}^c}{r^*}\frac{\partial}{\partial\theta} + \cos\dot{\varphi}^c\sin\dot{\theta}^c\frac{\partial}{\partial r^*}, \tag{38}$$

$$\frac{\partial}{\partial\widetilde{a}_2} = \frac{\cos\dot{\varphi}^c}{r^*\sin\dot{\theta}^c}\frac{\partial}{\partial\varphi} + \frac{\sin\dot{\varphi}^c\cos\dot{\theta}^c}{r^*}\frac{\partial}{\partial\theta} + \sin\dot{\varphi}^c\sin\dot{\theta}^c\frac{\partial}{\partial r^*}, \tag{39}$$

$$\frac{\partial}{\partial\widetilde{a}_3} = -\frac{\sin\dot{\theta}^c}{r^*}\frac{\partial}{\partial\theta} + \cos\dot{\theta}^c\frac{\partial}{\partial r^*}. \tag{40}$$

$$\frac{\partial}{\partial \widetilde{a}_1^*} = -i\frac{\partial}{\partial \widetilde{a}_1} = \frac{\sin\dot{\varphi}^c}{r^*\sin\dot{\theta}^c}\frac{\partial}{\partial\epsilon} - \frac{\cos\dot{\varphi}^c\cos\dot{\theta}^c}{r^*}\frac{\partial}{\partial\tau} - i\cos\dot{\varphi}^c\sin\dot{\theta}^c\frac{\partial}{\partial r^*}, \quad (41)$$

$$\frac{\partial}{\partial \widetilde{a}_2^*} = -i\frac{\partial}{\partial \widetilde{a}_2} = -\frac{\cos\dot{\varphi}^c}{r^*\sin\dot{\theta}^c}\frac{\partial}{\partial\epsilon} - \frac{\sin\dot{\varphi}^c\cos\dot{\theta}^c}{r^*}\frac{\partial}{\partial\tau} - i\sin\dot{\varphi}^c\sin\dot{\theta}^c\frac{\partial}{\partial r^*}, \quad (42)$$

$$\frac{\partial}{\partial \widetilde{a}_3^*} = -i\frac{\partial}{\partial \widetilde{a}_3} = \frac{\sin\dot{\theta}^c}{r^*}\frac{\partial}{\partial\tau} - i\cos\dot{\theta}^c\frac{\partial}{\partial r^*} \quad (43)$$

are derivatives defined on the spheres (10) and (11). Solutions of wave equations for extended object are found in the form of series in associated hyperspherical functions which defined on the spheres S_c^2 and $\dot{S}_c^2$:

$$\psi^k_{lm;\dot{l}\dot{m}} = \boldsymbol{f}^{l_0}_{lmk}(r)\mathfrak{M}^{l_0}_{mn}(\varphi,\epsilon,\theta,\tau,0,0),$$
$$\psi^{\dot{k}}_{\dot{l}\dot{m};lm} = \boldsymbol{f}^{\dot{l}_0}_{\dot{l}\dot{m}\dot{k}}(r^*)\mathfrak{M}^{\dot{l}_0}_{\dot{m}\dot{n}}(\varphi,\epsilon,\theta,\tau,0,0),$$

where $l_0 \geq l$, $-l_0 \leq m, n \leq l_0$ and $\dot{l}_0 \geq \dot{l}$, $-\dot{l}_0 \leq \dot{m}, \dot{n} \leq \dot{l}_0$.

We claim that a system of Dirac like wave equations

$$\left\{\begin{array}{rcl} \Gamma_i\dfrac{\partial\psi(x)}{\partial x_i} + \kappa\psi(x) & = & 0, \\ \Gamma_i^T\dfrac{\partial\overline{\psi}(x)}{\partial x_i} - \kappa\overline{\psi}(x) & = & 0, \quad (i=0,\ldots,3); \\ \Gamma_k\dfrac{\partial\psi(\mathfrak{g})}{\partial\mathfrak{g}_k} + \kappa\psi(\mathfrak{g}) & = & 0, \\ \Gamma_k^T\dfrac{\overline{\psi}(\mathfrak{g})}{\partial\mathfrak{g}_k} - \kappa\overline{\psi}(\mathfrak{g}) & = & 0 \quad (k=1,\ldots,6) \end{array}\right. \quad (44)$$

describes extended particles of any spin on the group manifold $\mathcal{M}_{10} = \mathbb{R}^{1,3}\times\mathfrak{L}_6$ *of the Poincaré group* $\mathcal{P}$ (in particular, on the homogeneous space $\mathcal{M}_8 = \mathbb{R}^{1,3}\times S_c^2$). In its turn, the Ginzburg-Tamm system of Klein-Gordon like wave equations (2) describes extended particles on the homogeneous space $\mathcal{M} = \mathbb{R}^{1,3}\times H^3$ of $\mathcal{P}$, where H^3 is a 3-dimensional one-sheeted hyperboloid. A relation between the systems (44) and (2) is similar to a relation between usual Dirac and Klein-Gordon wave equations for the point-like particles. Namely, equations (44) (differential equations of the first order) can be understood as 'square roots' of the equations (2) (differential equations of the second order).

3.1. The Dirac Field

In accordance with the general Fermi-scheme (Fig. 1) of the interlocking representations of $\mathfrak{G}_+$ the field $(1/2,0)\oplus(0,1/2)$ is defined within the following chain:

$$\left(\frac{1}{2},0\right) \longleftrightarrow \left(0,\frac{1}{2}\right)$$

We start with the Lagrangian (18) on the group manifold $\mathcal{M}_{10}$:

$$\mathcal{L}(\boldsymbol{\alpha}) = -\frac{1}{2}\left(\overline{\boldsymbol{\psi}}(\boldsymbol{\alpha})\Gamma_\mu\frac{\partial\boldsymbol{\psi}(\boldsymbol{\alpha})}{\partial x_\mu} - \frac{\partial\overline{\boldsymbol{\psi}}(\boldsymbol{\alpha})}{\partial x_\mu}\Gamma_\mu\boldsymbol{\psi}(\boldsymbol{\alpha})\right) - \\ -\frac{1}{2}\left(\overline{\boldsymbol{\psi}}(\boldsymbol{\alpha})\Upsilon_\nu\frac{\partial\boldsymbol{\psi}(\boldsymbol{\alpha})}{\partial \mathfrak{g}_\nu} - \frac{\partial\overline{\boldsymbol{\psi}}(\boldsymbol{\alpha})}{\partial \mathfrak{g}_\nu}\Upsilon_\nu\boldsymbol{\psi}(\boldsymbol{\alpha})\right) - \kappa\overline{\boldsymbol{\psi}}(\boldsymbol{\alpha})\boldsymbol{\psi}(\boldsymbol{\alpha}), \quad (45)$$

where $\boldsymbol{\psi}(\boldsymbol{\alpha}) = \psi(x)\psi(\mathfrak{g})$ ($\mu = 0,1,2,3,\ \nu = 1,\ldots,6$), and

$$\gamma_0 = \begin{pmatrix}\sigma_0 & 0\\ 0 & -\sigma_0\end{pmatrix},\ \gamma_1 = \begin{pmatrix}0 & \sigma_1\\ -\sigma_1 & 0\end{pmatrix},\ \gamma_2 = \begin{pmatrix}0 & \sigma_2\\ -\sigma_2 & 0\end{pmatrix},\ \gamma_3 = \begin{pmatrix}0 & \sigma_3\\ -\sigma_3 & 0\end{pmatrix}, \quad (46)$$

$$\Upsilon_1 = \begin{pmatrix}0 & \Lambda_1^*\\ \Lambda_1 & 0\end{pmatrix},\quad \Upsilon_2 = \begin{pmatrix}0 & \Lambda_2^*\\ \Lambda_2 & 0\end{pmatrix},\quad \Upsilon_3 = \begin{pmatrix}0 & \Lambda_3^*\\ \Lambda_3 & 0\end{pmatrix}, \quad (47)$$

$$\Upsilon_4 = \begin{pmatrix}0 & i\Lambda_1^*\\ i\Lambda_1 & 0\end{pmatrix},\quad \Upsilon_5 = \begin{pmatrix}0 & i\Lambda_2^*\\ i\Lambda_2 & 0\end{pmatrix},\quad \Upsilon_6 = \begin{pmatrix}0 & i\Lambda_3^*\\ i\Lambda_3 & 0\end{pmatrix}, \quad (48)$$

where σ_i are the Pauli matrices, and the matrices Λ_j and Λ_j^* are derived from (23)–(25) and (26)–(28) at $l = 1/2$:

$$\Lambda_1 = \frac{1}{2}c_{\frac{1}{2}\frac{1}{2}}\begin{pmatrix}0 & 1\\ 1 & 0\end{pmatrix},\quad \Lambda_2 = \frac{1}{2}c_{\frac{1}{2}\frac{1}{2}}\begin{pmatrix}0 & -i\\ i & 0\end{pmatrix},\quad \Lambda_3 = \frac{1}{2}c_{\frac{1}{2}\frac{1}{2}}\begin{pmatrix}1 & 0\\ 0 & -1\end{pmatrix},$$

$$\Lambda_1^* = \frac{1}{2}\dot{c}_{\frac{1}{2}\frac{1}{2}}\begin{pmatrix}0 & 1\\ 1 & 0\end{pmatrix},\quad \Lambda_2^* = \frac{1}{2}\dot{c}_{\frac{1}{2}\frac{1}{2}}\begin{pmatrix}0 & -i\\ i & 0\end{pmatrix},\quad \Lambda_3^* = \frac{1}{2}\dot{c}_{\frac{1}{2}\frac{1}{2}}\begin{pmatrix}1 & 0\\ 0 & -1\end{pmatrix}. \quad (49)$$

It is easy to see that these matrices coincide with the Pauli matrices σ_i when $c_{\frac{1}{2}\frac{1}{2}} = 2$. The Dirac bispinor $\psi = (\psi_1, \psi_2, \dot{\psi}_1, \dot{\psi}_2)^T$ is defined on $\mathcal{M}_8 = \mathbb{R}^{1,3}\times S_c^2$ by the following components [63]:

$$\psi_{1n}^l(\boldsymbol{\alpha}) = \psi_1^+(x)\psi_{1n}^l(\mathfrak{g}) = u_1(\mathbf{p})e^{-ipx}\boldsymbol{f}^l_{\frac{1}{2},\frac{1}{2}}(\operatorname{Re} r)\mathfrak{M}^l_{\frac{1}{2},n}(\varphi,\epsilon,\theta,\tau,0,0),$$
$$\psi_{2n}^l(\boldsymbol{\alpha}) = \psi_2^+(x)\psi_{2n}^l(\mathfrak{g}) = \pm u_2(\mathbf{p})e^{-ipx}\boldsymbol{f}^l_{\frac{1}{2},\frac{1}{2}}(\operatorname{Re} r)\mathfrak{M}^l_{-\frac{1}{2},n}(\varphi,\epsilon,\theta,\tau,0,0),$$
$$\dot{\psi}_{1\dot{n}}^{\dot{l}}(\boldsymbol{\alpha}) = \psi_1^-(x)\dot{\psi}_{1\dot{n}}^{\dot{l}}(\mathfrak{g}) = \mp v_1(\mathbf{p})e^{ipx}\boldsymbol{f}^{\dot{l}}_{\frac{1}{2},-\frac{1}{2}}(\operatorname{Re} r^*)\mathfrak{M}^{\dot{l}}_{\frac{1}{2},\dot{n}}(\varphi,\epsilon,\theta,\tau,0,0),$$
$$\dot{\psi}_{2\dot{n}}^{\dot{l}}(\boldsymbol{\alpha}) = \psi_2^-(x)\dot{\psi}_{2\dot{n}}^{\dot{l}}(\mathfrak{g}) = v_2(\mathbf{p})e^{ipx}\boldsymbol{f}^{\dot{l}}_{\frac{1}{2},-\frac{1}{2}}(\operatorname{Re} r^*)\mathfrak{M}^{\dot{l}}_{-\frac{1}{2},\dot{n}}(\varphi,\epsilon,\theta,\tau,0,0),$$

where

$$u_1(\mathbf{p}) = \left(\frac{E+m}{2m}\right)^{1/2}\begin{pmatrix}1\\ 0\\ \frac{p_z}{E+m}\\ \frac{p_+}{E+m}\end{pmatrix},\quad u_2(\mathbf{p}) = \left(\frac{E+m}{2m}\right)^{1/2}\begin{pmatrix}0\\ 1\\ \frac{p_-}{E+m}\\ \frac{-p_z}{E+m}\end{pmatrix},$$

$$v_1(\mathbf{p}) = \left(\frac{E+m}{2m}\right)^{1/2}\begin{pmatrix}\frac{p_z}{E+m}\\ \frac{p_+}{E+m}\\ 1\\ 0\end{pmatrix},\quad v_2(\mathbf{p}) = \left(\frac{E+m}{2m}\right)^{1/2}\begin{pmatrix}\frac{p_-}{E+m}\\ \frac{-p_z}{E+m}\\ 0\\ 1\end{pmatrix},$$

here $p_{\pm} = p_x \pm ip_y$. Radial functions have the form

$$\boldsymbol{f}^{l}_{\frac{1}{2},\frac{1}{2}}(\mathrm{Re}\,r) = C_1\sqrt{\kappa^c\dot{\kappa}^c}\mathrm{Re}\,r J_l\left(\sqrt{\kappa^c\dot{\kappa}^c}\mathrm{Re}\,r\right) + C_2\sqrt{\kappa^c\dot{\kappa}^c}\mathrm{Re}\,r J_{-l}\left(\sqrt{\kappa^c\dot{\kappa}^c}\mathrm{Re}\,r\right),$$

$$\boldsymbol{f}^{\dot{l}}_{\frac{1}{2},-\frac{1}{2}}(\mathrm{Re}\,r^*) = \frac{C_1}{2}\sqrt{\frac{\dot{\kappa}^c}{\kappa^c}}\mathrm{Re}\,r^* J_{l+1}\left(\sqrt{\kappa^c\dot{\kappa}^c}\mathrm{Re}\,r^*\right) - \\ - \frac{C_2}{2}\sqrt{\frac{\dot{\kappa}^c}{\kappa^c}}\mathrm{Re}\,r^* J_{-l-1}\left(\sqrt{\kappa^c\dot{\kappa}^c}\mathrm{Re}\,r^*\right),$$

where $J_l\left(\sqrt{\kappa^c\dot{\kappa}^c}\mathrm{Re}\,r\right)$ are the Bessel functions of half-integer order, and

$$l = \frac{1}{2}, \frac{3}{2}, \frac{5}{2}, \ldots;$$
$$\dot{l} = \frac{1}{2}, \frac{3}{2}, \frac{5}{2}, \ldots;$$

$$\mathfrak{M}^{\pm\frac{1}{2}}_{l}(\varphi, \epsilon, \theta, \tau, 0, 0) = e^{\mp\frac{1}{2}(\epsilon+i\varphi)}\mathfrak{Z}^{\pm\frac{1}{2}}_{l}(\theta, \tau),$$

$$\mathfrak{Z}^{\pm\frac{1}{2}}_{l}(\theta, \tau) = \cos^{2l}\frac{\theta}{2}\cosh^{2l}\frac{\tau}{2}\sum^{l}_{k=-l} i^{\pm\frac{1}{2}-k}\tan^{\pm\frac{1}{2}-k}\frac{\theta}{2}\tanh^{-k}\frac{\tau}{2}\times \\ {}_2F_1\left(\begin{matrix}\pm\frac{1}{2}-l+1, 1-l-k\\ \pm\frac{1}{2}-k+1\end{matrix}\,\middle|\, i^2\tan^2\frac{\theta}{2}\right){}_2F_1\left(\begin{matrix}-l+1, 1-l-k\\ -k+1\end{matrix}\,\middle|\,\tanh^2\frac{\tau}{2}\right),$$

$$\mathfrak{M}^{\pm\frac{1}{2}}_{\dot{l}}(\varphi, \epsilon, \theta, \tau, 0, 0) = e^{\mp\frac{1}{2}(\epsilon-i\varphi)}\mathfrak{Z}^{\pm\frac{1}{2}}_{\dot{l}}(\theta, \tau),$$

$$\mathfrak{Z}^{\pm\frac{1}{2}}_{\dot{l}}(\theta, \tau) = \cos^{2\dot{l}}\frac{\theta}{2}\cosh^{2\dot{l}}\frac{\tau}{2}\sum^{\dot{l}}_{\dot{k}=-\dot{l}} i^{\pm\frac{1}{2}-\dot{k}}\tan^{\pm\frac{1}{2}-\dot{k}}\frac{\theta}{2}\tanh^{-\dot{k}}\frac{\tau}{2}\times \\ {}_2F_1\left(\begin{matrix}\pm\frac{1}{2}-\dot{l}+1, 1-\dot{l}-\dot{k}\\ \pm\frac{1}{2}-\dot{k}+1\end{matrix}\,\middle|\, i^2\tan^2\frac{\theta}{2}\right){}_2F_1\left(\begin{matrix}-\dot{l}+1, 1-\dot{l}-\dot{k}\\ -\dot{k}+1\end{matrix}\,\middle|\,\tanh^2\frac{\tau}{2}\right)$$

are the associated hyperspherical functions defined on the spheres S^c_2 and $\dot{S}^c_2$. General solutions are found in the form of generalized Fourier integrals

$$\psi_1(\boldsymbol{\alpha}) = \sum_{l=\frac{1}{2}}^{\infty} \boldsymbol{f}^l_{\frac{1}{2},\frac{1}{2}}(\operatorname{Re} r) \sum_{n=-l}^{l} \int_{T_4} u_1(\mathbf{p}) e^{ipx} \alpha^{\frac{1}{2}}_{l,n} \mathfrak{M}^l_{\frac{1}{2},n}(\varphi,\epsilon,\theta,\tau,0,0) d^4x,$$

$$\psi_2(\boldsymbol{\alpha}) = \pm \sum_{l=\frac{1}{2}}^{\infty} \boldsymbol{f}^l_{\frac{1}{2},\frac{1}{2}}(\operatorname{Re} r) \sum_{n=-l}^{l} \int_{T_4} u_2(\mathbf{p}) e^{ipx} \alpha^{-\frac{1}{2}}_{l,n} \mathfrak{M}^l_{-\frac{1}{2},n}(\varphi,\epsilon,\theta,\tau,0,0) d^4x,$$

$$\dot{\psi}_1(\boldsymbol{\alpha}) = \mp \sum_{\dot{l}=\frac{1}{2}}^{\infty} \boldsymbol{f}^{\dot{l}}_{\frac{1}{2},-\frac{1}{2}}(\operatorname{Re} r^*) \sum_{\dot{n}=-\dot{l}}^{\dot{l}} \int_{T_4} v_1(\mathbf{p}) e^{-ipx} \alpha^{\frac{1}{2}}_{\dot{l},\dot{n}} \mathfrak{M}^{\dot{l}}_{\frac{1}{2},\dot{n}}(\varphi,\epsilon,\theta,\tau,0,0) d^4x,$$

$$\dot{\psi}_2(\boldsymbol{\alpha}) = \sum_{\dot{l}=\frac{1}{2}}^{\infty} \boldsymbol{f}^{\dot{l}}_{\frac{1}{2},-\frac{1}{2}}(\operatorname{Re} r^*) \sum_{\dot{n}=-\dot{l}}^{\dot{l}} \int_{T_4} v_2(\mathbf{p}) e^{-ipx} \alpha^{-\frac{1}{2}}_{\dot{l},\dot{n}} \mathfrak{M}^{\dot{l}}_{-\frac{1}{2},\dot{n}}(\varphi,\epsilon,\theta,\tau,0,0) d^4x,$$

where

$$\alpha^{\pm\frac{1}{2}}_{l,n} = \frac{(-1)^n(2l+1)(2\dot{l}+1)}{32\pi^4 \boldsymbol{f}^l_{\frac{1}{2},\frac{1}{2}}(\operatorname{Re} a)} \int_{S^2_c} \int_{T_4} F_{\pm\frac{1}{2}}(\boldsymbol{\alpha}) e^{-ipx} \mathfrak{M}^l_{\pm\frac{1}{2},n}(\varphi,\epsilon,\theta,\tau,0,0) d^4x d^4\mathfrak{g},$$

$$\alpha^{\pm\frac{1}{2}}_{\dot{l},\dot{n}} = \frac{(-1)^{\dot{n}}(2l+1)(2\dot{l}+1)}{32\pi^4 \boldsymbol{f}^{\dot{l}}_{\frac{1}{2},-\frac{1}{2}}(\operatorname{Re} a^*)} \int_{S^2_c} \int_{T_4} \dot{F}_{\pm\frac{1}{2}}(\boldsymbol{\alpha}) e^{ipx} \mathfrak{M}^{\dot{l}}_{\pm\frac{1}{2},\dot{n}}(\varphi,\epsilon,\theta,\tau,0,0) d^4x d^4\mathfrak{g}.$$

3.2. The Maxwell Field

In accordance with the general Bose-scheme of the interlocking representations of $\mathfrak{G}_+$ (see Fig. 1), the field $(1,0)\oplus(0,1)$ is defined within the following interlocking scheme:

$$(1,0) \longleftrightarrow \left(\frac{1}{2},\frac{1}{2}\right) \longleftrightarrow (0,1)$$

We start with the Lagrangian (18) on the group manifold $\mathcal{M}_{10}$. Let us rewrite (18) in the form

$$\mathcal{L}(\boldsymbol{\alpha}) = -\frac{1}{2}\left(\overline{\boldsymbol{\phi}}(\boldsymbol{\alpha})\Gamma_\mu \frac{\partial \boldsymbol{\phi}(\boldsymbol{\alpha})}{\partial x_\mu} - \frac{\partial \overline{\boldsymbol{\phi}}(\boldsymbol{\alpha})}{\partial x_\mu}\Gamma_\mu \boldsymbol{\phi}(\boldsymbol{\alpha})\right) - \\ -\frac{1}{2}\left(\overline{\boldsymbol{\phi}}(\boldsymbol{\alpha})\Upsilon_\nu \frac{\partial \boldsymbol{\phi}(\boldsymbol{\alpha})}{\partial \mathfrak{g}_\nu} - \frac{\partial \overline{\boldsymbol{\phi}}(\boldsymbol{\alpha})}{\partial \mathfrak{g}_\nu}\Upsilon_\nu \boldsymbol{\phi}(\boldsymbol{\alpha})\right), \quad (50)$$

where $\boldsymbol{\phi}(\boldsymbol{\alpha}) = \phi(x)\phi(\mathfrak{g})$ ($\mu = 0,1,2,3,\ \nu = 1,\ldots,6$), and

$$\Gamma_0 = \begin{pmatrix} 0 & I \\ I & 0 \end{pmatrix}, \quad \Gamma_1 = \begin{pmatrix} 0 & -\alpha_1 \\ \alpha_1 & 0 \end{pmatrix}, \quad \Gamma_2 = \begin{pmatrix} 0 & -\alpha_2 \\ \alpha_2 & 0 \end{pmatrix}, \quad \Gamma_3 = \begin{pmatrix} 0 & -\alpha_3 \\ \alpha_3 & 0 \end{pmatrix}, \quad (51)$$

$$\Upsilon_1 = \begin{pmatrix} 0 & \Lambda_1^* \\ \Lambda_1 & 0 \end{pmatrix}, \quad \Upsilon_2 = \begin{pmatrix} 0 & \Lambda_2^* \\ \Lambda_2 & 0 \end{pmatrix}, \quad \Upsilon_3 = \begin{pmatrix} 0 & \Lambda_3^* \\ \Lambda_3 & 0 \end{pmatrix}, \quad (52)$$

$$\Upsilon_4 = \begin{pmatrix} 0 & i\Lambda_1^* \\ i\Lambda_1 & 0 \end{pmatrix}, \quad \Upsilon_5 = \begin{pmatrix} 0 & i\Lambda_2^* \\ i\Lambda_2 & 0 \end{pmatrix}, \quad \Upsilon_6 = \begin{pmatrix} 0 & i\Lambda_3^* \\ i\Lambda_3 & 0 \end{pmatrix}, \tag{53}$$

where

$$\alpha_1 = \begin{pmatrix} 0 & 0 & 0 \\ 0 & 0 & i \\ 0 & -i & 0 \end{pmatrix}, \quad \alpha_2 = \begin{pmatrix} 0 & 0 & -i \\ 0 & 0 & 0 \\ i & 0 & 0 \end{pmatrix}, \quad \alpha_3 = \begin{pmatrix} 0 & i & 0 \\ -i & 0 & 0 \\ 0 & 0 & 0 \end{pmatrix}, \tag{54}$$

and the matrices Λ_j and Λ_j^* are derived from (23)–(25) and (26)–(28) at $l = 1$:

$$\Lambda_1 = \frac{c_{11}}{\sqrt{2}} \begin{pmatrix} 0 & 1 & 0 \\ 1 & 0 & 0 \\ 0 & 1 & 0 \end{pmatrix}, \quad \Lambda_2 = \frac{c_{11}}{\sqrt{2}} \begin{pmatrix} 0 & -i & 0 \\ i & 0 & -i \\ 0 & i & 0 \end{pmatrix}, \quad \Lambda_3 = c_{11} \begin{pmatrix} 1 & 0 & 0 \\ 0 & 0 & 0 \\ 0 & 0 & -1 \end{pmatrix}. \tag{55}$$

$$\Lambda_1^* = \frac{\sqrt{2}}{2}\dot{c}_{11} \begin{pmatrix} 0 & 1 & 0 \\ 1 & 0 & 1 \\ 0 & 1 & 0 \end{pmatrix}, \quad \Lambda_2^* = \frac{\sqrt{2}}{2}\dot{c}_{11} \begin{pmatrix} 0 & -i & 0 \\ i & 0 & -i \\ 0 & i & 0 \end{pmatrix}, \quad \Lambda_3^* = \dot{c}_{11} \begin{pmatrix} 1 & 0 & 0 \\ 0 & 0 & 0 \\ 0 & 0 & -1 \end{pmatrix}. \tag{56}$$

At this point, electromagnetic field should be defined in the Riemann-Silberstein representation [69, 57, 10]. The Riemann-Silberstein (Majorana-Oppenheimer) representation considered during long time by many authors [40, 46, 28, 42, 51, 41, 13, 23]. The interest to this formulation of electrodynamics has grown in recent years [32, 58, 22, 17]. One of the main advantages of this approach lies in the fact that Dirac and Maxwell fields are derived similarly from the Dirac-like Lagrangians. These fields have the analogous mathematical structure, namely, they are the functions on the Poincaré group. This circumstance allows us to consider the fields $(1/2, 0) \oplus (0, 1/2)$ and $(1, 0) \oplus (0, 1)$ on an equal footing, from the one group theoretical viewpoint.

In 1949, Newton and Wigner [44] showed that for the photon there exist no localized states. Therefore, a massless field $(1, 0) \oplus (0, 1)$ should be considered within the unitary infinite-dimensional representation of the Lorentz group. In accordance with the Naimark theorem [43], the representation S_k of the subgroup $\mathrm{SU}(2)$ is contained in $\mathfrak{S}_{\lambda,\rho}$ not more then one time. Matrix elements of the representation $\mathfrak{S}_{\lambda,\rho}$ are defined by the functions (6). *We suppose that the photon field is described within the infinite-dimensional representation* $\mathfrak{S}_{\lambda,\rho}$. This representation contains a finite-dimensional representation associated with the field $(1, 0) \oplus (0, 1)$.

Since longitudinal solutions $\phi_{0,n}^{-\frac{1}{2}+i\rho,l_0}(\boldsymbol{\alpha})$ and $\dot{\phi}_{0,\dot{n}}^{-\frac{1}{2}-i\rho,l_0}(\boldsymbol{\alpha})$ do not contribute to a real photon due to their transversality conditions, then particular solutions for the Majorana-Oppenheimer bispinor $\boldsymbol{\phi} = (\phi_1, \phi_2, \phi_3, \dot{\phi}_1, \dot{\phi}_2, \dot{\phi}_3)^T$ on $\mathcal{M}_8$ are defined by the following expressions [64]:

$$\phi_{\pm 1,n}^{-\frac{1}{2}+i\rho,l_0}(\boldsymbol{\alpha}) = \phi_\pm(\mathbf{k}; \mathbf{x}, t)\phi_{\pm 1,n}^{-\frac{1}{2}+i\rho,l_0}(\mathfrak{g}) =$$
$$\{2(2\pi)^3\}^{-\frac{1}{2}} \begin{pmatrix} \varepsilon_\pm(\mathbf{k}) \\ \varepsilon_\pm(\mathbf{k}) \end{pmatrix} \exp[i(\mathbf{k}\cdot\mathbf{x} - \omega t)] \boldsymbol{f}_{1,\pm 1}^{-\frac{1}{2}+i\rho,l_0}(r)\mathfrak{M}_{\pm 1,n}^{-\frac{1}{2}+i\rho,l_0}(\varphi, \epsilon, \theta, \tau, 0, 0),$$

$$\dot{\phi}^{-\frac{1}{2}-i\rho,\dot{l}_0}_{\pm 1,\dot{n}}(\boldsymbol{\alpha}) = \phi^*_{\pm}(\mathbf{k};\mathbf{x},t)\dot{\phi}^{-\frac{1}{2}-i\rho,\dot{l}_0}_{\pm 1,\dot{n}}(\mathfrak{g}) =$$
$$\{2(2\pi)^3\}^{-\frac{1}{2}}\begin{pmatrix}\varepsilon^*_{\pm}(\mathbf{k})\\ \varepsilon^*_{\pm}(\mathbf{k})\end{pmatrix}\exp[-i(\mathbf{k}\cdot\mathbf{x}-\omega t)]\boldsymbol{f}^{-\frac{1}{2}-i\rho,\dot{l}_0}_{1,\pm 1}(r^*)\mathfrak{M}^{-\frac{1}{2}-i\rho,\dot{l}_0}_{\pm 1,\dot{n}}(\varphi,\epsilon,\theta,\tau,0,0),$$

where

$$\varepsilon_{\pm}(\mathbf{k}) = \{2|\mathbf{k}|^2(k_1^2+k_2^2)\}^{-\frac{1}{2}}\begin{bmatrix}-k_1k_3 \pm ik_2|\mathbf{k}|\\ -k_2k_3 \mp ik_1|\mathbf{k}|\\ k_1^2+k_2^2\end{bmatrix}$$

are the polarization vectors of a photon. Radial functions have the form

$$\boldsymbol{f}^{-\frac{1}{2}+i\rho,l_0}_{1,1}(r) = C\sqrt{r} + \sqrt{\left(l_0+i\rho+\frac{1}{2}\right)\left(\frac{l_0}{2}+i\frac{\rho}{2}+\frac{5}{4}\right)}r,$$
$$\boldsymbol{f}^{-\frac{1}{2}-i\rho,\dot{l}_0}_{1,1}(r^*) = \dot{C}\sqrt{r^*} + \sqrt{\left(\dot{l}_0-i\rho+\frac{3}{2}\right)\left(\frac{\dot{l}_0}{2}-i\frac{\rho}{2}+\frac{7}{4}\right)}r^*,$$

and

$$\mathfrak{M}^{-\frac{1}{2}+i\rho,l_0}_{\pm 1,n}(\varphi,\epsilon,\theta,\tau,0,0) = e^{\mp(\epsilon+i\varphi)}\mathfrak{Z}^{-\frac{1}{2}+i\rho,l_0}_{\pm 1,n}(\theta,\tau),$$

$$\mathfrak{Z}^{-\frac{1}{2}+i\rho,l_0}_{\pm 1,n}(\theta,\tau) = \cos^{2l_0}\frac{\theta}{2}\cosh^{-1+2i\rho,l_0}\frac{\tau}{2}\sum_{k=-l_0}^{l_0} i^{\pm 1-k}\tan^{\pm 1-k}\frac{\theta}{2}\tanh^{n-k}\frac{\tau}{2}\times$$
$${}_2F_1\left(\begin{matrix}\pm 1-l_0+1, 1-l_0-k\\ \pm 1-k+1\end{matrix}\middle| i^2\tan^2\frac{\theta}{2}\right){}_2F_1\left(\begin{matrix}n-i\rho+\frac{1}{2}, \frac{1}{2}-i\rho-k\\ n-k+1\end{matrix}\middle|\tanh^2\frac{\tau}{2}\right),$$

$$\mathfrak{M}^{-\frac{1}{2}-i\rho,\dot{l}_0}_{\pm 1,\dot{n}}(\varphi,\epsilon,\theta,\tau,0,0) = e^{\mp(\epsilon-i\varphi)}\mathfrak{Z}^{-\frac{1}{2}-i\rho,\dot{l}_0}_{\pm 1,\dot{n}}(\theta,\tau),$$

$$\mathfrak{Z}^{-\frac{1}{2}-i\rho,\dot{l}_0}_{\pm 1,\dot{n}}(\theta,\tau) = \cos^{2\dot{l}_0}\frac{\theta}{2}\cosh^{-1-2i\rho,\dot{l}_0}\frac{\tau}{2}\sum_{\dot{k}=-\dot{l}_0}^{\dot{l}_0} i^{\pm 1-\dot{k}}\tan^{\pm 1-\dot{k}}\frac{\theta}{2}\tanh^{\dot{n}-\dot{k}}\frac{\tau}{2}\times$$
$${}_2F_1\left(\begin{matrix}\pm 1-\dot{l}_0+1, 1-\dot{l}_0-\dot{k}\\ \pm 1-\dot{k}+1\end{matrix}\middle| i^2\tan^2\frac{\theta}{2}\right){}_2F_1\left(\begin{matrix}\dot{n}+i\rho+\frac{1}{2}, \frac{1}{2}+i\rho-\dot{k}\\ \dot{n}-\dot{k}+1\end{matrix}\middle|\tanh^2\frac{\tau}{2}\right)$$

are associated hyperspherical functions defined on the spheres S^2_c and $\dot{S}^2_c$. General solutions are found via the following generalized Fourier integrals:

$$\phi_{\pm 1}(\boldsymbol{\alpha}) = \{2(2\pi)^3\}^{-\frac{1}{2}}\sum_{l_0=1}^{\infty}\boldsymbol{f}^{-\frac{1}{2}+i\rho,l_0}_{1,\pm 1}(r)\times$$
$$\sum_{n=-l_0}^{l_0}\int_{T_4}\begin{pmatrix}\varepsilon_{\pm}(\mathbf{k})\\ \varepsilon_{\pm}(\mathbf{k})\end{pmatrix}e^{ikx}\alpha^{\pm 1}_{l_0,n}\mathfrak{M}^{-\frac{1}{2}+i\rho,l_0}_{\pm 1,n}(\varphi,\epsilon,\theta,\tau,0,0)d^4x,$$

$$\dot{\phi}_{\pm 1}(\boldsymbol{\alpha}) = \left\{2(2\pi)^3\right\}^{-\frac{1}{2}} \sum_{l_0=1}^{\infty} \boldsymbol{f}_{1,\pm 1}^{-\frac{1}{2}-i\rho,\dot{l}_0}(r^*) \times$$
$$\sum_{\dot{n}=-l_0}^{l_0} \int\limits_{T_4} \begin{pmatrix} \varepsilon_{\pm}^*(\mathbf{k}) \\ \varepsilon_{\pm}^*(\mathbf{k}) \end{pmatrix} e^{-ikx} \alpha_{l_0,\dot{n}}^{\pm 1} \mathfrak{M}_{\pm 1,\dot{n}}^{-\frac{1}{2}-i\rho,l_0}(\varphi,\epsilon,\theta,\tau,0,0)d^4x,$$

where

$$\alpha_{l_0,n}^{\pm 1} = \frac{(-1)^n(2l_0+1)^2}{32\pi^4 \boldsymbol{f}_{1,\pm 1}^{-\frac{1}{2}+i\rho,l_0}(a)} \int\limits_{S_c^2}\int\limits_{T_4} F_{\pm 1}(\boldsymbol{\alpha}) e^{-ikx} \mathfrak{M}_{\pm 1,n}^{-\frac{1}{2}+i\rho,l_0}(\varphi,\epsilon,\theta,\tau,0,0)d^4x d^4\mathfrak{g},$$
$$\alpha_{l_0,\dot{n}}^{\pm 1} = \frac{(-1)^{\dot{n}}(2l_0+1)^2}{32\pi^4 \boldsymbol{f}_{1,\pm 1}^{-\frac{1}{2}-i\rho,\dot{l}_0}(a^*)} \int\limits_{S_c^2}\int\limits_{T_4} \dot{F}_{\pm 1}(\boldsymbol{\alpha}) e^{ikx} \mathfrak{M}_{\pm 1,\dot{n}}^{-\frac{1}{2}-i\rho,l_0}(\varphi,\epsilon,\theta,\tau,0,0)d^4x d^4\mathfrak{g}.$$

4. Interaction

Up to now we analyze free Dirac and Maxwell fields. Let us consider an interaction between these fields. As usual, interactions between the fields are described by an interaction Lagrangian $\mathcal{L}_I$. In our case we take the following Lagrangian [66]:

$$\mathcal{L}_I(\boldsymbol{\alpha}) = \mu(\overline{\boldsymbol{\psi}}(\boldsymbol{\alpha})\sigma_{(\mu\nu)k}^D \boldsymbol{\psi}(\boldsymbol{\alpha}))(\Xi_{(\rho)k}^M \boldsymbol{\phi}(\boldsymbol{\alpha})), \tag{57}$$

where

$$\sigma_{\mu\nu}^D = \frac{1}{2}(\Xi_\mu^D \Xi_\nu^D - \Xi_\nu^D \Xi_\mu^D)$$

and

$$\Xi^D = (\Gamma_0^D, \Gamma_1^D, \Gamma_2^D, \Gamma_3^D, \Upsilon_1^D, \Upsilon_2^D, \Upsilon_3^D, \Upsilon_4^D, \Upsilon_5^D, \Upsilon_6^D),$$
$$\Xi^M = (\Gamma_0^M, \Gamma_1^M, \Gamma_2^M, \Gamma_3^M, \Upsilon_1^M, \Upsilon_2^M, \Upsilon_3^M, \Upsilon_4^M, \Upsilon_5^M, \Upsilon_6^M),$$

here Γ^D and Υ^D are the matrices (46) and (47)–(48), and Γ^M and Υ^M are the matrices (51) and (52)–(53).

The full Lagrangian of interacting Dirac and Maxwell fields equals to a sum of the free field Lagrangians and the interaction Lagrangian:

$$\mathcal{L}(\boldsymbol{\alpha}) = \mathcal{L}_D(\boldsymbol{\alpha}) + \mathcal{L}_M(\boldsymbol{\alpha}) + \mathcal{L}_I(\boldsymbol{\alpha}),$$

where $\mathcal{L}_D(\boldsymbol{\alpha})$ and $\mathcal{L}_M(\boldsymbol{\alpha})$ are of the type (45) and (50), respectively. Or,

$$\mathcal{L}(\boldsymbol{\alpha}) = -\frac{1}{2}\left(\overline{\boldsymbol{\psi}}(\boldsymbol{\alpha})\Xi_\mu^D \frac{\partial \boldsymbol{\psi}(\boldsymbol{\alpha})}{\partial \boldsymbol{\alpha}_\mu} - \frac{\partial \overline{\boldsymbol{\psi}}(\boldsymbol{\alpha})}{\partial \boldsymbol{\alpha}_\mu}\Xi_\mu^D \boldsymbol{\psi}(\boldsymbol{\alpha})\right) -$$
$$-\frac{1}{2}\left(\overline{\boldsymbol{\phi}}(\boldsymbol{\alpha})\Xi_\mu^M \frac{\partial \boldsymbol{\phi}(\boldsymbol{\alpha})}{\partial \boldsymbol{\alpha}_\mu} - \frac{\partial \overline{\boldsymbol{\phi}}(\boldsymbol{\alpha})}{\partial \boldsymbol{\alpha}_\mu}\Xi_\mu^M \boldsymbol{\phi}(\boldsymbol{\alpha})\right) -$$
$$-\kappa\overline{\boldsymbol{\psi}}(\boldsymbol{\alpha})\boldsymbol{\psi}(\boldsymbol{\alpha}) + \mu(\overline{\boldsymbol{\psi}}(\boldsymbol{\alpha})\sigma_{(\mu\nu)k}^D \boldsymbol{\psi}(\boldsymbol{\alpha}))(\Xi_{(\rho)k}^M \boldsymbol{\phi}(\boldsymbol{\alpha})).$$

Since the Lagrangian (57) does not contain derivatives on the field functions, then for a Hamiltonian density we have $\mathcal{H}_I(\boldsymbol{\alpha}) = -\mathcal{L}_I(\boldsymbol{\alpha})$.

As is known, in the standard quantum field theory the S-matrix is expressed via the Dyson formula [53]

$$S = T\left[\exp\left(-\frac{i}{\hbar c}\int\limits_{-\infty}^{+\infty}\mathcal{H}_I(x)d^4x\right)\right], \tag{58}$$

where T is the time ordering operator.

In our case, the electron-positron and photon fields are defined on the space $\mathcal{M}_8 = \mathbb{R}^{1,3}\times S^2_c$ which larger then the Minkowski space $\mathbb{R}^{1,3}$. With a view to define a formula similar to the equation (58) it is necessarily to replace d^4x by the following invariant measure on $\mathcal{M}_8$:

$$d^8\mu = d^4x d^4\mathfrak{g},$$

where

$$d^4\mathfrak{g} = \sin\theta^c d\theta d\tau d\varphi d\epsilon.$$

Therefore, an analogue of the Dyson formula (58) on the manifold $\mathcal{M}_8$ can be written as follows

$$S = T\left[\exp\left(-\frac{i}{\hbar c}\int\limits_{T_4}\int\limits_{S^2_c}\mathcal{H}_I(\alpha)d^4x d^4\mathfrak{g}\right)\right].$$

Let us consider an integral on trilinear form, defining the interaction, over the two-dimensional complex sphere:

$$\int\limits_{S^2_c}\mathcal{H}_I(\alpha)dS^2_c \sim$$

$$\int\limits_{S^2_c}\mathfrak{M}^{\dot{m}_e}_{\dot{l}_e}(\varphi,\epsilon,\theta,\tau,0,0)\mathfrak{M}^{m_e}_{l_e}(\varphi,\epsilon,\theta,\tau,0,0)\mathfrak{M}^{m_f}_{-\frac{1}{2}+i\rho}(\varphi,\epsilon,\theta,\tau,0,0)\sin\theta^c d\theta^c d\varphi^c,$$

that is, to investigate convergence of the integral

$$I_1 = \int\limits_{-\infty}^{+\infty}\int\limits_{-\infty}^{+\infty} e^{-i(m_e\varphi^c - \dot{m}_e\dot{\varphi}^c - m_f\varphi^c)}\mathfrak{Z}^{\dot{m}_e}_{\dot{l}_e}(\cos\dot{\theta}^c)\mathfrak{Z}^{m_e}_{l_e}(\cos\theta^c)\mathfrak{Z}^{m_f}_{-\frac{1}{2}+i\rho}(\cos\theta^c)\sin\theta^c d\theta^c d\varphi^c.$$

Here the symbols l_e, m_e correspond to the electron field $\psi(\mathfrak{g})$, $\dot{l}_e$, $\dot{m}_e$ correspond to the positron field $\overline{\psi(\mathfrak{g})}$, and m_f corresponds to the photon field $\phi(\mathfrak{g})$. It is obvious that convergence of the integral

$$I_2 = \int\limits_{-\infty}^{+\infty} e^{-i(m_e\varphi^c - \dot{m}_e\dot{\varphi}^c - m_f\varphi^c)}d\varphi^c$$

is not difficult to investigate. Let us consider the following integral:

$$I_3 = \int\limits_{-\infty}^{+\infty}\mathfrak{Z}^{\dot{m}_e}_{\dot{l}_e}(\cos\dot{\theta}^c)\mathfrak{Z}^{m_e}_{l_e}(\cos\theta^c)\mathfrak{Z}^{m_f}_{-\frac{1}{2}+i\rho}(\cos\theta^c)\sin\theta^c d\theta^c.$$

Rewriting hyperspherical functions via the hypergeometric functions, we obtain

$$I_3 = \frac{i^{m_e+\dot{m}_e+m_f}}{\Gamma(m_e+1)\Gamma(\dot{m}_e+1)\Gamma(m_f+1)} \times$$

$$\sqrt{\frac{\Gamma(l_e+m_e+1)\Gamma(\dot{l}_e+\dot{m}_e+1)\Gamma(m_f+i\rho+\frac{1}{2})}{\Gamma(l_e-m_e+1)\Gamma(\dot{l}_e-\dot{m}_e+1)\Gamma(i\rho-m_f+\frac{1}{2})}} \times$$

$$\int\limits_{-\infty}^{+\infty} \sin^{\dot{m}_e+m_e+m_f}\frac{\theta^c}{2}\cos^{\dot{m}_e+m_e+m_f}\frac{\theta^c}{2}\, {}_2F_1\left(\begin{matrix} \dot{l}_e+\dot{m}_e+1, \dot{m}_e-\dot{l}_e \\ \dot{m}_e+1 \end{matrix} \middle| \sin^2\frac{\theta^c}{2}\right) \times$$

$${}_2F_1\left(\begin{matrix} l_e+m_e+1, m_e-l_e \\ m_e+1 \end{matrix} \middle| \sin^2\frac{\theta^c}{2}\right) \times$$

$${}_2F_1\left(\begin{matrix} i\rho+m_f+\frac{1}{2}, m_f-i\rho+\frac{1}{2} \\ m_f+1 \end{matrix} \middle| \sin^2\frac{\theta^c}{2}\right) \sin\theta^c d\theta^c. \quad (59)$$

Further, using an explicit expression for the associated hyperspherical function

$$\mathfrak{Z}_l^m(\cos\theta^c) = i^m\sqrt{\frac{\Gamma(l-m+1)}{\Gamma(l+m+1)}}\cos^m\frac{\theta^c}{2}\sin^m\frac{\theta^c}{2} \times$$

$$\sum_{j=0}^{l-m} \frac{(-1)^j\Gamma(l+m+j+1)}{\Gamma(j+1)\Gamma(m+j+1)\Gamma(l-m-j+1)}\sin^{2j}\frac{\theta^c}{2},$$

we rewrite the first two hypergeometric functions in the integral (59). Then

$$I_3 = \frac{i^{m_e+\dot{m}_e+m_f}}{\Gamma(m_f+1)}\sqrt{\frac{\Gamma(l_e-m_e+1)\Gamma(\dot{l}_e-\dot{m}_e+1)\Gamma(m_f+i\rho+\frac{1}{2})}{\Gamma(l_e+m_e+1)\Gamma(\dot{l}_e+\dot{m}_e+1)\Gamma(i\rho-m_f+\frac{1}{2})}} \times$$

$$\sum_{j_1=0}^{l_e-m_e}\sum_{j_2=0}^{\dot{l}_e-\dot{m}_e} \frac{(-1)^{j_1+j_2}\Gamma(l_e+m_e+j_1+1)}{\Gamma(j_1+1)\Gamma(j_2+1)\Gamma(m_e+j_1+1)\Gamma(\dot{m}_e+j_2+1)} \times$$

$$\int\limits_{-\infty}^{+\infty} \cos^{m_e+\dot{m}_e+m_f}\frac{\theta^c}{2}\sin^{m_e+\dot{m}_e+m_f+2j_1+2j_2}\frac{\theta^c}{2} \times$$

$${}_2F_1\left(\begin{matrix} i\rho+m_f+\frac{1}{2}, m_f-i\rho+\frac{1}{2} \\ m_f+1 \end{matrix} \middle| \sin^2\frac{\theta^c}{2}\right) \sin\theta^c d\theta^c.$$

Making the substitution $z = \cos\theta^c$ in the integral

$$I_4 = \frac{1}{2^{m_e+\dot{m}_e+m_f}}\int\limits_{-\infty}^{+\infty} \sin^{m_e+\dot{m}_e+m_f}\theta^c \sin^{2j_1+2j_2}\frac{\theta^c}{2} \times$$

$${}_2F_1\left(\begin{matrix} i\rho+m_f+\frac{1}{2}, m_f-i\rho+\frac{1}{2} \\ m_f+1 \end{matrix} \middle| \sin^2\frac{\theta^c}{2}\right) \sin\theta^c d\theta^c,$$

we obtain

$$I_4 = \frac{1}{2^{m_e+\dot{m}_e+m_f+j_1+j_2}} \int\limits_{-\infty}^{+\infty} (1-z^2)^{\frac{m_e+\dot{m}_e+m_f}{2}} \left(\frac{1-z}{2}\right)^{j_1+j_2} \times$$
$${}_2F_1\left(\begin{matrix} i\rho+m_f+\frac{1}{2}, m_f-i\rho+\frac{1}{2} \\ m_f+1 \end{matrix} \middle| \frac{1-z}{2} \right) dz.$$

Or,

$$I_4 = \frac{1}{2^{m_e+\dot{m}_e+m_f+j_1+j_2}} \sum_{q=0}^{m_e+\dot{m}_e+m_f} (-1)^q \frac{(m_e+\dot{m}_e+m_f)!}{q!(m_e+\dot{m}_e+m_f-q)!} \times$$
$$\int\limits_{-\infty}^{+\infty} z^k \left(\frac{1-z}{2}\right)^{j_1+j_2} {}_2F_1\left(\begin{matrix} i\rho+m_f+\frac{1}{2}, m_f-i\rho+\frac{1}{2} \\ m_f+1 \end{matrix} \middle| \frac{1-z}{2} \right) dz.$$

Introducing a new variable $t = (1-z)/2$, we find

$$I_4 = \frac{1}{2^{m_e+\dot{m}_e+m_f+j_1+j_2}} \sum_{q=0}^{m_e+\dot{m}_e+m_f} (-1)^{q+1} \frac{(m_e+\dot{m}_e+m_f)!}{q!(m_e+\dot{m}_e+m_f-q)!} \times$$
$$\int\limits_{-\infty}^{+\infty} (1-2t)^q t^{j_1+j_2} {}_2F_1\left(\begin{matrix} i\rho+m_f+\frac{1}{2}, m_f-i\rho+\frac{1}{2} \\ m_f+1 \end{matrix} \middle| t \right) dt.$$

Decomposing $(1-2t)^q$ via the Newton binomial, we obtain

$$I_4 = \frac{1}{2^{m_e+\dot{m}_e+m_f+j_1+j_2}} \sum_{q=0}^{m_e+\dot{m}_e+m_f} \sum_{p=0}^{q} (-1)^{q+p+1} \frac{(m_e+\dot{m}_e+m_f)!}{(m_e+\dot{m}_e+m_f-q)!} \times$$
$$\frac{2^p}{p!(k-p)!} \int\limits_{-\infty}^{+\infty} t^{j_1+j_2+p} {}_2F_1\left(\begin{matrix} i\rho+m_f+\frac{1}{2}, m_f-i\rho+\frac{1}{2} \\ m_f+1 \end{matrix} \middle| t \right) dt.$$

With the aim to calculate the latter integral we use the following formula [49]:

$$I_5 = \int t^n {}_2F_1\left(\begin{matrix} a, b \\ c \end{matrix} \middle| t \right) dt =$$
$$= n! \sum_{k=1}^{n+1} (-1)^{k+1} \frac{(c-k)_k t^{n-k+1}}{(a-k+1)!(a-k)_k(b-k)_k} {}_2F_1\left(\begin{matrix} a-k, b-k \\ c-k \end{matrix} \middle| t \right).$$

Then

$$I_4 = \frac{1}{2^{m_e+\dot{m}_e+m_f+j_1+j_2}} \sum_{q=0}^{m_e+\dot{m}_e+m_f} \sum_{p=0}^{q} (j_1+j_2+p)! \times$$

$$\sum_{k=1}^{j_1+j_2+p} \frac{(-1)^{q+p+k} 2^p (m_e+\dot{m}_e+m_f)!}{p!(k-p)!(m_e+\dot{m}_e+m_f-q)!(m_f+i\rho+\frac{3}{2}-k)!} \times$$

$$\frac{(m_f-k+1)_k t^{j_1+j_2+p-k+1}}{(m_f+i\rho-k+\frac{1}{2})_k (m_f-k+1)_k} \times$$

$${}_2F_1\left(\begin{matrix} i\rho+m_f-k+\frac{1}{2}, m_f-i\rho-k+\frac{1}{2} \\ m_f-k+1 \end{matrix} \middle| t \right) dt.$$

With the aim to investigate the convergence of (59) let us apply the following asymptotic expansion for the hypergeometric function [9]:

$${}_2F_1\left(\begin{matrix} a, b \\ c \end{matrix} \middle| t \right) = \frac{\Gamma(c)\Gamma(b-a)}{\Gamma(b)\Gamma(c-a)} (-t)^{-a} {}_2F_1\left(\begin{matrix} a, 1-c+a \\ 1-b+a \end{matrix} \middle| \frac{1}{t} \right) +$$

$$+ \frac{\Gamma(c)\Gamma(a-b)}{\Gamma(a)\Gamma(c-b)} (-t)^{-b} {}_2F_1\left(\begin{matrix} b, 1-c+b \\ 1-a+b \end{matrix} \middle| \frac{1}{t} \right).$$

Thus,

$$I_3 = \frac{i^{m_e+\dot{m}_e+m_f}}{\Gamma(m_f+1)} \sqrt{\frac{\Gamma(l_e-m_e+1)\Gamma(\dot{l}_e-\dot{m}_e+1)\Gamma(m_f+i\rho+\frac{1}{2})}{\Gamma(l_e+m_e+1)\Gamma(\dot{l}_e+\dot{m}_e+1)\Gamma(i\rho-m_f+\frac{1}{2})}} \times$$

$$\sum_{j_1=0}^{l_e-m_e} \sum_{j_2=0}^{\dot{l}_e-\dot{m}_e} \frac{(-1)^{j_1+j_2}\Gamma(l_e+m_e+j_1+1)}{2^{m_e+\dot{m}_e+m_f+j_1+j_2}\Gamma(j_1+1)\Gamma(j_2+1)\Gamma(m_e+j_1+1)\Gamma(\dot{m}_e+j_2+1)} \times$$

$$\frac{\Gamma(\dot{l}_e+\dot{m}_e+j_2+1)}{\Gamma(l_e-m_e-j_1+1)\Gamma(\dot{l}_e-\dot{m}_e-j_2+1)} \times$$

$$\sum_{q=0}^{m_e+\dot{m}_e+m_f} \sum_{p=0}^{q} (j_1+j_2+p)! \sum_{k=1}^{j_1+j_2+p} \frac{(-1)^{q+p+k} 2^p (m_e+\dot{m}_e+m_f)!}{p!(k-p)!(m_e+\dot{m}_e+m_f-q)!(m_f+i\rho+\frac{3}{2}-k)!} \times$$

$$\frac{(m_f-k+1)_k}{(m_f+i\rho-k+\frac{1}{2})_k (m_f-k+1)_k} \left[(-1)^{k-m_f-i\rho-\frac{1}{2}} \frac{\Gamma(m_f-k+1)\Gamma(-2i\rho)}{\Gamma(m_f-i\rho-k+\frac{1}{2})\Gamma(\frac{1}{2}-i\rho)} \times \right.$$

$$t^{j_1+j_2+p-m_f-i\rho+\frac{1}{2}} {}_2F_1\left(\begin{matrix} m_f+i\rho-k+\frac{1}{2}, i\rho+\frac{1}{2} \\ 2i\rho+1 \end{matrix} \middle| \frac{1}{t} \right) +$$

$$(-1)^{k-m_f+i\rho-\frac{1}{2}}\frac{\Gamma(m_f-k+1)\Gamma(2i\rho)}{\Gamma(m_f+i\rho-k+\frac{1}{2})\Gamma(\frac{1}{2}-i\rho)}\times$$
$$t^{j_1+j_2+p-m_f+i\rho+\frac{1}{2}}{}_2F_1\left(\begin{array}{c} m_f+i\rho-k+\frac{1}{2},\frac{1}{2}-i\rho \\ 1-2i\rho \end{array}\middle|\frac{1}{t}\right)\Bigg],$$

where $t=\sin^2\frac{\theta^c}{2}$. Since $I_3\sim 1/t^M$ (the hypergeometric function ${}_2F_1$ can be written as a power series in $1/t$) and $M\to\infty$, then I_3 converges at $M>j_1+j_2+p-m_f+i\rho+\frac{1}{2}$ and, therefore, in accordance with (58) the elements of S-matrix are defined by convergent expressions.

5. Summary

We have proved convergence of quantum electrodynamics on the Poincaré group in the case of homogeneous space $\mathcal{M}_8=\mathbb{R}^{1,3}\times S^2_c$. We considered Dirac like equations for extended objects on $\mathcal{M}_8$ and their particular solutions corresponding to the fields of spin 1/2 and 1. At this point, the spin-1 field (Maxwell field) is defined within an infinite-dimensional representation of the Lorentz group. We showed that an analogue of the Dyson formula for S-matrix in the case of $\mathcal{M}_8$ is defined by convergent integrals. It would be interesting to consider quantum field models and their convergence on other homogeneous spaces of the Poincaré group, such as $\mathcal{M}_6=\mathbb{R}^{1,3}\times S^2$, $\mathcal{M}_7=\mathbb{R}^{1,3}\times H^3$ and $\mathcal{M}_{10}=\mathbb{R}^{1,3}\times\mathfrak{L}_6$. It would be interesting also to consider wave equations and field models for the fields $\psi(\boldsymbol{\alpha})=\langle x,\mathfrak{q}\,|\psi\rangle$ on the de Sitter group, where $x\in T_5$ and $\mathfrak{q}\in\mathbf{Spin}_+(1,4)\simeq\mathrm{Sp}(1,1)$, and for the fields $\psi(\boldsymbol{\alpha})=\langle x,\mathfrak{c}\,|\psi\rangle$ on the conformal group, where $x\in T_6$ and $\mathfrak{c}\in\mathbf{Spin}_+(2,6)\simeq\mathrm{SU}(2,2)$. Our next paper will be devoted to these questions.

Appendix: Spinor Groups and Bivector Spaces

As is known, rotations of pseudo-Euclidean spaces $\mathbb{R}^{p,q}$ are defined by spinor groups [37]

$$\mathbf{Spin}(p,q)=\left\{s\in\boldsymbol{\Gamma}^+_{p,q}\mid N(s)=\pm 1\right\},$$

where $\boldsymbol{\Gamma}^+_{p,q}=\boldsymbol{\Gamma}_{p,q}\cap C\ell^+_{p,q}$ is a special Lipschitz group, $s\in C\ell^+_{p,q}$ is an even invertible element of the real Clifford algebra $C\ell_{p,q}$. In more detail, the element s is a linear combination of even basic elements, that is,

$$s=\sum_k a^{i_1\ldots i_{2k}}\mathbf{e}_{i_1\ldots i_{2k}}.$$

The condition $N(s)=\pm 1$ means that

$$\sum_k\sigma(i_1)\cdots\sigma(i_{2k})\left(a^{i_1\ldots i_{2k}}\right)^2=\pm 1. \tag{60}$$

On the other hand, from the fact that the elements $s\in\boldsymbol{\Gamma}^+_{p,q}$ are even products of $\nu=\sum\nu^i\mathbf{e}_i$, it is easy to derive that coordinates of $s\in\mathbf{Spin}(p,q)$ are related by the following conditions [50]:

$$a^{i_1i_2\ldots i_{2k}}(a)^{k-1}=(2k-1)!!a^{[i_1i_2}a^{i_2i_3}\cdots a^{i_{2k-1}i_{2k}]}. \tag{61}$$

Conditions (61) can be rewritten in the form

$$\left.\begin{array}{rcl} aa^{i_1i_2i_3i_4} &=& 3!!a^{[i_1i_2}a^{i_3i_4]}, \\ aa^{i_1i_2i_3i_4i_5i_6} &=& 5!!a^{[i_1i_2}a^{i_3i_4i_5i_6]}, \\ \multicolumn{3}{c}{\dots\dots\dots\dots\dots\dots\dots\dots\dots} \\ aa^{i_1i_2\dots i_{2k}} &=& (2k-1)!!a^{[i_1i_2\dots}a^{i_3i_4\dots i_{2k-1}i_{2k}]}. \end{array}\right\} \qquad (62)$$

Conditions (60) and (61) express all coordinates of the elements s, belonging to the spinor group $\mathbf{Spin}(p,q)$, via $n(n-1)/2$ coordinates a^{ij}, where $n = p+q$. The number of the coordinates a^{ij} coincides with the number of parameters of the rotation group of the space $\mathbb{R}^{p,q}$. This fact shows that expressions (60) and (61) form a full system of the conditions separating the spinor group $\mathbf{Spin}(p,q)$ from the algebra $C\ell_{p,q}$.

Further, parameters of the rotation group of the space $\mathbb{R}^{p,q}$ form a *bivector space* $\mathbb{R}^N$, where $N = \frac{n(n-1)}{2}$. Indeed, Let $\mathbb{R}^{p,q}$ be the n-dimensional pseudo-Euclidean space, $p+q=n$. Let us evolve in $\mathbb{R}^{p,q}$ all the tensors satisfying the following two conditions: 1) a rank of the tensors is even; 2) covariant and contravariant indexes are divided into separate skewsymmetric pairs. Such tensors can be exemplified by bivectors (skewsymmetric tensors of the second rank). The set of all bivector tensor fields in $\mathbb{R}^{p,q}$ is called *a bivector set*, and its representation in a given point of $\mathbb{R}^{p,q}$ is called *a local bivector set*. In any tensor from the bivector set we take the each skewsymmetric pair $\alpha\beta$ as one collective index. At this point, from the two possible pairs $\alpha\beta$ and $\beta\alpha$ we fix only one, for example, $\alpha\beta$. The number of all collective indexes is equal to $N = \frac{n(n-1)}{2}$. In a given point, the bivector set of the space $\mathbb{R}^{p,q}$ with the contravariant components defines in the collective indexes a vector set, and the each vector of this set has N components. Identifying these vectors with the points of the N-dimensional manifold, we come to an affine manifold E^N if and only if this manifold admits a Klein geometry with the group

$$\eta^{a'} = A^{a'}_{a}\eta^{a}, \quad \eta^{a} = A^{a}_{a'}\eta^{a'},$$
$$\det A^{a'}_{a} \neq 0, \quad A^{a}_{b'}A^{b'}_{c} = \delta^{a}_{c},$$

where

$$A^{a'}_{a} \longrightarrow A^{\alpha'}_{[\alpha}A^{\beta'}_{\beta]}.$$

Thus, any local bivector set of the space $\mathbb{R}^{p,q}$ $(p+q=n)$ can be mapped onto the affine space E^N . Therefore, E^N is related with the each point of the space $\mathbb{R}^{p,q}$. The space E^N is called *a bivector space*. It should be noted that the bivector space is a particular case of the most general mathematical construction called a Grassmannian manifold (a manifold of m-dimensional planes of the affine space). In the case $m = 2$ the manifold of two-dimensional planes is isometric to the bivector space, and the Grassmann coordinates in this case are called Pluecker coordinates.

The metrization of the bivector space E^N is given by the formula (see [47])

$$g_{ab} \longrightarrow g_{\alpha\beta\gamma\delta} \equiv g_{\alpha\gamma}g_{\beta\delta} - g_{\alpha\delta}g_{\beta\gamma}, \qquad (63)$$

where $g_{\alpha\beta}$ is a metric tensor of the space $\mathbb{R}^{p,q}$, and the collective indexes are skewsymmetric pairs $\alpha\beta \to a$, $\gamma\delta \to b$. After introduction of g_{ab}, the bivector affine space E^N is transformed to a metric space $\mathbb{R}^N$.

As is known, any transformation from the rotation group of the space $\mathbb{R}^{p,q}$ can be represented via $\frac{n(n-1)}{2}$ transformations in the planes (x_1, x_2), (x_1, x_3), ..., (x_{p+q}, x_1). The full number of the planes (x_i, x_j) is equal to $N = \frac{n(n-1)}{2}$, and the each plane (x_i, x_j) corresponds to the coordinate a^{ij} of the spinor group $\mathbf{Spin}(p, q)$.

References

[1] Altaisky, M. V.: Wavelet-Based Quantum Field Theory. *SIGMA* 3, 105–118 (2007); arXiv:0711.1671 [hep-th] (2007).

[2] Altaisky, M. V.: Quantum field theory without divergences. *Phys. Rev. D.* 81, 125003 (2010); arXiv:1002.2566 [hep-th] (2010).

[3] Amar, V., Dozzio, U.: Gel'fand-Yaglom Equations with Charge or Energy Density of Definite Sign. *Nuovo Cimento* A11, 87–99 (1972).

[4] Arnold, V. I.: *Mathematical Methods of Classical Mechanics*. Nauka, Moscow (1989) [in Russian].

[5] Arodź, H.: Metric tensors, Lagrangian formalism and Abelian gauge field on the Poincaré group. *Acta Phys. Pol. B.* 7, 177–190 (1976).

[6] Bacry, H., Kihlberg, A.: Wavefunctions on homogeneous spaces. *J. Math. Phys.* 10, 2132–2141 (1969).

[7] Bacry, H., Nuyts, J.: Mass-Spin Relation in a Lagrangian Model. *Phys. Rev.* 157, 1471–1472 (1967).

[8] Bargmann, V., Wigner, E. P.: Group theoretical discussion of relativistic wave equations. *Proc. Nat. Acad. USA* 34, 211–223 (1948).

[9] Bateman, H., Erdélyi, A.: *Higher Transcendental Functions.* vol. I. Mc Grow-Hill Book Company, New York (1953).

[10] Bialynicki-Birula, I.: Photon wave function. *Progress in Optics*, Vol. XXXVI, Ed. E. Wolf, Elsevier, Amsterdam (1996); arXiv:quant-th/0508202 (2005).

[11] Biedenharn, L. C., Braden, H. W, Truini, P., van Dam, H.: Relativistic wavefunctions on spinor spaces. *J. Phys. A: Math. Gen.* 21, 3593–3610 (1988).

[12] Boyer, C. P., Fleming, G. N.: Quantum field theory on a seven-dimensional homogeneous space of the Poincaré group. *J. Math. Phys.* 15, 1007–1024 (1974).

[13] Da Silveira.: Dirac-like equations for the photon. *Z. Naturforsh* A34, 646–647 (1979).

[14] de Broglie L. *Theorie Generale des Particules a Spin (Methode de Fusion).* Gauthier-Villars, Paris (1943).

[15] de Broglie, L.: *La théorie de la mesure en mécanique ondulatorie*. Gauthier-Villars, Paris (1957).

[16] Drechsler, W.: Geometro-stohastically quantized fields with internal spin variables. *J. Math. Phys.* 38, 5531–5558 (1997); arXiv:gr-qc/9610046 (1996).

[17] Esposito, S.: Covariant Majorana Formulation of Electrodynamics. *Found. Phys.* 28, 231–244 (1998); arXiv:hep-th/9704144 (1997).

[18] Finkelstein, D.: Internal Structure of Spinning Particles. *Phys. Rev.* 100, 924–931 (1955).

[19] Fierz, M., Pauli, W.: On Relativistic Wave Equations of Particles of Arbitrary Spin in an Electromagnetic Field. *Proc. Roy. Soc.* (London) A. 173, 211–232 (1939).

[20] Gel'fand, I. M., Minlos, R. A., Shapiro, Z. Ya.: *Representations of the Rotation and Lorentz Groups and their Applications*. Pergamon Press, Oxford (1963).

[21] Gel'fand, I. M., Graev, M. I., Vilenkin, N. Ya.: *Generalized Functions* Vol. 5. Integral Geometry and Representation Theory. Academic Press, New York, London (1986).

[22] Gersten, A.: Maxwell equations as one-photon quantum equation. *Found. Phys. Lett.* 12, 291–298 (1998); arXiv:quant-ph/9911049 (1999).

[23] Giannetto, E.: A Majorana-Oppenheimer Formulation of Quantum Electrodynamics. *Lettere al Nuovo Cimento* 44, 140–144 (1985).

[24] Ginzburg, V. L., Tamm, I. E.: On the theory of spin. *Zh. Ehksp. Teor. Fiz.* 17, 227–237 (1947).

[25] Gitman, D. M., Shelepin, A. L.: Fields on the Poincaré Group: Arbitrary Spin Description and Relativistic Wave Equations. *Int. J. Theor. Phys.* 40, 3, 603–684 (2001); arXiv:hep-th/0003146 (2000).

[26] Gitman, D. M., Shelepin, A. L.: Field on the Poincaré group and quantum description of orientable objects. *Eur. Phys. J. C.* 61, 111-139 (2009); arXiv:0901.2537 [hep-th] (2009).

[27] Gitman, D. M., Shelepin, A. L.: Classification of quantum relativistic orientable objects. *Phys. Scr.* 83, 015103 (2011);arXiv: 1001.5290 [hep-th] (2010).

[28] Good, R. H.: Particle aspect of the electromagnetic field equations. *Phys. Rev.* 105, 1914 (1957).

[29] Grandpeix, J.-Y., Lurçat, F.: Particle description of zero energy vacuum. *Found. Phys.* 32, 109–158 (2002); arXiv:hep-th/0106229 (2001).

[30] Huszar, M.: *Angular Momentum and Unitary Spinor Bases of the Lorentz Group*. Preprint JINR No. E2-5429, Dubna (1970).

[31] Huszar, M., Smorodinsky, J.: *Representations of the Lorentz Group on the Two-Dimensional Complex Sphere and Two-Particle States*. Preprint JINR No. E2-5020, Dubna (1970).

[32] Inagaki, T.: Quantum-mechanical approach to a free photon. *Phys. Rev.* A49, 2839–2843 (1994).

[33] Kaiser, G.: *Quantum Physics, Relativity, and Complex Spacetime: Towards a New Synthesis*. arXiv: 0910.0352 [math-ph] (2009).

[34] Kihlberg, A.: Internal Co-ordinates and Explicit Representations of the Poincaré Group. *Nuovo Cimento* A53, 592–609 (1968).

[35] Kihlberg, A.: Fields on a homogeneous space of the Poincaré group. *Ann. Inst. Henri Poincaré*. 13, 57–76 (1970).

[36] Kuzenko, S. M., Lyakhovich, S. L., Segal, A. Yu.: A geometric model of the arbitrary spin massive particle. *Int. J. Mod. Phys. A*. 10, 1529–1552 (1995); arXiv:hep-th/9403196 (1994).

[37] Lipschitz R.: *Untersuchungen über die Summen von Quadraten*. Max Cohen und Sohn, Bonn (1886).

[38] Lurçat, F.: Quantum field theory and the dynamical role of spin. *Physics*. 1, 95 (1964).

[39] Lyakhovich, S. L., Segal, A. Yu., Sharapov, A. A.: Universal model of a $D = 4$ spinning particles. *Phys. Rev. D*. 54, 5223–5238 (1996); arXiv:hep-th/9603174 (1996).

[40] Majorana, E.: Scientific Papers, unpublished, deposited at the “Domus Galileana”, *Pisa, quaderno* 2, p.101/1; 3, p.11, 160; 15, p.16; 17, p.83, 159.

[41] Mignani, R., Recami, E., Baldo, M.: About a Dirac-Like Equation for the Photon according to Ettore Majorana. *Lettere al Nuovo Cimento* 11, 568–572 (1974).

[42] Moses, H. E.: Solution of Maxwell’s Equations in Terms of a Spinor Notation: the Direct and Inverse Problem. *Phys. Rev.* 113, 1670–1679 (1959).

[43] Naimark, M. A.: *Linear Representations of the Lorentz Group*. Pergamon, London (1964).

[44] Newton, T. D., Wigner, E. P.: Localized states for elementary systems. *Rev. Mod. Phys.* 21, 400 (1949).

[45] Nilsson, J., Beskow, A.: The concept of wave function and irreducible representations of the Poincaré group. *Arkiv för Fysik* 34, 307–324 (1967).

[46] Oppenheimer, J. R.: Note on light quanta and the electromagnetic field. *Phys. Rev.* 38, 725 (1931).

[47] Petrov, A. Z.: *Einstein Spaces*. Pergamon Press, Oxford (1969).

[48] Pletyukhov, V. A., Strazhev, V. I.: On Dirac-like relativistic wave equations. *Russian J. Phys.* n.12, 38-41 (1983).

[49] Prudnikov, A. P., Brychkov, Yu. A., Marichev, O. I.: *Integrals and Series: Supplementary Chapters* [in Russian]. Nauka, Moscow (1981).

[50] Rozenfeld, B. A.: *Geometry of Lie groups*. Dordrecht-Boston-London (1997).

[51] Sachs, M., Schwebel, S. L.: On covariant formulation of the Maxwell-Lorentz theory of electromagnetism. *J. Math. Phys.* 3, 843–848 (1962).

[52] Sastry, R. R.: *Quantum Mechanics of Extended Objects*. arXiv:quant-ph/9903025 (1999).

[53] Schweber, S. S.: *An Introduction to Relativistic Quantum Field Theory*. Harper & Row, New York (1961).

[54] Segal, I. E., Zhou, Z.: Convergence of nonlinear massive quantum field theory in the Einstein universe. *Ann. Phys.* 218, 2, 279–292 (1992).

[55] Segal, I. E., Zhou, Z.: Convergence of Quantum Electrodynamics in a Curved Deformation of Minkowski Space. *Ann. Phys.* 232, 1, 61–87 (1994).

[56] Shirokov, Yu. M.: Relativistskaia teoria spina. *Zh. Ehksp. Teor. Fiz.* 21, 748–760 (1951).

[57] Silberstein, L.: Elektromagnetische Grundgleichungen in bivectorieller Behandlung. *Ann. d. Phys.* 22, 579 (1907).

[58] Sipe, J. E.: Photon wave functions. *Phys. Rev.* A52, 1875–1883 (1995).

[59] Smorodinsky, Ya. A., Huszar, M.: Representations of the Lorentz group and the generalization of helicity states. *Teor. Mat. Fiz.* 4, 3, 328–340 (1970).

[60] Todorov, N. S.: Extended Particles Part I: Reformulation and Reinterpretation of the Dirac and Klein-Gordon Theories. *Annales de la Fondation Louis de Broglie* 25, 41–66 (2000).

[61] Toller, M.: Free quantum fields on the Poincaré group. *J. Math. Phys.* 37, 2694–2730 (1996); arXiv:gr-qc/9602031 (1996).

[62] Varlamov, V. V.: General Solutions of Relativistic Wave Equations. *Int. J. Theor. Phys.* 42, 3, 583–633 (2003); arXiv:math-ph/0209036 (2002).

[63] Varlamov, V. V.: Relativistic wavefunctions on the Poincaré group. *J. Phys. A: Math. Gen.* 37, 5467–5476 (2004); arXiv:math-ph/0308038 (2003).

[64] Varlamov, V. V.: Maxwell field on the Poincaré group. *Int. J. Mod. Phys. A.* 20, 17, 4095–4112 (2005); arXiv:math-ph/0310051 (2003).

[65] Varlamov V. V.: Relativistic spherical functions on the Lorentz group. *J. Phys. A: Math. Gen.* 39, 805–822 (2006); arXiv:math-ph/0507056 (2005).

[66] Varlamov V. V.: Towards the Quantum Electrodynamics on the Poincaré Group. *New Topics in Mathematical Physics Research* (Ed. C. V. Benton) New York. – Nova Science Publishers, 2006. – P. 109–179; arXiv:hep-th/0403070 (2004).

[67] Varlamov, V. V.: General Solutions of Relativistic Wave Equations II: Arbitrary Spin Chains. *Int. J. Theor. Phys.* 46, 4, 741–805 (2007); arXiv:math-ph/0503058 (2005).

[68] Vilenkin, N. Ya.: *Special Functions and the Theory of Group Representations*. AMS, Providence (1968).

[69] Weber, H.: *Die partiellen Differential-Gleichungen der mathematischen Physik nach Riemann's Vorlesungen*. Friedrich Vieweg und Sohn, Braunschweig (1901).

[70] Yukawa, H.: Quantum theory of non-local fields. I. Free fields. *Phys. Rev.* 77, 219–226 (1950).

In: Einstein and Others: Unification
Editor: Valeriy V. Dvoeglazov

ISBN: 978-1-63463-276-8

Chapter 5

A New Track for Unifying General Relativity with Quantum Field Theories

C. Pierre*
Institut de Mathématique pure et appliquée,
Université de Louvain Chemin du Cyclotron, Louvain-la-Neuve, Belgium

Abstract

In the perspective of unifying quantum field theories with general relativity, the equations of the internal dynamics of the vacuum and mass structures of a set of interacting particles are proved to be in one-to-one correspondence with the equations of general relativity.

This leads to envisage a high value for the cosmological constant, as expected theoretically.

1. Introduction

It is generally believed that a convincing theory of quantum gravity [3], [2], [8], [38], [41] will emerge from some unification of quantum field theory with general relativity.

This objective motivated the creation of the standard perturbative approaches based on Feynman perturbation theory for graviton modes of the form $g_{\mu\nu} = \eta_{\mu\nu} + h_{\mu\nu}$ where $h_{\mu\nu}$ refers to a small excitation of a flat metric [26], [14], [21]. But, these models failed because they are perturbatively nonrenormalizable [25], [18], [9].

In order to overcome these problems, string theories [19], [34], [43] and loop quantum gravity [35], [37], [5] were created in such a way that point like models of elementary particles be replaced by one-dimensional models.

In spite of an intense activity in this field and evident successes, the contact with experiment still remain nebulous and, furthermore, the conceptual framework of (super)string theories has not reached the expected maturity degree.

So, as G. 't Hooft [20] pointed out, it might be that quantum gravity could not be solved without revising the principles of quantum theory and, especially, at the Planck scale.

*E-mail address: pierre@math.ucl.ac.be

As a matter of fact, it is commonly admitted that Lorentz invariance could be broken at the Planck length due to the strong fluctuations of the space-time at this scale [22], [42].

To go beyond these difficulties, it was envisaged in [31] and in [33] **to enlarge the conceptual framework of quantum field theories** in order that the new proposed quantum model takes into account the quantum Physics below the Planck scale and corresponds to quantum field theories at the scale for which they were developed [4].

But, to become acceptable, the enlargement of the conceptual frame of quantum field theories has to imply some unification of these with the equations of general relativity.

The problem is that general relativity is a classical theory having resisted until now to a convincing quantization.

It is thus the aim of this paper to reconsider the matter in question on new basis at the light of what can bring close together general relativity and quantum field theories.

What characterizes general relativity with respect to a possible quantization is:

1. The influence of the space-time curvature on the matter and vice versa.
2. The existence of the cosmological constant λ interpreted by Y. Zel'dovich [44] as a medium tied to the vacuum polarization of the QFT (quantum field theory).

Reciprocally, one of the main features of QFT which could be used to quantize GR (General relativity) consists in the creation of (pairs of) particles from the vacuum energy which reacts backwards on the curvature of the space-time: this means that the energy of the created particles is taken away from the space-time itself [8].

In this respect, as the quantum level is essentially that of elementary particles, it was envisaged [29] by the author to set up equations describing the (internal) dynamics of a set of elementary particles in such a way that these equations be isomorphic to the equations of general relativity with respect to a metric contraction given by a suitable compactification from the quantum level to the classical level and associated with the condition $\hbar \to 0$.

To fulfil these conditions **in order to bridge the gap between QFT and GR**, it was envisaged that **the expanding space-time**, to which the cosmological constant of GR may correspond, **could constitute the fundamental structure of the QFT vacuum shared out amongst the considered elementary particles.**

In this context, every elementary particle would then be characterized by an internal expanding space-time structure constituting its own vacuum from which its matter shell could be generated due to the strong fluctuations of the internal space-time of its internal vacuum.

Indeed, at this scale which likely corresponds to the Planck scale, the fluctuations of the space-time generate local strong curvatures which are responsible for the generation of degenerate singularities allowing by versal deformations and blowups of these to create two external contracting enveloping structures above the particle internal vacuum in such a way that the most external structure is interpreted as its mass shell.

So, an important part of the vacuum of QFT would be of space-time nature distributed amongst **the internal vacua of the elementary particles** constituting massless particles **potentially able to generate their mass shells** due to the strong fluctuations of their internal space-time vacua.

Thus, the attempt of unifying QFT with GR leads us to **envisage elementary particles endowed with internal structures which must be of space-time type and quantized.**

The first step, developed in chapter 1, then consists in **generating mathematically space and time internal fields that must be really quantized**: this can be only realized algebraically by assuming that **quanta must have an algebraic structure given by algebraic closed subsets** characterized by a Galois extension degree N , $N \in \mathbb{N}$.

As extensively developed in [33], the internal fields of elementary particles have **a twofold nature** due to the fact that **every elementary particle has to be viewed as an elementary bisemiparticle** composed of a left semiparticle, localized in the upper half space, and of a right symmetric (co)semiparticle, localized in the lower half space in such a way that:

a) the right semiparticle, dual to the left semiparticle, is observed as projected on the latter and is thus normally unobservable.

b) the product of the internal structure of the right semiparticle by its left equivalent gives rise to a "working interaction space" responsible for the electric charge and the magnetic moment.

In this respect, the internal space-time structure of the vacuum of a bisemiparticle will be composed of an internal time field and of an internal space field having a bilinear nature and being localized in orthogonal spaces.

The time internal field (as well as the space internal field) is then given mathematically by a (bisemi)sheaf of differentiable bifunctions over a compactified algebraic bilinear semigroup: a bifunction is defined by the product of a right function localized in the lower half space by its symmetric left equivalent over the considered algebraic bilinear semigroup $\mathrm{GL}_2(F_{\overline{v}} \times F_v)$ covering its linear correspondent [30] and defined over the product $(F_{\overline{v}} \times F_v)$ of the set of pairs $\{F_{\overline{v}_{\mu,m_\mu}}, F_{v_{\mu,m_\mu}}\}$ or right and left real symmetric pseudo-ramified algebraic subsets having a structure at μ quanta of degree

$$[F_{\overline{v}_{\mu,m_\mu}} : K] = [F_{v_{\mu,m_\mu}} : K] = \mu \, . \, N$$

over a number field K of characteristic 0 .

The representation space $\mathrm{Repsp}(\mathrm{GL}_2(F_{\overline{v}} \times F_v))$ of $\mathrm{GL}_2(F_{\overline{v}} \times F_v)$ then consists in the product, right by left, of **two symmetric towers of conjugacy class representatives** in such a way that the functional representation subspaces of the products of **the corresponding right and left compactified conjugacy class representatives behave like harmonic oscillators.** This leads us to consider that **the functional (modular) representation space of $\mathbf{GL_2}(\boldsymbol{L_{\overline{v}}} \times \boldsymbol{L_v})$**, where $L_{\overline{v}}$ (resp. L_v) results from the compactification of $F_{\overline{v}}$ (resp. F_v), **can be interpreted as a time internal field of an elementary bisemiparticle.**

The corresponding internal space field can be obtained from the internal time field throughout a smooth (bi)endomorphism based on Galois antiautomorphisms as developed in proposition 2.7. The end of chapter 1 deals with **the compactification of the conjugacy class representatives of the "time" and "space" bilinear algebraic semigroups** leading to continuous upper and lower 4-dimensional (semi)manifolds of space-time.

So, **the consideration of algebraic quanta at a sub-Planckian scale allows to generate time and space quantized internal fields of the internal vacua** of elementary particles likely at the Planck scale.

These time and space internal fields are products, right by left, of time and space semifields restricted to the lower and upper half spaces and described mathematically by (semi)sheaves of differentiable functions over the appropriate algebraic bilinear semi-groups.

Due to the strong fluctuations of space-time at this level, these differentiable functions become afflicted by degenerate singularities of which versal deformations and blowups allow **to generate two covering "middle-ground" and "mass" fields of space and time.**

The internal vacuum of an elementary (bisemi)particle is thus described mathematically by a space and time internal fields and by an enveloping "middle-ground" space and time fields. A process of versal deformations and blowups of the singularities on the "middle-ground" space and time fields is responsible for the creation or **generation of the enveloping "mass" fields of space and time, phenomenon corresponding to the creation of this (bisemi)particle from its own vacuum and bring an interesting alternative to the Higgs mechanism by the Higgs boson.**

The generation of the middle-ground and mass fields of an elementary particle from its internal space and time fields constitute the contents of chapter 3, which allows to formulate **the equations of the internal dynamics of a bisemifermion.**

On the other hand, the unification of quantum field theory with general relativity, leading to a coherent quantum gravity theory, requires **a new interpretation of the equations of general relativity at the Planck scale**.

These objectives will be reached if it is taken into account that:

1. an **initial state mathematically well-defined** must be introduced "at the beginning" in such a way that:
   (a) it corresponds to some **universal structure** in the perspective of the Langlands program by a one-to-one correspondence with its automorphic representation.
   (b) it is of (bilinear) space-time nature.
   (c) its **space-time structure is totally quantized** from an algebraic point of view.
   (d) it gives rise to the fundamental structure of the **internal vacua of the elementary particles**.
   (e) it is directly **connected to the (small) value of the cosmological constant Λ** [40].
2. **the interactions between elementary particles** must appear naturally (i.e. without divergences) in the model by considering:
   (a) bilinear interactions between pairs of semiparticles as it was developed in chapter 5 of [33].
   (b) a unification of electro-magnetism with gravitation as it was hoped by Einstein [12].

In this new context, a special importance is thus given to the structure of elementary particles whose space-time nature is quantized and less to the evolution of states of these particles as currently done in quantum field theories and (super)string theories.

The idea then consists in building up a quantum gravity theory at the level of elementary particles in such a way that:

1. **the space-time structure**, constituting the "initial" state **of each elementary particle**, corresponds to its internal vacuum structure of expanding nature from which its mass structure of contracting nature can be generated dynamically from the singularities on its internal vacuum structure submitted to strong fluctuations at the Planck scale.

2. **the set of equations of vacuum and mass structure** of elementary particles **is in one-to-one correspondence with the equations of general relativity** if:

   - the big points, describing the quanta, at the Planck scale are viewed as ordinary points at the classical level by means of a metric contraction based on a condition equivalent to $\hbar \to 0$.
   - the one-dimensional components of the time and space structures of the elementary particles, considered as flow lines, are compactified to give rise to a four dimensional manifold endowed with a Riemann geometry.

The equations of general relativity, generated in that manner at the Planck scale, are characterized by a **high value of the cosmological constant**, corresponding to the set of internal vacua of elementary particles, in contrast with the standard interpretation of these equations where the cosmological constant, being very small [10], is used to adjust the geometry in order that it be on a average flat.

Chapter 4 is then devoted to the set up of the equations of the internal dynamics of a set of interacting particles and to the proof of their equivalence with the equations of general relativity in the perspective of unifying quantum field theories with general relativity.

All developments of this paper refer to the preprint "Algebraic quantum theory" [33].

2. The Space-Time Structure of the Internal Vacua of Elementary (Bisemi)fermions

A first step will then consist in recalling the algebraic quantized space-time structure of the vacuum of elementary particles, under the circumstances the elementary (stable) fermions, i.e.

- the leptons e^- , μ^- , τ^- and their neutrinos.
- the quarks u^+ , d^- , s^- , c^+ , b^- , t^+ .

2.1. An Elementary Bisemifermion

As developed in [29] and [31], **an elementary fermion must be viewed as an elementary bisemifermion**:

- composed of (the product of) a left semifermion, localized in the upper half space, and of a right semifermion, localized in the symmetric lower half space.

- centred on an emergence point allowing the transfer of quanta from the time structures of a semifermion to their space structures and vice-versa.
- to which it can be associated a "working space" composed of (tensor) products between right and left (time and space) internal structures, respectively of a right and of a left semifermion.

2.2. "Initial State" of a Bisemifermion

The "initial state" of a bisemifermion, corresponding to its most internal vacuum structure, must be chosen to be:

a) of mathematical nature, and, consequently, of space-time type.

b) discrete, and, thus quantized.

The only way to satisfy these two conditions is that the internal vacuum structure of a bisemifermion be **of algebraic type**.

And, as the bilinear nature of a bisemifermion must be taken into account, an algebraic bilinear semigroup over a set of products, right by left, of symmetric algebraic subsets characterized by increasing Galois extension degrees was introduced in [29] in such a way that a natural automorphic representation corresponds to it by means of a Langlands global correspondence [30].

2.3. Algebraic Bilinear Semigroup over Symmetric Splitting Fields

- Let K be a number field of characteristic 0 and let $K[x]$ denote a polynomial ring composed of a set of pairs of polynomials $\{P(x), P(-x)\}$, x being a time or space variable.

  The real and complex algebraic extensions of K , noted respectively $F_{\overline{v}-v}$ and $F_{\overline{\omega}-\omega}$, are assumed to be **symmetric splitting fields** $F_{\overline{v}-v} = F_{\overline{v}} \cup F_v$ and $F_{\overline{\omega}-\omega} = F_{\overline{\omega}} \cup F_\omega$ composed of right extension semifields $F_{\overline{v}}$ and $F_{\overline{\omega}}$ and of left extensions semifields F_v and F_ω in one-to-one correspondence [30], [33].

- **The real pseudo-ramified algebraic extensions** are assumed to be generated from **irreducible** (one-dimensional) **algebraic closed subsets** characterized by a Galois extension degree $[F_{v^1_\mu} : K] \equiv [F_{\overline{v}^1_\mu} : K] = N$, $1 \leq \mu \leq q \leq \infty$, equal to N and **interpreted as space or time quanta**.

  The real pseudo-ramified algebraic extensions F_{v_μ} (resp. $F_{\overline{v}_\mu}$) are indexed in equivalence classes characterized by their ranks (i.e., their Galois extension degrees) which are integers modulo N :

$$[F_{v_\mu} : K] = * + \mu \centerdot N \qquad \text{(resp.} \quad [F_{\overline{v}_\mu} : K] = * + \mu \centerdot N \text{)}$$

  where:

  - $F_{v_\mu} \in F_v$, $1 \leq \mu \leq q \leq \infty$ (resp. $F_{\overline{v}_\mu} \in F_{\overline{v}}$, $1 \leq \mu \leq q \leq \infty$).

– $*$ denotes an integer inferior to N .

The degrees will be generally chosen to be equal to 0 modulo N . So, we get **a tower of classes of real pseudo-ramified algebraic subsets**:

$$\begin{aligned} & F_{v_1} \subset \cdots \subset F_{v_{\mu,m_\mu}} \subset \cdots \subset F_{v_{q,m_q}} \\ \text{(resp.} \quad & F_{\overline{v}_1} \subset \cdots \subset F_{\overline{v}_{\mu,m_\mu}} \subset \cdots \subset F_{\overline{v}_{q,m_q}} \text{),} \end{aligned}$$

where $F_{v_{\mu,m_\mu}}$ (resp. $F_{\overline{v}_{\mu,m_\mu}}$) is the m_μ-th representative of the μ-th equivalence class,
in such a way that **all representatives of the μ-th equivalence class have a structure at μ quanta** of degree N , since their extension degrees (or ranks) are

$$[F_{v_{\mu,m_\mu}} : K] = \mu \centerdot N \qquad \text{(resp.} \quad [F_{\overline{v}_{\mu,m_\mu}} : K] = \mu \centerdot N \text{).}$$

It must be noted that the real pseudo-ramified algebraic subsets are included, in the sense of [30], into the corresponding complex pseudo-ramified algebraic subsets.

- The next step consists in considering an algebraic group over the product, right by left, of corresponding equivalent representatives of

$$\begin{aligned} F_v &= \{F_{v_1}, \cdots, F_{v_{\mu,m_\mu}}, \cdots, F_{v_{q,m_q}}\} \\ \text{and of} \quad F_{\overline{v}} &= \{F_{\overline{v}_1}, \cdots, F_{\overline{v}_{\mu,m_\mu}}, \cdots, F_{\overline{v}_{q,m_q}}\} \end{aligned}$$

to take into account the bilinear (or twofold) structure of the elementary bisemiparticles as noticed in section 2.2.

Let then $T_2(F_v)$ (resp. $T_2^t(F_{\overline{v}})$) denote the (semi)group of upper (resp. lower) triangular matrices of order 2 over F_v (resp. $F_{\overline{v}}$).

An **algebraic bilinear general semigroup**

$$\mathrm{GL}_2(F_{\overline{v}} \times F_v) \equiv T_2^t(F_{\overline{v}}) \times T_2(F_v)$$

can be introduced so that:

a) the product $(F_{\overline{v}} \times F_v)$ is taken over the set of corresponding pairs $\{F_{\overline{v}_{\mu,m_\mu}}, F_{v_{\mu,m_\mu}}\}_{v_{\mu,m_\mu}}$ of right and left real pseudo-ramified algebraic subsets.

b) $\mathrm{GL}_2(F_{\overline{v}} \times F_v)$ has the Gauss bilinear decomposition:

$$\mathrm{GL}_2(F_{\overline{v}} \times F_v) = [D_2(F_{\overline{v}}) \times D_2(F_v)][UT_2^t(F_{\overline{v}}) \times UT_2(F_v)]$$

where:

– $D_2(\centerdot)$ is the subgroup of diagonal matrices.

– $UT_2(\centerdot)$ (resp. $UT_2^t(\centerdot)$) is the subgroup of upper (resp. lower) unitriangular matrices.

c) $\mathrm{GL}_2(F_{\overline{v}} \times F_v)$ covers its linear equivalent $\mathrm{GL}_2(F_{\overline{v}} - F_v)$.

d) the representation space $\text{Repsp}(GL_2(F_{\overline{v}} \times F_v))$ of $GL_2(F_{\overline{v}} \times F_v)$ is given by the tensor product $M_R(F_{\overline{v}}) \otimes M_L(F_v)$ of a right $T_2^t(F_{\overline{v}})$-semimodule $M_R(F_{\overline{v}})$ by a left $T_2(F_v)$-semimodule $M_L(F_v)$.

e) the μ-th conjugacy class representative $GL_2(F_{\overline{v}_{\mu,m_\mu}} \times F_{v_{\mu,m_\mu}})$ of $GL_2(F_{\overline{v}} \times F_v)$, with respect to the product, right by left, $(F_{\overline{v}^1_\mu} \times F_{v^1_\mu})$ of irreducible algebraic closed subsets of rank N , has for representation the $GL_2(F_{\overline{v}_{\mu,m_\mu}} \times F_{v_{\mu,m_\mu}})$-subbisemimodule $(M_{F_{\overline{v}_{\mu,m_\mu}}} \otimes M_{F_{v_{\mu,m_\mu}}})$.

2.4. The Compactification of the Algebraic Bilinear Semigroup $GL_2(F_{\overline{v}} \times F_v)$

- The problem is that the conjugacy class representatives of $GL_2(F_{\overline{v}} \times F_v)$ are not locally compact.

  Now, the two elements $M_{F_{\overline{v}_{\mu,m_\mu}}}$ and $M_{F_{v_{\mu,m_\mu}}}$ of each conjugacy class representative $GL_2(F_{\overline{v}_{\mu,m_\mu}} \times F_{v_{\mu,m_\mu}})$ of $GL_2(F_{\overline{v}} \times F_v)$ rotate in opposite senses giving them a spin orientation according to [33].

  The rotation of the elements $M_{F_{\overline{v}_{\mu,m_\mu}}}$ and $M_{F_{v_{\mu,m_\mu}}}$ of the conjugacy class representatives leads to a compactification of them as developed by W. Fulton and R. McPherson [15] and adapted in the present case in [30].

- **The compactification of $M_{F_{v_{\mu,m_\mu}}}$ (resp. $M_{F_{\overline{v}_{\mu,m_\mu}}}$**) given by the map:

$$\gamma^c_{M_{F_{v_{\mu,m_\mu}}}} : M_{F_{v_{\mu,m_\mu}}} \longrightarrow M_{L_{v_{\mu,m_\mu}}} (\text{resp. } \gamma^c_{M_{F_{\overline{v}_{\mu,m_\mu}}}} : M_{F_{\overline{v}_{\mu,m_\mu}}} \longrightarrow M_{L_{v_{\mu,m_\mu}}})$$

results from:

a) a compactification of the μ irreducible algebraic closed subsets, i.e. of the μ quanta, of $M_{F_{v_{\mu,m_\mu}}}$ (resp. $M_{F_{\overline{v}_{\mu,m_\mu}}}$) by a sequence of blowups on the algebraic points of each irreducible subset transforming it into an irreducible completion (or transcendental extension) centred on a point in the upper (resp. lower) half space according to the maps:

$$\gamma^c_{v^1_\mu} : F_{v^1_\mu} \longrightarrow L_{v^1_\mu} \qquad (\text{resp. } \gamma^c_{\overline{v}^1_\mu} : F_{\overline{v}^1_\mu} \longrightarrow L_{\overline{v}^1_\mu})$$

where:

- the closed irreducible algebraic subset $F_{v^1_\mu}$ (resp. $F_{\overline{v}^1_\mu}$) is one of the μ quanta of $M_{F_{\overline{v}_{\mu,m_\mu}}}$ (resp. $M_{F_{v_{\mu,m_\mu}}}$).
- $L_{v^1_\mu}$ (resp. $L_{\overline{v}^1_\mu}$) is an irreducible transcendental extension of degree N , corresponding to a compactified quantum.

b) The connectedness of the set of the μ irreducible transcendental extensions put from end to end alter their compactifications in such a way that they generate a one-dimensional closed string $M_{L_{v_{\mu,m_\mu}}}$ (resp. $M_{L_{\overline{v}_{\mu,m_\mu}}}$).

- By this way, each compactified quantum $L_{v^1_\mu}$ (resp. $L_{\overline{v}^1_\mu}$) on the conjugacy class representative $M_{L_{v_{\mu,m_\mu}}}$ (resp. $M_{L_{\overline{v}_{\mu,m_\mu}}}$), which is a closed string, can be viewed as one of its μ **quantum "big points"** centred on the compactification points of the blowups as described in (a).

2.5. The Compactified Algebraic Bilinear Semigroup $\mathrm{GL}_2(L_{\overline{v}} \times L_v)$

- So, the compactification of all the conjugacy class representatives $(M_{F_{\overline{v}_{\mu,m_\mu}}} \otimes M_{F_{v_{\mu,m_\mu}}})$ of the bilinear algebraic semigroup $\mathrm{GL}_2(F_{\overline{v}} \times F_v)$ transforms it into the bilinear compactified algebraic semigroup $\mathrm{GL}_2(L_{\overline{v}} \times L_v)$ where

$$
\begin{aligned}
L_v &= \{L_{v_1}, \cdots, L_{v_{\mu,m_\mu}}, \cdots, L_{v_{q,m_q}}\} \\
(\text{resp.}\quad L_{\overline{v}} &= \{L_{\overline{v}_1}, \cdots, L_{\overline{v}_{\mu,m_\mu}}, \cdots, L_{\overline{v}_{q,m_q}}\}\,)
\end{aligned}
$$

  is the set of completions corresponding to the set F_v (resp. $F_{\overline{v}}$) of real pseudo-ramified algebraic subsets.

- The μ-th conjugacy class representative $\mathrm{GL}_2(L_{\overline{v}_{\mu,m_\mu}} \times L_{v_{\mu,m_\mu}})$ of $\mathrm{GL}_2(L_{\overline{v}} \times L_v)$, with respect to the product, right by left, $(L_{\overline{v}^1_\mu} \times L_{v^1_\mu})$ of irreducible transcendental extensions of rank N , has for representation the $\mathrm{GL}_2(L_{\overline{v}_{\mu,m_\mu}} \times L_{v_{\mu,m_\mu}})$-subbisemimodule $(M_{L_{\overline{v}_{\mu,m_\mu}}} \otimes M_{L_{v_{\mu,m_\mu}}})$.

- But, as the elements $M_{L_{\overline{v}_{\mu,m_\mu}}}$ and $M_{L_{v_{\mu,m_\mu}}}$ of $\mathrm{GL}_2(L_{\overline{v}} \times L_v)$ rotate in opposite senses as noticed in section 2.4, it is more exactly the Lie algebra $\mathfrak{gl}_2(L_{\overline{v}} \times L_v)$ of the bilinear compactified algebraic semigroup $\mathrm{GL}_2(L_{\overline{v}} \times L_v)$ which would be considered.

2.6. Proposition

The discrete structure, generated by the algebraic bilinear semigroup $\mathrm{GL}_2(F_{\overline{v}} \times F_v)$ *:*

- *corresponds to the time structure of the internal vacuum of an elementary (bisemi)-fermion.*

- *has been transformed by the compactification map:*

$$
\gamma^c_{F_{\overline{v}} \times F_v} : \quad \mathrm{Repsp}(\mathrm{GL}_2(F_{\overline{v}} \times F_v)) \longrightarrow \mathrm{Repsp}(\mathrm{GL}_2(L_{\overline{v}} \times L_v))
$$

  into a corresponding **locally compact quantized time structure**.

Proof.

1. It was proved in [33] and in [31] that $\mathrm{Repsp}(\mathrm{GL}_2(F_{\overline{v}} \times F_v))$ corresponds to the time structure of the internal vacuum of an elementary bisemifermion because its conjugacy class representatives $(M_{F_{\overline{v}_{\mu,m_\mu}}} \otimes M_{F_{v_{\mu,m_\mu}}})$ are isomorphic, under the compact-

ifications $\gamma^c_{M_{F_{\overline{v}_\mu}}} \times \gamma^c_{M_{F_{v_\mu}}}$, to products of pairs of strings $(M_{L_{\overline{v}_{\mu},m_\mu}} \otimes M_{L_{v_{\mu},m_\mu}})$ behaving like harmonic oscillators. And, a (bisemi)sheaf of differentiable (bi)functions on it is a (time) field as it will be seen.

2. As $M_{F_{v_{\mu},m_\mu}}$ (resp. $M_{F_{\overline{v}_{\mu},m_\mu}}$) is composed of μ quanta, its compactified equivalent $M_{L_{v_{\mu},m_\mu}}$ (resp. $M_{L_{\overline{v}_{\mu},m_\mu}}$) will also have a structure at μ quanta set out in a continuous way. ■

2.7. Proposition

Under the composition of maps $\gamma_{t_{R\times L}\to r_{R\times L}} \circ E_{R\times L}$ *, where:*

- $E_{R\times L} : \text{Repsp}(\text{GL}_2(L_{\overline{v}} \times L_v)_t) \to \text{Repsp}(\text{GL}_2^*(L^*_{\overline{v}} \times L^*_v)_t) \oplus \text{Repsp}(\text{GL}_2^I(L^I_{\overline{v}} \times L^I_v)_t)$ *is a smooth biendomorphism*
- $\gamma_{t_{R\times L}\to r_{R\times L}}$ *sends "time" complementary bistructures into orthogonal "space" bistructures,*

the "time" representation space $\text{Repsp}(\text{GL}_2(L_{\overline{v}} \times L_v)_t)$ *can be transformed into:*

a) a reduced "time" representation space $\text{Repsp}(\text{GL}_2^*(L^*_{\overline{v}} \times L^*_v)_t)$ *over a set of reduced completions* $(L^*_{\overline{v}} \times L^*_v)$ *.*

b) a complementary "space" representation space $\text{Repsp}(\text{GL}_2^I(L^I_{\overline{v}} \times L^I_v)_r)$ *over a complementary set of completions* $(L^I_{\overline{v}} \times L^I_v)$ *, localized in a space perpendicular to* $\text{Repsp}(\text{GL}_2^*(L^*_{\overline{v}} \times L^*_v)_t)$ *.*

Proof.

1. It was seen in [33] that the "time" representation space $\text{Repsp}(\text{GL}_2(L_{\overline{v}} \times L_v)_t)$, submitted to Galois antiautomorphisms on its right and left parts, can be decomposed into the direct sum of two non connected "time" representation spaces of bilinear algebraic semigroups:

$$E_{R\times L} : \quad \text{Repsp}(\text{GL}_2(L_{\overline{v}} \times L_v)_t) \longrightarrow$$
$$\text{Repsp}(\text{GL}_2^*(L^*_{\overline{v}} \times L^*_v)_t) \oplus \text{Repsp}(\text{GL}_2^I(L^I_{\overline{v}} \times L^I_v)_t)$$

corresponding to a smooth biendomorphsim $E_{R\times L}$ in such a way that the biquanta (i.e. products of pairs of corresponding quanta on the same conjugacy class representatives $(M_{L_{\overline{v}_{\mu},m_\mu}} \otimes M_{L_{v_{\mu},m_\mu}})$), extracted from $\text{Repsp}(\text{GL}_2^*(L^*_{\overline{v}} \times L^*_v)_t)$ by Galois antiautomorphisms, be used to build up a new non connected complementary "time" representation space $\text{Repsp}(\text{GL}_2^I(L^I_{\overline{v}} \times L^I_v)_t)$.

2. Under the map

$$\gamma_{t_{R\times L}\to r_{R\times L}} : \quad \text{Repsp}(\text{GL}_2^I(L^I_{\overline{v}} \times L^I_v)_t) \longrightarrow \text{Repsp}(\text{GL}_2^I(L^I_{\overline{v}} \times L^I_v)_r)$$

the complementary " time" representation space $\text{Repsp}(\text{GL}_2^I(L^I_{\overline{v}} \times L^I_v)_r)$ is sent, out of the origin, biquantum by biquantum, into the orthogonal space generating by this way the complementary representation space $\text{Repsp}(\text{GL}_2^I(L^I_{\overline{v}} \times L^I_v)_r)$ of "space".

3. Thus, the composition of maps:

$$\gamma_{t_{R\times L}\to r_{R\times L}} \circ E_{R\times L} : \quad \text{Repsp}(\text{GL}_2(L_{\overline{v}} \times L_v)_t) \\ \longrightarrow \text{Repsp}(\text{GL}_2^*(L_{\overline{v}}^* \times L_v^*)_t) \oplus \text{Repsp}(\text{GL}_2^I(L_{\overline{v}}^I \times L_v^I)_r)$$

transforms the "time" representation space $\text{Repsp}(\text{GL}_2(L_{\overline{v}} \times L_v)_t)$ into the reduced "time" representation space $\text{Repsp}(\text{GL}_2^*(L_{\overline{v}}^* \times L_v^*)_t)$ and into the complementary "space" representation space $\text{Repsp}(\text{GL}_2^I(L_{\overline{v}}^I \times L_v^I)_r)$.

As a consequence, **the "time" structure of the internal vacuum** of an elementary (bisemi)fermion, given by the "time" representation space $\text{Repsp}(\text{GL}_2(L_{\overline{v}} \times L_v)_t)$, **can be transformed** partially, or completely, **into a "space" structure**, given by the complementary "space" representation space $\text{Repsp}(\text{GL}_2^I(L_{\overline{v}}^I \times L_v^I)_r)$ **and this "space-time" structure is quantized**. ■

2.8. Bisemisheaf of Differentiable (Bi)functions on $\text{GL}_2(L_{\overline{v}} \times L_v)$

- Let $\phi_L(M_{v_\mu})$ (resp. $\phi_R(M_{\overline{v}_\mu})$) denote a complex-valued differentiable function (resp. cofunction) over the μ-th real conjugacy class representative $M_{L_{v_\mu}}$ (resp. $M_{L_{\overline{v}_\mu}}$) of $T_2(L_v)$ (resp. $T_2^t(L_{\overline{v}})$) $\subset \text{GL}_2(L_{\overline{v}} \times L_v)$ and let $\phi_R(M_{\overline{v}_\mu}) \otimes \phi_L(M_{v_\mu})$ be the corresponding bifunction on the conjugacy class representative $(M_{L_{\overline{v}_\mu}} \otimes M_{L_{v_\mu}})$ of $\text{GL}_2(L_{\overline{v}} \times L_v)$.

- The set $\{\phi_L(M_{v_{\mu,m_\mu}})\}_{\mu,m_\mu}$ (resp. $\{\phi_R(M_{\overline{v}_{\mu,m_\mu}})\}_{\mu,m_\mu}$) of $\mathbb{C}$ −valued differentiable functions, localized in the upper (resp. lower) half space and defined over the $T_2(L_v)$ (resp. $T_2^t(L_{\overline{v}})$)-semimodule $M_L(L_v)$ (resp. $M_R(L_{\overline{v}})$), constitutes the set $\Gamma(\phi_L(M_L(v)))$ (resp. $\Gamma(\phi_R(M_R(L_{\overline{v}})))$) of sections of **the semisheaf of rings $\boldsymbol{\phi_L(M_L(L_v))}$ (resp. $\boldsymbol{\phi_R(M_R(L_{\overline{v}}))}$)**, also noted $\widetilde{M}_L(L_v)$ (resp. $\widetilde{M}_R(L_{\overline{v}})$).

- And the set $\{\phi_R(M_{\overline{v}_{\mu,m_\mu}}) \otimes \phi_L(M_{v_{\mu,m_\mu}})\}_{\mu,m_\mu}$ of differentiable bifunctions over the $\text{GL}_2(L_{\overline{v}} \times L_v)$-bisemimodule $M_R(L_{\overline{v}}) \otimes M_L(L_v)$ constitutes the set of bisections of **the bisemisheaf of rings $\boldsymbol{\phi_R(M_R(L_{\overline{v}})) \otimes \phi_L(_L(L_v))}$.** ■

2.9. Proposition

*The bisemisheaf $\phi_R(M_R(L_{\overline{v}})) \otimes \phi_L(M_L(L_v))$ of differentiable (bi)functions is a physical **string field**.*

Proof.

- The bisections $\phi_R(M_{\overline{v}_{\mu,m_\mu}}) \otimes \phi_L(M_{v_{\mu,m_\mu}})$ of the bisemisheaf $\phi_R(M_R(L_{\overline{v}})) \otimes \phi_L(M_L(L_v))$ are $\mathbb{C}$-valued differentiable bifunctions on the conjugacy class representatives $(M_{\overline{v}_{\mu,m_\mu}} \otimes M_{v_{\mu,m_\mu}})$ which are (tensor) products of two symmetric (closed) strings at μ quanta in such a way that $(M_{\overline{v}_{\mu,m_\mu}} \otimes M_{v_{\mu,m_\mu}})$ (and also $\phi_R(M_{\overline{v}_{\mu,m_\mu}}) \otimes \phi_L(M_{v_{\mu,m_\mu}})$) behave like harmonic oscillators [29].

- So, the set $\{\phi_R(M_{\overline{v}_{\mu,m_\mu}}) \otimes \phi_L(M_{v_{\mu,m_\mu}})\}_{\mu,m_\mu}$ of bisections of the bisemisheaf $\phi_R(M_R(L_{\overline{v}})) \otimes \phi_L(M_L(L_v))$ is a physical string field composed of packets of products of symmetric (closed) strings characterized by increasing numbers of quanta and behaving like harmonic oscillators. ■

2.10. Getting a Compact 4-Dimensional Semimanifold of Space-Time

1. A smooth linear **compact 3-dimensional semispace** M^I_τ **(resp.** $M^I_{\overline{\tau}}$), restricted to the upper (resp. lower) half space, can be constructed by considering in the first place the compactification C^τ_L (resp. C^τ_R) of the conjugacy class representatives $M^I_{v_{\mu,m_\mu}}$ (resp. $M^I_{\overline{v}_{\mu,m_\mu}}$) of the "space" bilinear algebraic semigroup $\mathrm{GL}^I_2(L^I_{\overline{v}} \times L^I_v)_r$ in the following way:

   (a) The real conjugacy clas representatives $\{M^I_{v_{\mu,m_\mu}}\}_{m_\mu}$ (resp. $\{M^I_{\overline{v}_{\mu,m_\mu}}\}_{m_\mu}$), m_μ varying, for each class μ , are compactified into a surface in such a way that they coincide with the complex equivalent $M^I_{\omega_\mu}$ (resp. $M^I_{\overline{\omega}_\mu}$):

   $$\bigcup_c M^I_{v_{\mu,m_\mu}} \subseteq M^I_{\omega_\mu} \qquad (\text{resp. } \bigcup_c M^I_{\overline{v}_{\mu,m_\mu}} \subseteq M^I_{\overline{\omega}_\mu})$$

   as described in [29], where $\bigcup_c$ denotes the "compact unions" or compactification.

   (b) The set of these surfaces $\{M^I_{\omega_\mu}\}_\mu$ (resp. $\{M^I_{\overline{\omega}_\mu}\}_\mu$), μ varying, $1 \leq \mu \leq q \leq \infty$, are also compactified into a 3-dimensional volume $M^I_{\tau_L}$ (resp. $M^I_{\tau_R}$), which is assumed to be of "space type", in such a way that $M^I_{\tau_L}$ (resp. $M^I_{\tau_R}$) be foliated by the surfaces $M^I_{\omega_\mu}$ (resp. $M^I_{\overline{\omega}_\mu}$).

2. Consider then the complementary **"time" bilinear algebraic semigroup** $\boldsymbol{\mathrm{GL}_2(L_{\overline{v}} \times L_v)_t}$ from which $\mathrm{GL}^I_2(L^I_{\overline{v}} \times L^I_v)_r$ was generated according to proposition 2.7.

   As for $\mathrm{GL}^I_2(L^I_{\overline{v}} \times L^I_v)_r$, the left (resp. right) linear conjugacy class representatives $M_{v_{\mu,m_\mu}}$ (resp. $M_{\overline{v}_{\mu,m_\mu}}$) of the "time" bilinear algebraic semigroup $\mathrm{GL}^*_2(L^*_{\overline{v}} \times L^*_v)_t$ are one-dimensional, corresponding then physically to "time" strings.

3. If the set $\{M_{v_{\mu,m_\mu}}\}_{m_\mu,\mu}$ (resp. $\{M_{\overline{v}_{\mu,m_\mu}}\}_{m_\mu,\mu}$ of **the "time" conjugacy class representatives are glued together** under the map C^{t-r}_L (resp. C^{t-r}_R) **"above" the 3-dimensional volume $M^I_{\tau_L}$ (resp. $M^I_{\tau_R}$) of space**, then we get a compactified 4-dimensional semispace of space-time $M^{T-S}_{\tau_L}$ (resp. $M^{T-S}_{\tau_R}$).

4. **In summary**, let:

   - $$C^\tau_L : \{M^I_{v_{\mu,m_\mu}}\}_{\mu,m_\mu} \to M^I_{\tau_L} = \bigcup_C M^I_{\omega_\mu}$$
     $$(\text{resp. } C^\tau_R : \{M^I_{\overline{v}_{\mu,m_\mu}}\}_{\mu,m_\mu} \to M^I_{\tau_R} = \bigcup_C M^I_{\overline{\omega}_\mu})$$

     be the compactification of the one-dimensional conjugacy class representatives of space into a 3-dimensional semispace.

- $$\begin{aligned} & C_L^{t-r} : M^I_{\tau_L} \cup \{M_{v_{\mu,m_\mu}}\}_{\mu,m_\mu} \to M^{T-S}_{\tau_L} = M^I_{\tau_L} \underset{C}{\cup} \{M_{v_{\mu,m_\mu}}\}_{\mu,m_\mu} \\ \text{(resp. } & C_R^{t-r} : M^I_{\tau_R} \cup \{M_{\overline{v}_{\mu,m_\mu}}\}_{\mu,m_\mu} \to M^{T-S}_{\tau_R} = M^I_{\tau_R} \underset{C}{\cup} \{M_{\overline{v}_{\mu,m_\mu}}\}_{\mu,m_\mu} \text{)} \end{aligned}$$

  be the compactification of the one-dimensional conjugacy class representatives of time with the 3-dimensional compact semispace $M^I_{\tau_L}$ (resp. $M^I_{\tau_R}$).

5. A **4-dimensional left (resp. right) semimanifold $\mathcal{M}_L^{T-S}$ (resp. $\mathcal{M}_R^{T-S}$)**) is then defined from the 4-dimensional semispace $M^{T-S}_{\tau_L}$ (resp. $M^{T-S}_{\tau_R}$) as a collection of charts $(M^{T-S}_{\tau_\mu}, \phi_\mu)$ (resp. $(M^{T-S}_{\overline{\tau}_\mu}, \overline{\phi}_\mu)$), where:

   - $M^{T-S}_{\tau_\mu}$ (resp. $M^{T-S}_{\overline{\tau}_\mu}$) are the compactified 4-dimensional semispaces obtained from the compactification of the linear conjugacy class representatives of $\mathrm{GL}_2(L_{\overline{v}} \times L_v)_t$ and of $\mathrm{GL}_2^I(L^I_{\overline{v}} \times L^I_v)_r$.
   - ϕ_μ (resp. $\overline{\phi}_\mu$) are one-to-one maps of $M^{T-S}_{\tau_\mu}$ (resp. $M^{T-S}_{\overline{\tau}_\mu}$) to open sets in the upper (resp. lower) half of $\mathbb{R}^4$,

   such that

   (a) $\mathcal{M}_L^{T-S} = \bigcup\limits_\mu M^{T-S}_{\tau_\mu}$ (resp. $\mathcal{M}_R^{T-S} = \bigcup\limits_\mu M^{T-S}_{\overline{\tau}_\mu}$).

   (b) if $M^{T-S}_{\tau_\mu} \cap M^{T-S}_{\tau_\nu}$ (resp. $M^{T-S}_{\overline{\tau}_\mu} \cap M^{T-S}_{\overline{\tau}_\nu}$) is non-empty, then the map

   $$\begin{aligned} & \phi_\mu \circ \phi_\nu^{-1} : \quad \phi_\nu(M^{T-S}_{\tau_\mu} \cap M^{T-S}_{\tau_\nu}) \longrightarrow \phi_\mu(M^{T-S}_{\tau_\mu} \cap M^{T-S}_{\tau_\nu}) \\ \text{(resp. } & \overline{\phi}_\mu \circ \overline{\phi}_\nu^{-1} : \quad \overline{\phi}_\nu(M^{T-S}_{\overline{\tau}_\mu} \cap M^{T-S}_{\overline{\tau}_\nu}) \longrightarrow \overline{\phi}_\mu(M^{T-S}_{\overline{\tau}_\mu} \cap M^{T-S}_{\overline{\tau}_\nu}) \text{)} \end{aligned}$$

   is a map of open upper (resp. lower) half subset of $\mathbb{R}^4$ to an open (resp. lower) half subset of $\mathbb{R}^4$.

6. **The composition of maps**:

   $$\begin{aligned} & C_L^{t-r} \circ C_L^r : \quad \{M_{v_{\mu,m_\mu}}\}_{\mu,m_\mu} \cup \{M^I_{v_{\mu,m_\mu}}\}_{\mu,m_\mu} \longrightarrow M^{T-S}_{\tau_L} \\ \text{(resp. } & C_R^{t-r} \circ C_R^r : \quad \{M_{\overline{v}_{\mu,m_\mu}}\}_{\mu,m_\mu} \cup \{M^I_{\overline{v}_{\mu,m_\mu}}\}_{\mu,m_\mu} \longrightarrow M^{T-S}_{\tau_R} \text{),} \end{aligned}$$

   compactifying one-dimensional conjugacy class representatives of space with one-dimensional conjugacy class representatives of time into a compact 4-dimensional upper (resp. lower) semispace $M^{T-S}_{\tau_L}$ (resp. $M^{T-S}_{\tau_R}$) of space-time, **allows to get a classical continuous (Riemann) 4-dimensional semispace from locally compact one-dimensional conjugacy class representatives of space-time at the Planck scale, the compactification map $C_L^{t-r} \circ C_L^r$ (resp. $C_R^{t-r} \circ C_R^r$) corresponding to a change of scale**.

2.11. Proposition

The product $(C_R^{t-r} \circ C_R^r) \times (C_L^{t-r} \circ C_L^r)$ of the 4-dimensional compactifications introduced in section 2.10.6. sends the direct sum of the representation space $\mathrm{Repsp}(\mathrm{GL}_2^*(L^*_{\overline{v}} \times$

$L_v^*)_t$) of the reduced "time" bilinear algebraic semigroup $\mathrm{GL}_2^*(L_{\overline{v}}^* \times L_v^*)_t$ and of the representation space $\mathrm{Repsp}(\mathrm{GL}_2^I(L_{\overline{v}}^I \times L_v^I)_r)$ of the complementary "space" bilinear algebraic semigroup $\mathrm{GL}_2^I(L_{\overline{v}}^I \times L_v^I)$ into the product $(\boldsymbol{M}_{\tau_R}^{T-S} \otimes \boldsymbol{M}_{\tau_L}^{T-S})$ of the 4-dimensional compact lower semispace $\boldsymbol{M}_{\tau_R}^{T-S}$ by its upper equivalent $\boldsymbol{M}_{\tau_L}^{T-S}$ according to:

$$\begin{aligned}(C_R^{t-r} \circ C_R^r) \times (C_L^{t-r} \circ C_L^r) : \quad & \mathrm{Repsp}(\mathrm{GL}_2^*(L_{\overline{v}}^* \times L_v^*)_t) \oplus \mathrm{Repsp}(\mathrm{GL}_2^I(L_{\overline{v}}^I \times L_v^I))_r) \\ & \longrightarrow (\boldsymbol{M}_{\tau_R}^{T-S} \otimes \boldsymbol{M}_{\tau_L}^{T-S})\end{aligned}$$

in such a way that:

- *each point of $(\boldsymbol{M}_{\tau_R}^{T-S} \otimes \boldsymbol{M}_{\tau_L}^{T-S})$ is in fact a bipoint characterized by a metric tensor g.*
- *each bipoint of $(\boldsymbol{M}_{\tau_R}^{T-S} \otimes \boldsymbol{M}_{\tau_L}^{T-S})$ is in one-to-one correspondence with a bipoint of $\mathrm{Repsp}(\mathrm{GL}_2^*(L_{\overline{v}}^* \times L_v^*)_t) \oplus \mathrm{Repsp}(\mathrm{GL}_2^I(L_{\overline{v}}^I \times L_v^I))_r)$ so that $(C_R^{t-r} \circ C_R^r) \times (C_L^{t-r} \circ C_L^r)$ is a homeomorphism.*

Proof.

1. the fact that $(C_R^{t-r} \circ C_R^r) \times (C_L^{t-r} \circ C_L^r)$ is a (bi)isomorphism results from its compactifying nature introduced in section 2.10.
2. Each point $\mathcal{P}$ of $(\boldsymbol{M}_{\tau_R}^{T-S} \otimes \boldsymbol{M}_{\tau_L}^{T-S})$ is a bipoint $\mathcal{P}_R \times \mathcal{P}_L$, product of a point $\mathcal{P}_L \in M_{\tau_L}^{T-S}$ and of a point $\mathcal{P}_R \in M_{\tau_R}^{T-S}$.

   A metric tensor g at $\mathcal{P} \equiv \mathcal{P}_R \times \mathcal{P}_L$ has for components $g_a^b = g(E_a, E^b)$ (or g_{a_b}) with respect to basis vectors $E_a \in \boldsymbol{M}_{\tau_R}^{T-S}$and $E_b \in \boldsymbol{M}_{\tau_L}^{T-S}$, $a = t, x, y, z$, $b = t, x, y, z$, in such a way that $g(E_a, E^b)$ is a scalar product between basis vectors.
3. The compactified bisemispace $\boldsymbol{M}_{\tau_R}^{T-S} \otimes \boldsymbol{M}_{\tau_L}^{T-S}$ can be understood if it is realized that each bipoint $(\mathcal{P}_R \times \mathcal{P}_L)$ of it is in one-to-one correspondence with a bipoint $(p_R \times p_L)$ belonging to the product, right by left, of two **one-dimensional** symmetric conjugacy class representatives of $\mathrm{Repsp}(\mathrm{GL}_2^*(L_{\overline{v}}^* \times L_v^*)_t)$ or of $\mathrm{Repsp}(\mathrm{GL}_2^I(L_{\overline{v}}^I \times L_v^I))_r)$, if it is referred to section 2.10. ∎

3. The Equations of the Internal Dynamics of a Bisemifermion

3.1. Singularizations on the Internal Vacuum Semisheaves

- It was seen in propositions 2.7 and 2.11 that **the internal vacuum space-time structure** of an elementary (bisemi)fermion is given by the direct sum of the representation spaces of the bilinear algebraic semigroups $\mathrm{GL}_2(L_{\overline{v}} \times L_v)_t$ and $\mathrm{GL}_2^I(L_{\overline{v}}^I \times L_v^I)_r$ according to:

$$\begin{gathered}(M_{ST_R}^T \otimes M_{ST_L}^T) \oplus (M_{ST_R}^S \otimes M_{ST_L}^S) \simeq \\ \mathrm{Repsp}(\mathrm{GL}_2(L_{\overline{v}} \times L_v)_t) \oplus \mathrm{Repsp}(\mathrm{GL}_2^I(L_{\overline{v}}^I \times L_v^I)_r)\end{gathered}$$

  where:

- $M^T_{ST_L} \simeq \text{Repsp}(T_2(L_{\overline{v}})_t)$ (resp. $M^T_{ST_R} \simeq \text{Repsp}(T^t_2(L_{\overline{v}})_t)$) represents the algebraic time structure of the internal vacuum of a left (resp. right) semifermion.
- $M^S_{ST_L} \simeq \text{Repsp}(T^I_2(L^I_v))$ (resp. $M^S_{ST_R} \simeq \text{Repsp}(T^{t,I}_2(L^I_{\overline{v}}))$) represents the algebraic space structure of the internal vacuum of a left (resp. right) semifermion.

- **The semisheaf of differentiable functions** on $M^T_{ST_L}$ (resp. $M^T_{ST_R}$) and on $M^S_{ST_L}$ (resp. $M^S_{ST_R}$) is noted respectively $\widetilde{M}^T_{ST_L}$ (resp. $\widetilde{M}^T_{ST_R}$) and $\widetilde{M}^S_{ST_L}$ (resp. $\widetilde{M}^S_{ST_R}$).

  The bisemisheaves $(\widetilde{M}^T_{ST_L} \otimes \widetilde{M}^T_{ST_R})$ and $(\widetilde{M}^S_{ST_L} \otimes \widetilde{M}^S_{ST_R})$ are, respectively, a "time" and a "space" string field of the internal vacuum of an elementary bisemifermion.

- As these internal vacuum string fields have a spatial extension of the order of the Planck length, they are submitted to strong fluctuations generating **degenerate singularities** on the sections (or strings) of these, as described mathematically in [32].

  Taking into account that the time (and space) left and right semisheaves are symmetric by construction and localized in small open balls, it is reasonable to assume that the singularities are generated symmetrically on the corresponding sections respectively in the upper and lower half spaces.

- By this way, it is easy to understand that **the metric tensor** g^b_a at each bipoint $(\mathcal{P}_R \times \mathcal{P}_L)$ of a bisection $(\widetilde{M}^T_{\overline{v}_{\mu,m_\mu}} \otimes \widetilde{M}^T_{v_{\mu,m_\mu}})$ (also noted $\phi_R(M_{\overline{v}_{\mu,m_\mu}}) \otimes \phi_L(M_{v_{\mu,m_\mu}})$ in proposition 2.9) of $(\widetilde{M}^T_{ST_R} \otimes \widetilde{M}^T_{ST_L})$ or of a bisection $(\widetilde{M}^S_{\overline{v}_{\nu,m_\nu}} \otimes \widetilde{M}^S_{v_{\nu,m_\nu}})$ of $(\widetilde{M}^S_{ST_R} \otimes \widetilde{M}^S_{ST_L})$ is not constant but varies according to the singularities on the sections.

3.2. Versal Deformation and Its Blowup

- Under a strong external perturbation, **a degenerate singularity** of corank 1 and multiplicity inferior or equal to 3 is assumed to be generated on each section $\widetilde{M}^S_{v_{\nu,m_\nu}}$ (resp. $\widetilde{M}^S_{\overline{v}_{\nu,m_\nu}}$) of the "space" semisheaf $\widetilde{M}^S_{ST_L}$ (resp. $\widetilde{M}^S_{ST_R}$).

- Then, **a versal deformation** of $\widetilde{M}^S_{ST_L}$ (resp. $\widetilde{M}^S_{ST_R}$) can be envisaged as given by the fibre bundle:

$$
\begin{aligned}
D_{S_L} : \quad & \widetilde{M}^S_{ST_L} \times \theta_{S_L} \longrightarrow \widetilde{M}^S_{ST_L} \\
\text{(resp.} \quad D_{S_R} : \quad & \widetilde{M}^S_{ST_R} \times \theta_{S_R} \longrightarrow \widetilde{M}^S_{ST_R} \text{)}
\end{aligned}
$$

where the fibre $\theta_{S_L} = \{\theta_1(\omega^1_L), \theta_2(\omega^2_L), \theta_3(\omega^3_L)\}$ (resp. $\theta_{S_R} = \{\theta_1(\omega^1_R), \theta_2(\omega^2_R), \theta_3(\omega^3_R)\}$) is composed of the set of three sheaves of the base S_L (resp. S_R) of the versal deformation in such a way that the ω^i_L (resp. ω^i_R , $1 \leq i \leq 3$, are the generators of the base of the quotient algebra.

- We refer to [32] for more complete developments of the versal deformation and of its blowup succinctly recalled in the following.

- **The blowup of the versal deformation** allows to generate a semisheaf $\widetilde{M}^S_{MG_L}$ (resp. $\widetilde{M}^S_{MG_R}$) covering the space internal vacuum semisheaf $\widetilde{M}^S_{ST_L}$ (resp. $\widetilde{M}^S_{ST_R}$) by means of the spreading-out isomorphism:

$$SOT_L = \tau_{v_{\omega_L}} \circ \pi_{s_L} \qquad \text{(resp.} \quad SOT_R = \tau_{v_{\omega_R}} \circ \pi_{s_R} \text{)}$$

where:

  - $$\begin{array}{lll} & \pi_{s_L} : & \widetilde{M}^S_{ST_L} \times \theta_{S_L} \longrightarrow \widetilde{M}^S_{ST_L} \oplus \theta_{S_L} \\ \text{(resp.} & \pi_{s_R} : & \widetilde{M}^S_{ST_R} \times \theta_{S_R} \longrightarrow \widetilde{M}^S_{ST_R} \oplus \theta_{S_R} \text{)} \end{array}$$
    is an endomorphism, based on Galois antiautomorphisms [33], which disconnect the fibre θ_{S_L} (resp. θ_{S_R}) from $\widetilde{M}^S_{ST_L}$ (resp. $M^S_{ST_R}$).

  - $\tau_{v_{\omega_L}}$ (resp. $\tau_{v_{\omega_R}}$) is the projective map:

$$\tau_{v_{\omega_L}} : \quad \text{TAN}(\theta_{S_L}) \longrightarrow \theta_{S_L} \qquad \text{(resp.} \quad \tau_{v_{\omega_R}} : \quad \text{TAN}(\theta_{S_R}) \longrightarrow \theta_{S_R} \text{)}$$

    of the vertical tangent bundle $T_{v_{\omega_L}}$ (resp. $T_{v_{\omega_R}}$) sending θ_{S_L} (resp. θ_{S_R}) into the total tangent space $\text{TAN}(\theta_{S_L})$ (resp. $\text{TAN}(\theta_{S_R})$).

By this way, the three functions $\omega^i_L(v_{\nu,m_\nu})$ (resp. $\omega^i_R(\overline{v}_{\nu,m_\nu})$) of the base of the versal deformation are projected above each section $\widetilde{M}^S_{v_{\nu,m_\nu}}$ (resp. $\widetilde{M}^S_{\overline{v}_{\nu,m_\nu}}$) in the vertical tangent space.

Being glued together, these three functions generate the sections $\widetilde{M}^{S_p}_{MG_{v_{\nu,m_\nu}}}$ (resp. $\widetilde{M}^{S_p}_{MG_{\overline{v}_{\nu,m_\nu}}}$) of a semisheaf $\widetilde{M}^{S_p}_{MG_L}$ (resp. $\widetilde{M}^{S_p}_{MG_R}$) (called middle ground) which covers the internal vacuum semisheaf.

If the numbers of quanta on the sections $\widetilde{M}^{S_p}_{MG_{v_{\nu,m_\nu}}}$ (resp. $\widetilde{M}^{S_p}_{MG_{\overline{v}_{\nu,m_\nu}}}$) of the **middle ground semisheaf $\widetilde{M}^{S_p}_{MG_L}$ (resp. $\widetilde{M}^{S_p}_{MG_R}$**) are equal to the numbers of quanta on the sections $\widetilde{M}^S_{v_{\nu,m_\nu}}$ (resp. $\widetilde{M}^S_{\overline{v}_{\nu,m_\nu}}$) (rewritten according to $\widetilde{M}^S_{ST_{v_{\nu,m_\nu}}}$ (resp. $\widetilde{M}^S_{ST_{\overline{v}_{\nu,m_\nu}}}$)) of the internal vacuum semisheaf $\widetilde{M}^S_{ST_L}$ (resp. $\widetilde{M}^S_{ST_R}$),then these sections $\widetilde{M}^{S_p}_{MG_{v_{\nu,m_\nu}}}$ (resp. $\widetilde{M}^{S_p}_{MG_{\overline{v}_{\nu,m_\nu}}}$) are open strings covering the closed strings $\widetilde{M}^S_{ST_{v_{\nu,m_\nu}}}$ (resp. $\widetilde{M}^S_{ST_{\overline{v}_{\nu,m_\nu}}}$) of $\widetilde{M}^S_{ST_L}$ (resp. $\widetilde{M}^S_{ST_R}$).

3.3. Proposition

The inverse of the projective map $\tau_{v_{\omega_L}}$ (resp. $\tau_{v_{\omega_R}}$) of the tangent bundle $T_{v_{\omega_L}}$ (resp. $T_{v_{\omega_R}}$) of the spreading-out isomorphism is the elliptic operator:

$$DT^S_{L;MG} = \left\{ i\,\frac{\hbar_{MG}}{c_{t\to r;MG}}\,\frac{\partial}{\partial x}, i\,\frac{\hbar_{MG}}{c_{t\to r;MG}}\,\frac{\partial}{\partial y}, i\,\frac{\hbar_{MG}}{c_{t\to r;MG}}\,\frac{\partial}{\partial z} \right\}$$

$$\left(\textit{resp.} \quad DT^S_{R;MG} = \left\{ -i\,\frac{\hbar_{MG}}{c_{t\to r;MG}}\,\frac{\partial}{\partial x}, -i\,\frac{\hbar_{MG}}{c_{t\to r;MG}}\,\frac{\partial}{\partial y}, -i\,\frac{\hbar_{MG}}{c_{t\to r;MG}}\,\frac{\partial}{\partial z} \right\} \right),$$

where the constants $\hbar_{MG}$ and $c_{t\to r;MG}$ are defined in chapter 3 of [33], sending the semisheaf $\widetilde{M}^S_{MG_L}$ (resp. $\widetilde{M}^S_{MG_R}$) into the perverse semisheaf $\widetilde{M}^{S_p}_{MG_L}$ (resp. $\widetilde{M}^{S_p}_{MG_R}$) according to:

$$DT^S_{L;MG}: \quad \widetilde{M}^S_{MG_L} \longrightarrow \widetilde{M}^{S_p}_{MG_L} \qquad (\textit{resp.} \quad DT^S_{R;MG}: \quad \widetilde{M}^S_{MG_R} \longrightarrow \widetilde{M}^{S_p}_{MG_R}).$$

Proof. The elliptic operator $DT^S_{L;MG}$ (resp. $DT^S_{R;MG}$) maps the semisheaf $\widetilde{M}^S_{MG_L}$ (resp. $\widetilde{M}^S_{MG_R}$) into its perverse equivalent $\widetilde{M}^{S_p}_{MG_L}$ (resp. $\widetilde{M}^{S_p}_{MG_R}$) since this latter belongs to the derived category of string semifields shifted in the three geometrical dimensions of space, a string semifield being given by a semisheaf, for example $\widetilde{M}^S_{MG_L}$ (resp. $\widetilde{M}^S_{MG_R}$). ■

3.4. Generation of Perverse Mass Semisheaves $\widetilde{M}^{S_p}_{M_L}$ and $\widetilde{M}^{S_p}_{M_R}$

- As the singularities on the sections of the internal vacuum semisheaf $\widetilde{M}^S_{ST_L}$ (resp. $\widetilde{M}^S_{ST_R}$) have a multiplicity inferior or equal to 3 , the functions $\omega^i_L(v_{\nu,m_\nu})$ (resp. $\omega^i_r(\overline{v}_{\nu,m_\nu})$) of the quotient algebra of the versal deformation of $\widetilde{M}^S_{ST_L}$ (resp. $\widetilde{M}^S_{ST_R}$) can have degenerate singularities of corank 1 and multiplicity one.

- So, the middle ground semisheaf $\widetilde{M}^S_{MG_L}$ (resp. $\widetilde{M}^S_{MG_R}$), of which sections $\widetilde{M}^S_{MG_{v_{\nu,m_\nu}}}$ (resp. $\widetilde{M}^S_{MG_{\overline{v}_{\nu,m_\nu}}}$) are the functions $\omega^i_L(v_{\nu,m_\nu})$ (resp. $\omega^i_R(\overline{v}_{\nu,m_\nu})$), $1 \le i \le 3$, glued together, can undergo a versal deformation and a blowup of it according to:

$$SOT^{(MG)}_L \circ D^{(MG)}_{S_L}: \quad \widetilde{M}^S_{MG_L} \longrightarrow \widetilde{M}^S_{MG_L} \oplus \widetilde{M}^{S_p}_{M_L}$$
$$(\text{resp.} \quad SOT^{(MG)}_R \circ D^{(MG)}_{S_R}: \quad \widetilde{M}^S_{MG_R} \longrightarrow \widetilde{M}^S_{MG_R} \oplus \widetilde{M}^{S_p}_{M_R})$$

  where:

  - $$D^{(MG)}_{S_L} : \widetilde{M}^S_{MG_L} \times \theta^{(MG)}_{S_L} \to \widetilde{M}^S_{MG_L}$$
    $$(\text{resp.} \quad D^{(MG)}_{S_R} : \widetilde{M}^S_{MG_R} \times \theta^{(MG)}_{S_R} \to \widetilde{M}^S_{MG_R})$$
    is the versal deformation of $\widetilde{M}^S_{MG_L}$ (resp. $\widetilde{M}^S_{MG_R}$).
  - $$SOT^{(MG)}_L : \widetilde{M}^S_{MG_L} \times \theta^{(MG)}_{S_L} \to \widetilde{M}^S_{MG_L} \oplus \widetilde{M}^{S_p}_{M_L}$$
    $$(\text{resp.} \quad SOT^{(MG)}_R : \widetilde{M}^S_{MG_R} \times \theta^{(MG)}_{S_R} \to \widetilde{M}^S_{MG_R} \oplus \widetilde{M}^{S_p}_{M_R})$$
    in such a way that $\widetilde{M}^{S_p}_{M_L}$ (resp. $\widetilde{M}^{S_p}_{M_R}$) is the perverse mass semisheaf generated from $\widetilde{M}^S_{M_L} \equiv \theta^{(MG)}_{S_L}$ (resp. $\widetilde{M}^S_{M_R} \equiv \theta^{(MG)}_{S_R}$) by the map:

    $$DT^S_{L;M}: \quad \widetilde{M}^S_{M_L} \longrightarrow \widetilde{M}^{S_p}_{M_L} \qquad (\text{resp.} \quad DT^S_{R;M}: \quad \widetilde{M}^S_{M_R} \longrightarrow \widetilde{M}^{S_p}_{M_R})$$

so that

$$DT^S_{L;M} = \left\{ i\,\frac{\hbar}{c_{t\to r;M}}\,\frac{\partial}{\partial x}, i\,\frac{\hbar}{c_{t\to r;M}}\,\frac{\partial}{\partial y}, i\,\frac{\hbar}{c_{t\to r;M}}\,\frac{\partial}{\partial z} \right\}$$

$$\left(\text{resp.}\quad DT^S_{R;M} = \left\{ -i\,\frac{\hbar}{c_{t\to r;M}}\,\frac{\partial}{\partial x}, -i\,\frac{\hbar}{c_{t\to r;M}}\,\frac{\partial}{\partial y}, -i\,\frac{\hbar}{c_{t\to r;M}}\,\frac{\partial}{\partial z} \right\}\right)$$

corresponds to the inverse of the projective map

$$\tau^{(M)}_{v_{\omega_L}} : \quad \text{TAN}(\theta^{(MG)}_{S_L}) \longrightarrow \theta^{(MG)}_{S_L}$$

$$(\text{resp.}\quad \tau^{(M)}_{v_{\omega_R}} : \quad \text{TAN}(\theta^{(MG)}_{S_R}) \longrightarrow \theta^{(MG)}_{S_R}\ )$$

of the vertical tangent bundle of the blowup of the versal deformation of the middle ground semisheaf $\widetilde{M}^S_{MG_L}$ (resp. $\widetilde{M}^S_{MG_R}$).

- **The sections $\widetilde{M}^{S_p}_{M_{v_\nu,m_\nu}}$ (resp. $\widetilde{M}^{S_p}_{M_{\overline{v}_\nu,m_\nu}}$) of the perverse "mass" semisheaf $\widetilde{M}^{S_p}_{M_L}$ (resp. $\widetilde{M}^{S_p}_{M_R}$) cover the corresponding sections of the "middle ground" and "internal vacuum" semisheaves** $\widetilde{M}^{S_p}_{MG_L}$ (resp. $\widetilde{M}^{S_p}_{MG_L}$) and $\widetilde{M}^S_{ST_L}$ (resp. $\widetilde{M}^S_{ST_R}$): they are open strings if the numbers of quanta on their sections are inferior or equal to the number of quanta on the sections of the "middle ground" semisheaves $\widetilde{M}^{S_p}_{MG_L}$ (resp. $\widetilde{M}^{S_p}_{MG_R}$).

3.5. Embedding of "Internal Vacuum", "Middle Ground" and "Mass" Semisheaves of Space

- So, the "middle ground" and "mass" semisheaves of space $\widetilde{M}^S_{MG_L}$ (resp. $\widetilde{M}^S_{MG_R}$) and $\widetilde{M}^S_{M_L}$ (resp. $\widetilde{M}^S_{M_R}$) can be generated by versal deformations and blowups of these from the "internal vacuum" semisheaf $\widetilde{M}^S_{ST_L}$ (resp. $\widetilde{M}^S_{ST_R}$) of space leading to the **embedding**:

$$\widetilde{M}^S_{ST_L} \subset \widetilde{M}^S_{MG_L} \subset \widetilde{M}^S_{M_L} \qquad (\text{resp.}\quad \widetilde{M}^S_{ST_R} \subset \widetilde{M}^S_{MG_R} \subset \widetilde{M}^S_{M_R}\ ).$$

- **The perverse "middle ground' and "mass" semisheaves $\widetilde{M}^{S_p}_{MG_L}$ (resp. $\widetilde{M}^{S_p}_{MG_R}$) and $\widetilde{M}^{S_p}_{M_L}$ (resp. $\widetilde{M}^{S_p}_{M_R}$)**, belonging to the derived category of string semifields, **are of contracting nature**, while the **"internal vacuum" semisheaf of space $\widetilde{M}^S_{ST_L}$ (resp. $\widetilde{M}^S_{ST_R}$) is of expanding nature**: so, it was assumed in [33] that the $3D$-differential operator

$$T^S_{L;ST} = \left\{ i\,\frac{\hbar_{ST}}{c_{t\to r;ST}}\,dx, i\,\frac{\hbar_{ST}}{c_{t\to r;ST}}\,dy, i\,\frac{\hbar_{ST}}{c_{t\to r;ST}}\,dz \right\}$$

$$\left(\text{resp.}\quad T^S_{R;ST} = \left\{ -i\,\frac{\hbar_{ST}}{c_{t\to r;ST}}\,dx, -i\,\frac{\hbar_{ST}}{c_{t\to r;ST}}\,dy, -i\,\frac{\hbar_{ST}}{c_{t\to r;ST}}\,dz \right\}\right)$$

applies to all the sections of $\widetilde{M}^S_{ST_L}$ (resp. $\widetilde{M}^S_{ST_R}$) according to:

$$T^S_{L;ST} : \quad \widetilde{M}^S_{ST_L} \longrightarrow \widetilde{M}^{S_p}_{ST_L} \qquad (\text{resp.} \quad T^S_{R;ST} : \quad \widetilde{M}^S_{ST_R} \longrightarrow \widetilde{M}^{S_p}_{ST_R} \,)$$

where $\widetilde{M}^{S_p}_{ST_L}$ (resp. $\widetilde{M}^{S_p}_{ST_R}$) was written abusively as a perverse semisheaf.

Similarly, we have the following embedding

$$\widetilde{M}^{S_p}_{ST_L} \subset \widetilde{M}^{S_p}_{MG_L} \subset \widetilde{M}^{S_p}_{M_L} \qquad (\text{resp.} \quad \widetilde{M}^{S_p}_{ST_R} \subset \widetilde{M}^{S_p}_{MG_R} \subset \widetilde{M}^{S_p}_{M_R} \,)$$

between "perverse" semisheaves.

3.6. Middle Ground and Mass Semisheaves of Time

- Similarly, as the "middle ground" and "mass" semisheaves of space are generated by versal deformations and blowups from the "internal vacuum" semisheaf of space $\widetilde{M}^S_{ST_L}$ (resp. $\widetilde{M}^S_{ST_R}$), the "middle ground" and "mass" semisheaves of time $\widetilde{M}^T_{MG_L}$ (resp. $\widetilde{M}^T_{MG_R}$) and $\widetilde{M}^T_{M_L}$ (resp. $\widetilde{M}^T_{M_R}$) can be generated:
  - either by versal deformations and blowups from the "internal vaccuum" semisheaf of time $\widetilde{M}^T_{ST_L}$ (resp. $\widetilde{M}^T_{ST_R}$).
  - or from the respective semisheaves of space by the composition of maps:

$$\begin{aligned} \gamma^{(MG)}_{r_L\to t_L} \circ E^{(MG)}_L : &\quad \widetilde{M}^S_{MG_L} \longrightarrow \widetilde{M}^T_{MG_L} \\ \text{and} \quad \gamma^{(M)}_{r_L\to t_L} \circ E^{(M)}_L : &\quad \widetilde{M}^S_{M_L} \longrightarrow \widetilde{M}^T_{M_L} \end{aligned}$$

$$\begin{aligned} (\text{resp.} \quad \gamma^{(MG)}_{r_R\to t_R} \circ E^{(MG)}_R : &\quad \widetilde{M}^S_{MG_R} \longrightarrow \widetilde{M}^T_{MG_R} \\ \text{and} \quad \gamma^{(M)}_{r_R\to t_R} \circ E^{(M)}_R : &\quad \widetilde{M}^S_{M_R} \longrightarrow \widetilde{M}^T_{M_R} \,), \end{aligned}$$

  as described in proposition 2.7, where $E^{(MG)}_L$ and $E^{(M)}_L$ (resp. $E^{(MG)}_R$ and $E^{(M)}_R$) are endomorphisms based on Galois antiautomorphisms.

As for the semisheaves of space, we have **for the semisheaves of time the embedding**:

$$\widetilde{M}^T_{ST_L} \subset \widetilde{M}^T_{MG_L} \subset \widetilde{M}^T_{M_L} \qquad (\text{resp.} \quad \widetilde{M}^T_{ST_R} \subset \widetilde{M}^T_{MG_R} \subset \widetilde{M}^T_{M_R} \,)$$

as well as for the perverse semisheaves of time:

$$\widetilde{M}^{T_p}_{ST_L} \subset \widetilde{M}^{T_p}_{MG_L} \subset \widetilde{M}^{T_p}_{M_L} \qquad (\text{resp.} \quad \widetilde{M}^{T_p}_{ST_R} \subset \widetilde{M}^{T_p}_{MG_R} \subset \widetilde{M}^{T_p}_{M_R} \,).$$

These perverse semisheaves of time result form the morphisms:

$$\begin{aligned} T^T_{L;ST} : &\quad \widetilde{M}^T_{ST_L} \longrightarrow \widetilde{M}^{T_p}_{ST_L} \,, \\ DT^T_{L;MG} : &\quad \widetilde{M}^T_{MG_L} \longrightarrow \widetilde{M}^{T_p}_{MG_L} \,, \\ DT^T_{L;M} : &\quad \widetilde{M}^T_{M_L} \longrightarrow \widetilde{M}^{T_p}_{M_L} \,, \end{aligned}$$

where

$$T^T_{L;ST}=i\,\frac{\hbar_{ST}}{c_{t\to r;ST}}\,dt_0\;,\quad DT^T_{L;MG}=i\,\frac{\hbar_{MG}}{c_{t\to r;MG}}\,\frac{\partial}{\partial t_0}\;,\quad DT^T_{L;M}=i\,\frac{\hbar_{M}}{c_{t\to r;M}}\,\frac{\partial}{\partial t_0}$$

(the right cases are handled similarly).

3.7. Proposition

Let $(\widetilde M^{T-S}_{ST_R}\otimes \widetilde M^{T-S}_{ST_L})$ be the "internal vacuum" bisemisheaf or string field of an elementary bisemifermion.

Then, by versal deformations and blowups of these, ***the "internal vacuum" string field can generate the two covering "middle ground" and "mass" string fields of space-time***

$$(\widetilde M^{T-S}_{MG_R}\otimes \widetilde M^{T-S}_{MG_L})\qquad\qquad (\widetilde M^{T-S}_{M_R}\otimes \widetilde M^{T-S}_{M_L})$$

leading to the embedding:

$$(\widetilde M^{T-S}_{ST_R}\otimes \widetilde M^{T-S}_{ST_L})\subset(\widetilde M^{T-S}_{MG_R}\otimes \widetilde M^{T-S}_{MG_L})\subset(\widetilde M^{T-S}_{M_R}\otimes \widetilde M^{T-S}_{M_L}).$$

Proof.

- According to section 3.1, the "internal vacuum" string field is given by:

$$(\widetilde M^{T}_{ST_R}\otimes \widetilde M^{T}_{ST_L})\oplus(\widetilde M^{S}_{ST_R}\otimes \widetilde M^{S}_{ST_L})$$

  where $(\widetilde M^{T}_{ST_R}\otimes \widetilde M^{T}_{ST_L})$ (resp. $(\widetilde M^{S}_{ST_R}\otimes \widetilde M^{S}_{ST_L})$) is the time (resp. space) string field.

  But, this "internal vacuum" string field of space time is not complete, because it does not allow interactions between the time string field and the space string field.

- A more general approach would consider the following "internal vacuum" string field:

$$\begin{aligned}&(\widetilde M^{T-S}_{ST_R}\otimes \widetilde M^{T-S}_{ST_L})\equiv(\widetilde M^{T}_{ST_R}\oplus \widetilde M^{S}_{ST_R})\otimes(\widetilde M^{T}_{ST_L}\oplus \widetilde M^{S}_{ST_L})\\ &=(\widetilde M^{T}_{ST_R}\otimes \widetilde M^{T}_{ST_L})\oplus(\widetilde M^{S}_{ST_R}\otimes \widetilde M^{S}_{ST_L})\oplus(\widetilde M^{T}_{ST_R}\otimes \widetilde M^{S}_{ST_L})\oplus(\widetilde M^{S}_{ST_R}\otimes \widetilde M^{T}_{ST_L})\end{aligned}$$

  constituting the **completely reducible non orthogonal bilinear representation space in bisemisheaves** [32] of $\mathrm{GL}_{2(T+S)}(L_{\overline v}\times L_v))$ according to:

$$\begin{aligned}&\mathrm{Repsp}\,\theta(\mathrm{GL}_{2(T+S)}(L_{\overline v}\times L_v))\\ &\quad=\mathrm{Repsp}\,\theta(\mathrm{GL}_{2(T)}(L_{\overline v}\times L_v))\oplus\mathrm{Repsp}\,\theta(\mathrm{GL}_{2(S)}(L^I_{\overline v}\times L^I_v))\\ &\quad\oplus\mathrm{Repsp}\,\theta(T^t_{2(T)}(L_{\overline v})\times T_{2(S)}(L^I_v))\oplus\mathrm{Repsp}\,\theta(T^t_{2(S)}(L^I_{\overline v})\times T_{2(T)}(L_v))\end{aligned}$$

where

$$(\widetilde{M}^{T}_{ST_R} \otimes \widetilde{M}^{S}_{ST_L}) \equiv \text{Repsp}\,\theta(T^t_{2(T)}(L_{\overline{v}}) \times T_{2(S)}(L^I_v))$$
$$(\widetilde{M}^{S}_{ST_R} \otimes \widetilde{M}^{T}_{ST_L}) \equiv \text{Repsp}\,\theta(T^t_{2(S)}(L^I_{\overline{v}}) \times T_{2(T)}(L_v))$$

are mixed bisemisheaves over off diagonal bilinear representation spaces of "time-space" and "space-time" responsible for the electric charge of the considered (bisemi)- fermion as developed in [33].

- Referring to the preceding sections, the "internal vacuum" bisemisheaf $(\widetilde{M}^{T-S}_{ST_R} \otimes \widetilde{M}^{T-S}_{ST_L})$ generates by versal deformations and blowups of these the "middle ground" and "mass" bisemisheaves $(\widetilde{M}^{T-S}_{MG_R} \otimes \widetilde{M}^{T-S}_{MG_L})$ and $(\widetilde{M}^{T-S}_{M_R} \otimes \widetilde{M}^{T-S}_{M_L})$ which are embedded as announced in this proposition. ∎

3.8. Proposition

1. *The corresponding perverse bisemisheaves (or fields)* $(\widetilde{M}^{T_p-S_p}_{ST_R} \otimes \widetilde{M}^{T_p-S_p}_{ST_L})$, $(\widetilde{M}^{T_p-S_p}_{MG_R} \otimes \widetilde{M}^{T_p-S_p}_{MG_L})$ *and* $(\widetilde{M}^{T_p-S_p}_{M_R} \otimes \widetilde{M}^{T_p-S_p}_{M_L})$ *of the internal structure of an elementary bisemifermion give rise to* **the set of equations of its internal dynamics***:*

$$\begin{aligned}
&(\widetilde{M}^{T_p-S_p}_{ST_R} \oplus \widetilde{M}^{T_p-S_p}_{MG_R} \oplus \widetilde{M}^{T_p-S_p}_{M_R}) \otimes (\widetilde{M}^{T_p-S_p}_{ST_L} \oplus \widetilde{M}^{T_p-S_p}_{MG_L} \oplus \widetilde{M}^{T_p-S_p}_{M_L}) \\
&\quad = (\widetilde{M}^{T_p-S_p}_{ST_R} \otimes \widetilde{M}^{T_p-S_p}_{ST_L}) \oplus (\widetilde{M}^{T_p-S_p}_{MG_R} \otimes \widetilde{M}^{T_p-S_p}_{MG_L}) \oplus (\widetilde{M}^{T_p-S_p}_{M_R} \otimes \widetilde{M}^{T_p-S_p}_{M_L}) \\
&\qquad \oplus (\widetilde{M}^{T_p-S_p}_{ST_R} \otimes \widetilde{M}^{T_p-S_p}_{MG_L}) \oplus (\widetilde{M}^{T_p-S_p}_{ST_R} \otimes \widetilde{M}^{T_p-S_p}_{M_L}) \oplus (\widetilde{M}^{T_p-S_p}_{MG_R} \otimes \widetilde{M}^{T_p-S_p}_{ST_L}) \\
&\qquad \oplus (\widetilde{M}^{T_p-S_p}_{MG_R} \otimes \widetilde{M}^{T_p-S_p}_{M_L}) \oplus (\widetilde{M}^{T_p-S_p}_{M_R} \otimes \widetilde{M}^{T_p-S_p}_{ST_L}) \oplus (\widetilde{M}^{T_p-S_p}_{M_R} \otimes \widetilde{M}^{T_p-S_p}_{MG_L}) \\
&\qquad = 0
\end{aligned}$$

where the set of six mixed bisemisheaves $(\widetilde{M}^{T_p-S_p}_{ST_R} \otimes \widetilde{M}^{T_p-S_p}_{MG_L}) \cdots (\widetilde{M}^{T_p-S_p}_{M_R} \otimes \widetilde{M}^{T_p-S_p}_{MG_L})$ *generates the interactions between the right and left semisheaves of different levels "* ST *", "* MG *" and "* M *".*

2. ***If the interactions between the right and left internal semistructures "*** ST ***", "*** MG ***" and "*** M ***" are assumed to be negligible, then the equations of the internal dynamics of a bisemifermion are:***

$$(\widetilde{M}^{T_p-S_p}_{ST_R} \otimes \widetilde{M}^{T_p-S_p}_{ST_L}) \oplus (\widetilde{M}^{T_p-S_p}_{MG_R} \otimes \widetilde{M}^{T_p-S_p}_{MG_L}) = -(\widetilde{M}^{T_p-S_p}_{M_R} \otimes \widetilde{M}^{T_p-S_p}_{M_L}) \qquad (*)$$

Proof.

- This proposition is a direct consequence of the preceding sections.
- The interactions between the right and left internal semisheaves " ST ", " MG " and " M " are especially responsible for the internal angular momentum of the considered bisemifermion as developed in [33].

- If the interactions between the right and left internal semisheaves are assumed to be negligible, then **the equations of the internal dynamics**:

$$(\widetilde{M}^{T_p-S_p}_{ST_R}\otimes\widetilde{M}^{T_p-S_p}_{ST_L})\oplus(\widetilde{M}^{T_p-S_p}_{MG_R}\otimes\widetilde{M}^{T_p-S_p}_{MG_L})=-(\widetilde{M}^{T_p-S_p}_{M_R}\otimes\widetilde{M}^{T_p-S_p}_{M_L})\qquad(*)$$

describe the generation of the matter field $(\widetilde{M}^{T_p-S_p}_{M_R}\otimes\widetilde{M}^{T_p-S_p}_{M_L})$ of a bisemifemion from its own vacuum fields given the lefthand side of $(*)$. ■

3.9. Extended Bilinear Hilbert Spaces of Internal Structures

According to [33], each " ST ", " MG " or " M " field in $(*)$ is an **operator valued string field** on the corresponding string field which defines an extended bilinear Hilbert space.

Indeed, if we apply a $(B_L\circ p_L)$ map on the string fields $(\widetilde{M}^{T-S}_{ST_R}\otimes\widetilde{M}^{T-S}_{ST_L})$, $(\widetilde{M}^{T-S}_{MG_R}\otimes\widetilde{M}^{T-S}_{MG_L})$ and $(\widetilde{M}^{T-S}_{M_R}\otimes\widetilde{M}^{T-S}_{M_L})$ in such a way that:

- p_L be a projective map projecting each right semifield on its corresponding left equivalent,
- B_L be a bijective isometric map sending each covariant element into its contravariant equivalent,

then they are transformed into:

$$\begin{aligned}
(B^{(ST)}_L\circ p^{(ST)}_L):&\quad \widetilde{M}^{T-S}_{ST_R}\otimes\widetilde{M}^{T-S}_{ST_L}\longrightarrow H^+_{ST}=\widetilde{M}^{T-S}_{ST_{L_R}}\otimes\widetilde{M}^{T-S}_{ST_L}\\
(B^{(MG)}_L\circ p^{(MG)}_L):&\quad \widetilde{M}^{T-S}_{MG_R}\otimes\widetilde{M}^{T-S}_{MG_L}\longrightarrow H^+_{MG}=\widetilde{M}^{T-S}_{MG_{L_R}}\otimes\widetilde{M}^{T-S}_{MG_L}\\
(B^{(M)}_L\circ p^{(M)}_L):&\quad \widetilde{M}^{T-S}_{M_R}\otimes\widetilde{M}^{T-S}_{M_L}\longrightarrow H^+_{M}=\widetilde{M}^{T-S}_{M_{L_R}}\otimes\widetilde{M}^{T-S}_{M_L}
\end{aligned}$$

where H^+_{ST}, H^+_{MG} and H^+_M are the extended bilinear Hilbert spaces of the internal vacuum, middle ground and mass structures of a bisemifermion if complete internal bilinear forms are defined on them.

3.10. Actions of Bioperators on H^+_{ST}, H^+_{MG} and H^+_M

On the extended bilinear Hilbert spaces H^+_{ST}, H^+_{MG} and H^+_M, we have the actions of the bioperators:

$$\begin{aligned}
T^{T-S}_{R;ST}\otimes T^{T-S}_{L;ST}:&\quad \widetilde{M}^{T-S}_{ST_{L_R}}\otimes\widetilde{M}^{T-S}_{ST_L}\longrightarrow\widetilde{M}^{T_p-S_p}_{ST_{L_R}}\otimes\widetilde{M}^{T_p-S_p}_{ST_L}\\
DT^{T-S}_{R;MG}\otimes DT^{T-S}_{L;MG}:&\quad \widetilde{M}^{T-S}_{MG_{L_R}}\otimes\widetilde{M}^{T-S}_{MG_L}\longrightarrow\widetilde{M}^{T_p-S_p}_{MG_{L_R}}\otimes\widetilde{M}^{T_p-S_p}_{MG_L}\\
DT^{T-S}_{R;M}\otimes DT^{T-S}_{L;M}:&\quad \widetilde{M}^{T-S}_{M_{L_R}}\otimes\widetilde{M}^{T-S}_{M_L}\longrightarrow\widetilde{M}^{T_p-S_p}_{M_{L_R}}\otimes\widetilde{M}^{T_p-S_p}_{M_L}
\end{aligned}$$

sending the internal vacuum, middle ground and mass bisemisheaves $\widetilde{M}^{T-S}_{ST_{L_R}}\otimes\widetilde{M}^{T-S}_{ST_L}$, $\widetilde{M}^{T-S}_{MG_{L_R}}\otimes\widetilde{M}^{T-S}_{MG_L}$ and $\widetilde{M}^{T-S}_{M_{L_R}}\otimes\widetilde{M}^{T-S}_{M_L}$ into their perverse equivalents.

The bioperators are explicitly given by:

$$T_{R;ST}^{T-S} \otimes T_{L;ST}^{T-S} = \left(-i \, \frac{\hbar_{ST}}{c_{t \to r;ST}} \, \{ s_{0_R} \, dt_0; s_{x_R} \, dx, s_{y_R} \, dy, s_{z_R} \, dz \} \right) \otimes \left(-i \, \frac{\hbar_{ST}}{c_{t \to r;ST}} \, \{ s_{0_L} \, dt_0; s_{x_L} \, dx, s_{y_L} \, dy, s_{z_L} \, dz \} \right)$$

$$DT_{R;MG}^{T-S} \otimes DT_{L;MG}^{T-S} = \left(-i \, \frac{\hbar_{MG}}{c_{t \to r;MG}} \left\{ s_{0_R} \, \frac{\partial}{\partial t_0}; s_{x_R} \, \frac{\partial}{\partial x}, s_{y_R} \, \frac{\partial}{\partial y}, s_{z_R} \, \frac{\partial}{\partial z} \right\} \right) \otimes \left(-i \, \frac{\hbar_{MG}}{c_{t \to r;MG}} \left\{ s_{0_L} \, \frac{\partial}{\partial t_0}; s_{x_L} \, \frac{\partial}{\partial x}, s_{y_L} \, \frac{\partial}{\partial y}, s_{z_L} \, \frac{\partial}{\partial z} \right\} \right)$$

$$DT_{R;M}^{T-S} \otimes DT_{L;M}^{T-S} = \left(-i \, \frac{\hbar_{M}}{c_{t \to r;M}} \left\{ s_{0_R} \, \frac{\partial}{\partial t_0}; s_{x_R} \, \frac{\partial}{\partial x}, s_{y_R} \, \frac{\partial}{\partial y}, s_{z_R} \, \frac{\partial}{\partial z} \right\} \right) \otimes \left(-i \, \frac{\hbar_{M}}{c_{t \to r;M}} \left\{ s_{0_L} \, \frac{\partial}{\partial t_0}; s_{x_L} \, \frac{\partial}{\partial x}, s_{y_L} \, \frac{\partial}{\partial y}, s_{z_L} \, \frac{\partial}{\partial z} \right\} \right)$$

where s_0 and (s_x, s_y, s_z) are the direction cosines of the unit vectors $\vec{s}_{t_0}$ and $\vec{s}_r$ referring to the spin and allowing to defined directional gradients $\vec{s}_{t_0} \frac{\partial}{\partial t_0}$ and $\vec{s}_r \, \nabla$.
(More concretely, the function $\phi(t_0)$ is said to be derivable at t_0 in the direction s if $\lim\limits_{\varepsilon \to 0} \frac{\phi(t_0+\varepsilon s)-\phi(t_0)}{\varepsilon}$ exists).

3.11. Proposition

- *Let $r = q + p$ be the number of algebraic conjugacy (or equivalence) classes of time and space, the indices of time and space varying separately according to $1 \leq \mu \leq q$ and $1 \leq \nu \leq p$ and commonly according to $1 \leq \sigma \leq r$.*

- *Let*

  - $$\left\{ \widetilde{M}_{ST_{v_\sigma, m_\sigma}} = \widetilde{M}^{T}_{ST_{v_\mu, m_\mu}} + \widetilde{M}^{S}_{ST_{v_\nu, m_\nu}} \right\}_{\sigma=1}^{r} \subset \widetilde{M}^{T-S}_{ST_L}$$

    $$\left(\textit{resp.} \quad \left\{ \widetilde{M}_{ST_{\overline{v}_\sigma, m_\sigma}} = \widetilde{M}^{T}_{ST_{\overline{v}_\mu, m_\mu}} + \widetilde{M}^{S}_{ST_{\overline{v}_\nu, m_\nu}} \right\}_{\sigma=1}^{r} \subset \widetilde{M}^{T-S}_{ST_R} \right)$$

  - $$\left\{ \widetilde{M}_{MG_{v_\sigma, m_\sigma}} = \widetilde{M}^{T}_{MG_{v_\mu, m_\mu}} + \widetilde{M}^{S}_{MG_{v_\nu, m_\nu}} \right\}_{\sigma=1}^{r} \subset \widetilde{M}^{T-S}_{MG_L}$$

    $$\left(\textit{resp.} \quad \left\{ \widetilde{M}_{MG_{\overline{v}_\sigma, m_\sigma}} = \widetilde{M}^{T}_{MG_{\overline{v}_\mu, m_\mu}} + \widetilde{M}^{S}_{MG_{\overline{v}_\nu, m_\nu}} \right\}_{\sigma=1}^{r} \subset \widetilde{M}^{T-S}_{MG_R} \right)$$

$$\bullet \qquad \left\{\widetilde{M}_{M_{v_{\sigma,m_\sigma}}} = \widetilde{M}^T_{M_{v_{\mu,m_\mu}}} + \widetilde{M}^S_{M_{v_{\nu,m_\nu}}}\right\}_{\sigma=1}^r \subset \widetilde{M}^{T-S}_{M_L}$$

$$\left(\textit{resp.} \qquad \left\{\widetilde{M}_{M_{\overline{v}_{\sigma,m_\sigma}}} = \widetilde{M}^T_{M_{\overline{v}_{\mu,m_\mu}}} + \widetilde{M}^S_{M_{\overline{v}_{\nu,m_\nu}}}\right\}_{\sigma=1}^r \subset \widetilde{M}^{T-S}_{M_R}\right)$$

be the set of sections of the space-time left (resp. right) semisheaves of the internal vacuum (" ST "), middle ground (" MG ") and mas (" M ") structures of a left (resp. right) semifermion.

- *Let*

$$\bullet \qquad T_{ST_L} = -i\,\frac{\hbar_{ST}}{c_{t\to r;ST}}\,(\vec{s}_{t_0}\,dt_0 + \vec{s}_{r_x}\,dx + \vec{s}_{r_y}\,dy + \vec{s}_{r_z}\,dz)$$

$$\left(\textit{resp.} \qquad T_{ST_R} = +i\,\frac{\hbar_{ST}}{c_{t\to r;ST}}\,(\vec{s}_{t_0}\,dt_0 + \vec{s}_{r_x}\,dx + \vec{s}_{r_y}\,dy + \vec{s}_{r_z}\,dz)\right)$$

$$\bullet \qquad DT_{MG_L} = -i\,\frac{\hbar_{MG}}{c_{t\to r;MG}}\left(\vec{s}_{t_0}\,\frac{\partial}{\partial t_0} + \vec{s}_{r_x}\,\frac{\partial}{\partial x} + \vec{s}_{r_y}\,\frac{\partial}{\partial y} + \vec{s}_{r_z}\,\frac{\partial}{\partial z}\right)$$

$$\left(\textit{resp.} \quad DT_{MG_R} = +i\,\frac{\hbar_{MG}}{c_{t\to r;MG}}\left(\vec{s}_{t_0}\,\frac{\partial}{\partial t_0} + \vec{s}_{r_x}\,\frac{\partial}{\partial x} + \vec{s}_{r_y}\,\frac{\partial}{\partial y} + \vec{s}_{r_z}\,\frac{\partial}{\partial z}\right)\right)$$

$$\bullet \qquad DT_{M_L} = -i\,\frac{\hbar_{M}}{c_{t\to r;M}}\left(\vec{s}_{t_0}\,\frac{\partial}{\partial t_0} + \vec{s}_{r_x}\,\frac{\partial}{\partial x} + \vec{s}_{r_y}\,\frac{\partial}{\partial y} + \vec{s}_{r_z}\,\frac{\partial}{\partial z}\right)$$

$$\left(\textit{resp.} \qquad DT_{M_R} = +i\,\frac{\hbar_{M}}{c_{t\to r;M}}\left(\vec{s}_{t_0}\,\frac{\partial}{\partial t_0} + \vec{s}_{r_x}\,\frac{\partial}{\partial x} + \vec{s}_{r_y}\,\frac{\partial}{\partial y} + \vec{s}_{r_z}\,\frac{\partial}{\partial z}\right)\right)$$

be the corresponding differential operators acting on these sets of sections.

Then, ***the equations*** $(*)$ ***of the internal dynamics of a bisemifermion can be put in the equivalent form:***

$$\begin{aligned}
&\left[T_{ST_R}(\widetilde{M}_{ST_{\overline{v}_{\sigma,m_\sigma}}}) \otimes T_{ST_L}(\widetilde{M}_{ST_{v_{\sigma,m_\sigma}}})\right] \\
&\qquad + \left[DT_{MG_R}(\widetilde{M}_{MG_{\overline{v}_{\sigma,m_\sigma}}}) \otimes DT_{MG_L}(\widetilde{M}_{MG_{v_{\sigma,m_\sigma}}})\right] \\
&\qquad = -\left[DT_{M_R}(\widetilde{M}_{M_{\overline{v}_{\sigma,m_\sigma}}}) \otimes DT_{M_L}(\widetilde{M}_{M_{v_{\sigma,m_\sigma}}})\right]\,, \quad \forall\,\sigma, m_\sigma\,,\ 1 \le \sigma \le r\,.
\end{aligned}$$

Proof. Indeed, it appears from the preceding developments that the actions of the bioperators $(T_{ST_R} \otimes T_{ST_L})$, $(DT_{MG_R} \otimes DT_{MG_L})$ and $(DT_{M_R} \otimes DT_{M_L})$ on the corresponding bisections, as developed above, transform these into bisections of the corresponding perverse bisemisheaves and give rise to the equations $(*)$ of the internal dynamics of a bisemifermion. ∎

3.12. Corollary

Let $T^{\dagger}_{ST_{L_R}}$, $DT^{\dagger}_{MG_{L_R}}$ and $DT^{\dagger}_{M_{L_R}}$ be the adjoint operators of T_{ST_R}, DT_{MG_R} and DT_{M_R} respectively:

$$T_{ST_{L_R}} = T^{\dagger}_{ST_R} , \quad DT_{MG_{L_R}} = DT^{\dagger}_{MG_R} \qquad DT_{M_{L_R}} = DT^{\dagger}_{M_R} .$$

Let $\widetilde{M}_{\overline{ST}_{\overline{v}\sigma,m_\sigma}}$, $\widetilde{M}_{\overline{MG}_{\overline{v}\sigma,m_\sigma}}$ and $\widetilde{M}_{\overline{M}_{\overline{v}\sigma,m_\sigma}}$ be the sections of the semisheaves $\widetilde{M}^{T-S}_{ST_R}$, $\widetilde{M}^{T-S}_{MG_R}$ and $\widetilde{M}^{T-S}_{M_R}$ projected onto the sections of the corresponding left semisheaves according to:

$$\begin{aligned} B_L^{(ST)} \circ p_L^{(ST)} : \quad & \widetilde{M}_{ST_{\overline{v}\sigma,m_\sigma}} \longrightarrow \widetilde{M}_{\overline{ST}_{\overline{v}\sigma,m_\sigma}} , \\ B_L^{(MG)} \circ p_L^{(MG)} : \quad & \widetilde{M}_{MG_{\overline{v}\sigma,m_\sigma}} \longrightarrow \widetilde{M}_{\overline{MG}_{\overline{v}\sigma,m_\sigma}} , \\ B_L^{(M)} \circ p_L^{(M)} : \quad & \widetilde{M}_{M_{\overline{v}\sigma,m_\sigma}} \longrightarrow \widetilde{M}_{\overline{M}_{\overline{v}\sigma,m_\sigma}} . \end{aligned}$$

Then, the equations on the sections of the internal dynamics of a bisemifermion are transformed under the isomorphism $ID_{R\to L_R}$ into:

$$\begin{aligned} ID_{R\to L_R} : \quad & [T_{ST_R}(\widetilde{M}_{ST_{\overline{v}\sigma,m_\sigma}}) \otimes T_{ST_L}(\widetilde{M}_{ST_{v\sigma,m_\sigma}})] \\ & + [DT_{MG_R}(\widetilde{M}_{MG_{\overline{v}\sigma,m_\sigma}}) \otimes DT_{MG_L}(\widetilde{M}_{MG_{v\sigma,m_\sigma}})] \\ & = -[DT_{M_R}(\widetilde{M}_{M_{\overline{v}\sigma,m_\sigma}}) \otimes DT_{M_L}(\widetilde{M}_{M_{v\sigma,m_\sigma}})] \\ & \longrightarrow [(T_{ST_{L_R}} \otimes T_{ST_L})(\widetilde{M}_{\overline{ST}_{\overline{v}\sigma,m_\sigma}} \otimes \widetilde{M}_{ST_{v\sigma,m_\sigma}})] \\ & \quad + [(DT_{MG_{L_R}} \otimes DT_{MG_L})(\widetilde{M}_{\overline{MG}_{\overline{v}\sigma,m_\sigma}} \otimes \widetilde{M}_{MG_{v\sigma,m_\sigma}})] \\ & \quad = -[(DT_{M_{L_R}} \otimes DT_{M_L})(\widetilde{M}_{\overline{M}_{\overline{v}\sigma,m_\sigma}} \otimes \widetilde{M}_{M_{v\sigma,m_\sigma}})] . \end{aligned}$$

Proof. This isomorphism $ID_{R\to L_R}$ transforms the (tensor) product of the right action of DT_{M_R} on $\widetilde{M}_{M_{\overline{v}\sigma,m_\sigma}}$ by the left action of DT_{M_L} on $\widetilde{M}_{M_{v\sigma,m_\sigma}}$ into the biaction of $(DT_{M_{L_R}} \otimes DT_{M_L})$ onto the product of the sections $(\widetilde{M}_{\overline{M}_{\overline{v}\sigma,m_\sigma}} \otimes \widetilde{M}_{M_{v\sigma,m_\sigma}})$, and so on for the two other levels " ST " and " MG ".

And, the products of the following sections belong to the extended bilinear Hilbert spaces:

$$\widetilde{M}_{\overline{ST}_{\overline{v}\sigma,m_\sigma}} \otimes \widetilde{M}_{ST_{v\sigma,m_\sigma}} \in H^+_{ST} , \qquad \widetilde{M}_{\overline{MG}_{\overline{v}\sigma,m_\sigma}} \otimes \widetilde{M}_{MG_{v\sigma,m_\sigma}} \in H^+_{MG} ,$$
$$\widetilde{M}_{\overline{M}_{\overline{v}\sigma,m_\sigma}} \otimes \widetilde{M}_{M_{v\sigma,m_\sigma}} \in H^+_{M} ,$$

according to section 3.9. ■

4. Equivalence between the Equations of General Relativity and the Equations of the Internal Dynamics of Bisemiparticles

The aim of this chapter consists in generalizing the equations of the internal dynamics to a set of J (bisemi)particles, and, more particularly, to a set of J elementary (bisemi)fermions and in pointing out that they are in one-to-one correspondence with the equations of general relativity.

4.1. The Internal Structure of a Set of Bisemifermions

- It was seen in chapter 3 that the time or space string field of the internal vacuum, middle ground or mass structure of a bisemifermion is given by the representation space (in bisemisheaf) of the compact bilinear algebraic semigroup $\mathrm{GL}_2(L_{\overline{v}}\times L_v)$.

- Generalizing to a set of J bisemifermions, we have to take into account the partition $2J=2_1+\cdots+2_i+\cdots+2_J$ of $2J$ in such a way that the " ST ", " MG " or " M " string field of time or space of them be given by **the completely reducible non orthogonal representation space $\mathbf{Repsp(GL_2(}L_{\overline{v}}\times L_v\mathbf{))}$** of the bilinear algebraic semigroup $\mathrm{GL}_{2J}(L_{\overline{v}}\times L_v)$ of dimension $2J$, as introduced in [30].

  This representation space $\mathrm{Repsp}(\mathrm{GL}_{2J}(L_{\overline{v}}\times L_v))$ **decomposes non orthogonally according to**:
$$\mathrm{Repsp}(\mathrm{GL}_{2J}(L_{\overline{v}}\times L_v))=$$
$$\boxplus_{i=1}^{J}\mathrm{Repsp}(\mathrm{GL}_{2_i}(L_{\overline{v}}\times L_v))\boxplus_{i\neq j=1}^{J}\mathrm{Repsp}(T^t_{2_i}(L_{\overline{v}})\times T_{2_j}(L_v))$$
  where the **off-diagonal representation spaces** $\mathrm{Repsp}(T^t_{2_i}(L_{\overline{v}})\times T_{2_j}(L_v))$ **are responsible for the generation of gravito-electro-magnetic fields of interaction** between bisemifermions as developed in [33].

- Referring to Proposition 3.7, we see that the internal vacuum (" ST ") field structure of space-time of J interacting bisemifermions is given by the bisemisheaves:
$$\begin{aligned}(\widetilde{M}^{T-S}_{ST_{R_J}}\otimes\widetilde{M}^{T-S}_{ST_{L_J}})&=\mathrm{Repsp}\,\theta(\mathrm{GL}_{2J(T+S)}(L_{\overline{v}}\times L_v))\\&=\boxplus_{i=1}^{J}(\widetilde{M}^{T-S}_{ST_{R_i}}\otimes\widetilde{M}^{T-S}_{ST_{L_i}})\boxplus_{i\neq j=1}^{J}(\widetilde{M}^{T-S}_{ST_{R_i}}\otimes\widetilde{M}^{T-S}_{ST_{L_j}})\,.\end{aligned}$$

- By versal deformations and blowups of these, the "internal vacuum" string fields $(\widetilde{M}^{T-S}_{ST_{R_J}}\otimes\widetilde{M}^{T-S}_{ST_{L_J}})$ of a set of J interacting bisemifermions generate **the two covering "middle ground" and "mass" string fields of space-time**:
$$(\widetilde{M}^{T-S}_{MG_{R_J}}\otimes\widetilde{M}^{T-S}_{MG_{L_J}})=\boxplus_{i=1}^{J}(\widetilde{M}^{T-S}_{MG_{R_i}}\otimes\widetilde{M}^{T-S}_{MG_{L_i}})\boxplus_{i\neq j=1}^{J}(\widetilde{M}^{T-S}_{MG_{R_i}}\otimes\widetilde{M}^{T-S}_{MG_{L_j}})$$
  and

$$(\widetilde{M}^{T-S}_{M_{R_J}} \otimes \widetilde{M}^{T-S}_{M_{L_J}}) = \mathop{\boxplus}_{i=1}^{J} (\widetilde{M}^{T-S}_{M_{R_i}} \otimes \widetilde{M}^{T-S}_{M_{L_i}}) \mathop{\boxplus}_{i\neq j=1}^{J} (\widetilde{M}^{T-S}_{M_{R_i}} \otimes \widetilde{M}^{T-S}_{M_{L_j}})$$

in such a way that we have the embeddings:

$$(\widetilde{M}^{T-S}_{ST_{R_J}} \otimes \widetilde{M}^{T-S}_{ST_{L_J}}) \subset (\widetilde{M}^{T-S}_{MG_{R_J}} \otimes \widetilde{M}^{T-S}_{MG_{L_J}}) \subset (\widetilde{M}^{T-S}_{M_{R_J}} \otimes \widetilde{M}^{T-S}_{M_{L_J}})\ .$$

4.2. Proposition

Let $(\widetilde{M}^{T_p-S_p}_{ST_{R_J}} \otimes \widetilde{M}^{T_p-S_p}_{ST_{L_J}})$, $(\widetilde{M}^{T_p-S_p}_{MG_{R_J}} \otimes \widetilde{M}^{T_p-S_p}_{MG_{L_J}})$ *and* $(\widetilde{M}^{T_p-S_p}_{M_{R_J}} \otimes \widetilde{M}^{T-S}_{M_{L_J}})$ *be the perverse bisemisheaves of the internal structures " ST ", " MG " and " M " of a set of J interacting bisemifermions.*

If the interactions between the right and left internal semifields " ST ", " MG " and " M " are negligible, then ***the equations of the internal dynnamics of this set of J bisermifermions are****:*

$$(\widetilde{M}^{T_p-S_p}_{ST_{R_J}} \otimes \widetilde{M}^{T_p-S_p}_{ST_{L_J}}) \oplus (\widetilde{M}^{T_p-S_p}_{MG_{R_J}} \otimes \widetilde{M}^{T_p-S_p}_{MG_{L_J}}) = -(\widetilde{M}^{T_p-S_p}_{M_{R_J}} \otimes \widetilde{M}^{T-S}_{M_{L_J}})\ .$$

Proof.

- The perverse bisemisheaves $(\widetilde{M}^{T_p-S_p}_{ST_{R_J}} \otimes \widetilde{M}^{T_p-S_p}_{ST_{L_J}})$, $(\widetilde{M}^{T_p-S_p}_{MG_{R_J}} \otimes \widetilde{M}^{T_p-S_p}_{MG_{L_J}})$ and $(\widetilde{M}^{T_p-S_p}_{M_{R_J}} \otimes \widetilde{M}^{T-S}_{M_{L_J}})$ are obtained from the corresponding bisemisheaves $(\widetilde{M}^{T-S}_{ST_{R_J}} \otimes \widetilde{M}^{T-S}_{ST_{L_J}})$, $(\widetilde{M}^{T-S}_{MG_{R_J}} \otimes \widetilde{M}^{T-S}_{MG_{L_J}})$ and $(\widetilde{M}^{T-S}_{M_{R_J}} \otimes \widetilde{M}^{T-S}_{M_{L_J}})$ by the respective actions of the biopearators $(T_{ST_L} \otimes T_{ST_R})$, $(DT_{MG_L} \otimes DT_{MG_R})$ and $(DT_{M_R} \otimes DT_{M_L})$ introduced in section 3.11.

- The equations of the internal dynamics of a set of J bisemifermions are a generalization of the equations $(*)$ of proposiiton 3.8. ■

4.3. Proposition

- *Let* $r_i = q_i + p_i$, $1 \leq i \leq J$, *be the numbers of algebraic conjugacy classes of time and space varying commonly according to* $1 \leq \sigma_i \leq r_i$.

- *Let* $\{\widetilde{M}_{ST_{v_{\sigma_i},m_{\sigma_i}}}\}_{\sigma_i=1}^{r_i} \subset \widetilde{M}^{T-S}_{ST_{L_i}}$ *(resp.* $\{\widetilde{M}_{ST_{\overline{v}_{\sigma_i},m_{\sigma_i}}}\}_{\sigma_i=1}^{r_i} \subset \widetilde{M}^{T-S}_{ST_{R_i}}$ *)*
$\{\widetilde{M}_{MG_{v_{\sigma_i},m_{\sigma_i}}}\}_{\sigma_i=1}^{r_i} \subset \widetilde{M}^{T-S}_{MG_{L_i}}$ *(resp.* $\{\widetilde{M}_{MG_{\overline{v}_{\sigma_i},m_{\sigma_i}}}\}_{\sigma_i=1}^{r_i} \subset \widetilde{M}^{T-S}_{MG_{R_i}}$ *)*
and $\{\widetilde{M}_{M_{v_{\sigma_i},m_{\sigma_i}}}\}_{\sigma_i=1}^{r_i} \subset \widetilde{M}^{T-S}_{M_{L_i}}$ *(resp.* $\{\widetilde{M}_{M_{\overline{v}_{\sigma_i},m_{\sigma_i}}}\}_{\sigma_i=1}^{r_i} \subset \widetilde{M}^{T-S}_{M_{R_i}}$ *)*

  be the set of sections of the space-time left (resp. right) semisheaves of the "internal vacuum" (ST), "middle ground" (MG) and "mass" (M) structures of the J , $1 \leq i \leq J$, left (resp. right) considered semifermions.

*Then, **the equations***

$$(\widetilde{M}^{T_p-S_p}_{ST_{R_J}}\otimes \widetilde{M}^{T_p-S_p}_{ST_{L_J}})\oplus(\widetilde{M}^{T_p-S_p}_{MG_{R_J}}\otimes \widetilde{M}^{T_p-S_p}_{MG_{L_J}})=-(\widetilde{M}^{T_p-S_p}_{M_{R_J}}\otimes \widetilde{M}^{T_p-S_p}_{M_{L_J}})$$

of the internal dynamics of a set of J bisemifermions are in one-to-one correspondence with the equations

$$\begin{aligned}
&[T_{ST_{R_i}}(\widetilde{M}_{ST_{\overline{v}_{\sigma_i},m_{\sigma_i}}})\otimes T_{ST_{L_j}}(\widetilde{M}_{ST_{v_{\sigma_j},m_{\sigma_j}}})]\\
&\qquad +[DT_{MG_{R_i}}(\widetilde{M}_{MG_{\overline{v}_{\sigma_i},m_{\sigma_i}}})\otimes DT_{MG_{L_j}}(\widetilde{M}_{MG_{v_{\sigma_j},m_{\sigma_j}}})]\\
&\qquad\qquad =-[DT_{M_{R_i}}(\widetilde{M}_{M_{\overline{v}_{\sigma_i},m_{\sigma_i}}})\otimes DT_{M_{L_j}}(\widetilde{M}_{M_{v_{\sigma_j},m_{\sigma_j}}})]\\
&\qquad\qquad\qquad \forall\,\sigma_i,\sigma_j\;,\;1\le\sigma_i,\sigma_j\le r_i,r_j\;\text{and}\;\forall\,i,j\;\;1\le i,j\le J\le\infty
\end{aligned}$$

referring to the sections of the considered bisemisheaves.

Proof. The one-to-one correspondence between the two types of equations results from:

- a generalization of proposition 3.11 to a set of J bisemifermions.
- a homomorphism

$$\begin{aligned}
H_{ST}:\quad (\widetilde{M}^{T_p-S_p}_{ST_{R_J}}\otimes\widetilde{M}^{T_p-S_p}_{ST_{L_J}}) &= \boxplus_{i=1}^{J}(\widetilde{M}^{T_p-S_p}_{ST_{R_i}}\otimes\widetilde{M}^{T-S}_{ST_{L_i}})\ \boxplus_{i\neq j=1}^{J}(\widetilde{M}^{T_p-S_p}_{ST_{R_i}}\otimes\widetilde{M}^{T-S}_{ST_{L_j}})\\
&\longrightarrow [T_{ST_{R_i}}(\widetilde{M}_{ST_{\overline{v}_{\sigma_i},m_{\sigma_i}}})\otimes T_{ST_{L_j}}(\widetilde{M}_{ST_{v_{\sigma_j},m_{\sigma_j}}})]
\end{aligned}$$

  sending the sums $\boxplus_{i=1}^{J}$ and $\boxplus_{i\neq j=1}^{J}$ of bisemisheaves, referring respectively to diagonal and off diagonal interactions between right and left " ST " semistructures of the J bisemifermions, into the set $[T_{ST_{R_i}}(\widetilde{M}_{ST_{\overline{v}_{\sigma_i},m_{\sigma_i}}})\otimes T_{ST_{L_j}}(\widetilde{M}_{ST_{v_{\sigma_j},m_{\sigma_j}}})]$ of the corresponding bisections of these bisemishaves.

- two similar homomorphisms H_{MG} and H_M referring to the " MG " and " M " structures of the J bisemifermions. ■

4.4. Proposition

- *Let*

$$\begin{aligned}
&B^{(ST)}_L\circ p^{(ST)}_L: && \widetilde{M}_{ST_{\overline{v}_{\sigma_i},m_{\sigma_i}}} \longrightarrow \widetilde{M}_{\overline{ST}_{\overline{v}_{\sigma_i},m_{\sigma_i}}}\\
&B^{(MG)}_L\circ p^{(MG)}_L: && \widetilde{M}_{MG_{\overline{v}_{\sigma_i},m_{\sigma_i}}} \longrightarrow \widetilde{M}_{\overline{MG}_{\overline{v}_{\sigma_i},m_{\sigma_i}}}\\
\text{and}\ &B^{(M)}_L\circ p^{(M)}_L: && \widetilde{M}_{M_{\overline{v}_{\sigma_i},m_{\sigma_i}}} \longrightarrow \widetilde{M}_{\overline{M}_{\overline{v}_{\sigma_i},m_{\sigma_i}}}
\end{aligned}$$

be the maps projecting the sections $\widetilde{M}_{ST_{\overline{v}_{\sigma_i},m_{\sigma_i}}}$, $\widetilde{M}_{MG_{\overline{v}_{\sigma_i},m_{\sigma_i}}}$ and $\widetilde{M}_{M_{\overline{v}_{\sigma_i},m_{\sigma_i}}}$ onto their left equivalents.

- *Let $T^{\dagger}_{ST_{L_{R_i}}}$, $DT^{\dagger}_{MG_{L_{R_i}}}$ and $DT^{\dagger}_{M_{L_{R_i}}}$ be the adjoint operators of $T_{ST_{R_i}}$, $DT_{MG_{R_i}}$ and $DT_{M_{R_i}}$ respectively:*

$$T_{ST_{L_{R_i}}} = T^{\dagger}_{ST_{R_i}} , \quad T_{MG_{L_{R_i}}} = T^{\dagger}_{MG_{R_i}} \qquad T_{M_{L_{R_i}}} = T^{\dagger}_{M_{R_i}} .$$

Then, ***the equations on the sections of the internal dynamics of a set of*** *J* ***bisemifermions are transformed under the isomorphisms $ID_{R_J\to L_{R_J}}$*** *into:*

$$\begin{aligned}
ID_{R_J\to L_{R_J}} : \quad & [T_{ST_{R_i}}(\widetilde{M}_{ST_{\overline{v}_{\sigma_i},m_{\sigma_i}}}) \otimes T_{ST_{L_j}}(\widetilde{M}_{ST_{v_{\sigma_j},m_{\sigma_j}}})] \\
& +[DT_{MG_{R_i}}(\widetilde{M}_{MG_{\overline{v}_{\sigma_i},m_{\sigma_i}}}) \otimes DT_{MG_{L_j}}(\widetilde{M}_{MG_{v_{\sigma_j},m_{\sigma_j}}})] \\
& = -[DT_{M_{R_i}}(\widetilde{M}_{M_{\overline{v}_{\sigma_i},m_{\sigma_i}}}) \otimes DT_{M_{L_j}}(\widetilde{M}_{M_{v_{\sigma_j},m_{\sigma_j}}})] \\
& \longrightarrow [(T_{ST_{L_{R_i}}} \otimes T_{ST_{L_j}})(\widetilde{M}_{\overline{ST}_{\overline{v}_{\sigma_i},m_{\sigma_i}}} \otimes \widetilde{M}_{ST_{v_{\sigma_i},m_{\sigma_i}}})] \\
& +[(DT_{MG_{L_{R_i}}} \otimes DT_{MG_{L_j}})(\widetilde{M}_{\overline{MG}_{\overline{v}_{\sigma_i},m_{\sigma_i}}} \otimes \widetilde{M}_{MG_{v_{\sigma_j},m_{\sigma_j}}})] \\
& = -[(DT_{M_{L_{R_i}}} \otimes DT_{M_{L_j}})(\widetilde{M}_{\overline{M}_{\overline{v}_{\sigma_i},m_{\sigma_i}}} \otimes \widetilde{M}_{M_{v_{\sigma_j},m_{\sigma_j}}})] .
\end{aligned}$$

Proof. This proposition is a generalization of corollary 3.12 to a set of J bisemifermions. ■

4.5. The Small Value of the Cosmological Constant

The equations obtained under the isomorphism $ID_{R_J\to L_{R_J}}$ and describing the internal dynamics of a set of (bisemi)fermions are rather close to the equations of general relativity as it will be seen in the following sections.

But, one of the big problems of the equations of general relativity in connection with the phenomenology of quantum field theories consists in the small value given to the cosmological constant. Indeed, in order to avoid a static solution to the equations:

$$G_{\mu\nu} = 8\pi G T_{\mu\nu}$$

where:

a) $G_{\mu\nu} = R_{\mu\nu} - \frac{1}{2}\, g_{\mu\nu} R$ with:

- $R_{\mu\nu}$ the Ricci tensor which is the contracted form of the Riemann-Christoffel curvature tensor $R^{\lambda}_{\mu\lambda\nu}$ by $R_{\mu\nu} = R^{\lambda}_{\mu\lambda\nu}$,
- $R = g^{\mu\nu} R_{\mu\nu}$ the curvature scalar,
- $g_{\mu\nu}$ the metric tensor of space-time,

b) $T_{\mu\nu}$ is the symmetric energy-momentum tensor and G is the gravitational constant,

Einstein introduced in these a new term $\lambda g_{\mu\nu}$, where **$\boldsymbol{\lambda}$ is the cosmological constant**, leading to [12], [13]:

$$\boldsymbol{\lambda g_{\mu\nu} + G_{\mu\nu} = 8\pi G T_{\mu\nu}} \ .$$

The equations $G_{\mu\nu} = 8\pi G T_{\mu\nu}$ describe how matter, given by $T_{\mu\nu}$, generates gravitational forces, characterized by the tensor $G_{\mu\nu}$, by means of the curvature of the space-time; indeed, the gravitational field is assumed to be represented by the metric tensor itself.

In this context, Zel'dovich [44], [45] envisaged **the possible connection between the vacuum energy density of quantum field theories and the Einstein's cosmological constant $\boldsymbol{\lambda}$** in such a way that the author proposed in [33] to describe the vacuum energy density $\rho_{\rm vac}$ by

$$\rho_{\rm vac} \simeq \rho^{(ST)}_{\rm vac} + \rho^{(MG)}_{\rm vac} \simeq \lambda_{\rm eff}/8\pi G$$

where:

- the internal vacuum energy density $\rho^{(ST)}_{\rm vac}$ could correspond to $\lambda/8\pi G$,
- $\rho^{(MG)}_{\rm vac}$ would correspond to the middle ground energy density " MG ".

The revised equations of general relativity can then take the form:

$$G_{\mu\nu} = 8\pi G \left(T_{\mu\nu} - \tfrac{\lambda}{8\pi G}\, g_{\mu\nu}\right)$$

in such a way that $\lambda g_{\mu\nu}$ could correspond to the internal vacuum energy density $\rho^{(ST)}_{\rm vac}$ which would behave like an ideal fluid with negative pressure $\rho^{(ST)}_{\rm vac} = -p^{(ST)}$ [28], [23].

Considerations on the expansion of the universe allowed Weinberg [39] to show that the effective cosmological constant $\lambda_{\rm eff}$ can be related to the Hubble constant H_0 by:

$$|\lambda_{\rm eff}| \leq H_0^2$$

and, also, that:

$$|\rho_{\rm vac}| \leq 10^{-29}{\rm g/cm}^3 \simeq 10^{-47}{\rm Gev} \ .$$

So, the present expansion rate of the universe viewed throughout the curvature of space-time by means of the Friedmann's model [16] leads to envisage very small values for the vacuum energy density and the related cosmological constant [39].

This expectation small value of the cosmological constant λ must be understood in the frame of general relativity describing gravity by the curvature of space-time by noticing that **the square root of $\boldsymbol{\lambda^{-1}}$ is a "distance" referring to the domain where the vacuum energy density $\boldsymbol{\rho^{(ST)}_{\rm vac}}$ alters the geometry of space-time by its gravitational effects** [1] in such a way that the curvature be on an average null.

4.6. New Interpretation of the General Relativity Equations

Thus, the only way to go beyond this problem of the small value of λ is to take into account a new interpretation of the equations of general relativity [11]

$$\lambda g_{\mu\nu} + G_{\mu\nu} = 8\pi G T_{\mu\nu}$$

as developed in [33] and [31].

This new context is especially characterized by:

a) the fact that **the gravitational potential is no more** assumed to be **described by the metric tensor $g_{\mu\nu}$** and the gravity is thus not explicitly described by the tensor $G_{\mu\nu}$ when the context of curved space-time geometry remains.

b) **the composition of "matter" given by the three embedded structures $ST \subset MG \subset M$** at the elementary particle level, as developed from the beginning of chapter 3, in such a way that:

- the terms $\lambda g_{\mu\nu}+G_{\mu\nu}$ refer to the vacuum structure of matter as it clearly appears from section 4.5 that, when the energy stress tensor of matter $T_{\mu\nu}$ (corresponding to the level " M ") is null, then the revised equations of GR in the vacuum are:

$$\lambda g_{\mu\nu} + G_{\mu\nu} = 0 \; .$$

-
  - the term $\lambda g_{\mu\nu}$ would be associated to the " ST " internal vacuum substructures;
  - the term $G_{\mu\nu}$ would correspond to the " MG " middle ground vacuum substructures;
  - the term $8\pi G T_{\mu\nu}$ would correspond to the " M " mass boundary structures.

It then appears that the vacuum, considered at the elementary particle level, must be composed of:

1. the "internal vacuum" substructures " ST " of which nature is of space-time type having a dynamical and expansive aspect.

2. the "middle ground" substructures generated from the corresponding internal " ST " substructures and having a contracting aspect allowing to confine " ST " substructures inside the elementary particles.

This elementary particle vacuum then corresponds to the vacuum of the quantum field theories allowing to generate the particle masses [27] (i.e. the " M " boundary structures, in the context of AQT, from the degenerate singularities on the "middle ground" substructures) and would be associated with the map:

$$G_M : \quad \lambda g_{\mu\nu} + G_{\mu\nu} = 0 \longrightarrow \lambda g_{\mu\nu} + G_{\mu\nu} = 8\pi G T_{\mu\nu}$$

in the frame of general relativity equations.

This new interpretation of general relativity, connecting its vacuum energy with that of quantum field theories, was foreboded by Sakharov [36] who claimed that gravity is not a fundamental quantum field but an induced quantum effect caused by an interaction of quantum vacuum fluctuations with space-time curvature.

In other terms, the inhomogeneity of vacuum fluctuations induces the Riemann space-time geometry [17], [6] which does not describe ontologically gravity.

4.7. Compactification of the " ST ", " MG " and " M " Structures

It remains to prove that a one-to-one correspondence, given by the isomorphism $I_{QT\to GR}$, exists between the equations of general relativity and the string field equations on the sections of the internal dynamics of a set of (bisemi)fermions (or, more generally, of a set of (bisemi)particles) as given in proposition 4.4:

$$\begin{aligned} I_{QT\to GR}: \quad & [(T_{ST_{L_{R_i}}} \otimes T_{ST_{L_j}})(\widetilde{M}_{\overline{ST}_{\overline{v}_{\sigma_i},m_{\sigma_i}}} \otimes \widetilde{M}_{ST_{\overline{v}_{\sigma_j},m_{\sigma_j}}})] \\ & +[(DT_{MG_{L_{R_i}}} \otimes DT_{MG_{L_j}})(\widetilde{M}_{\overline{MG}_{\overline{v}_{\sigma_i},m_{\sigma_i}}} \otimes \widetilde{M}_{MG_{\overline{v}_{\sigma_j},m_{\sigma_j}}})] \\ & = -[(DT_{M_{L_{R_i}}} \otimes DT_{M_{L_j}})(\widetilde{M}_{\overline{M}_{\overline{v}_{\sigma_i},m_{\sigma_i}}} \otimes \widetilde{M}_{M_{\overline{v}_{\sigma_j},m_{\sigma_j}}})] \\ & \longrightarrow \lambda g_{\mu\nu} + G_{\mu\nu} = 8\pi G T_{\mu\nu}\ . \end{aligned}$$

To reach this objective, it is necessary to:

1. compactify the diagonal perverse semisheaves $(T_{ST_{L_{R_i}}} \widetilde{M}_{\overline{ST}_{\overline{v}_{\sigma_i},m_{\sigma_i}}})$ and $(T_{ST_{L_i}} \widetilde{M}_{ST_{\overline{v}_{\sigma_i},m_{\sigma_i}}})$ of the internal vacua " ST " of the " J " considered bisemifermions as it was done in section 2.10 in order to obtain four-dimensional (1 dimension of time and 3 dimensions of space) perverse right (resp. left) compactified semisheaves according to the compactification map:

$$\begin{aligned} c^{t-r}_{ST_R}: \quad & T_{ST_{L_{R_i}}} \widetilde{M}_{\overline{ST}_{\overline{v}_{\sigma_i},m_{\sigma_i}}} \longrightarrow T_{ST_{L_{R_i}}} \widetilde{M}^c_{\overline{ST}_{\overline{v}_{\sigma_i},m_{\sigma_i}}} \\ \text{(resp. } c^{t-r}_{ST_L}: \quad & T_{ST_{L_i}} \widetilde{M}_{ST_{v_{\sigma_i},m_{\sigma_i}}} \longrightarrow T_{ST_{L_i}} \widetilde{M}^c_{ST_{v_{\sigma_i},m_{\sigma_i}}}\ )\ . \end{aligned}$$

2. compactify similarly the diagonal perverse semisheaves of the right and left "middle ground" (MG) and "mass" (M) structures of the considered bisemifermions.

3. bring together the " J " bisemifermions in such a way that the off-diagonal bisemisheaves

$$(T_{ST_{L_{R_i}}} \widetilde{M}^c_{\overline{ST}_{\overline{v}_{\sigma_i},m_{\sigma_i}}} \otimes T_{ST_{L_j}} \widetilde{M}^c_{ST_{v_{\sigma_j},m_{\sigma_j}}})\ , \quad \forall\, i \neq j\ ,$$

   of interaction of the " ST " structures, but also of the " MG " and " M " structures, have a continuous character.

4.8. Proposition

An isomorphism $I^c_{QT\to GR}$ exists between the string field equations of the internal dynamics of a set of compactified bisemiparticles and the equations of general relativity:

$$\begin{aligned} I^c_{QT\to GR}: \quad & [(T_{ST_{L_{R_i}}} \otimes T_{ST_{L_j}})(\widetilde{M}^c_{\overline{ST}_{\overline{v}_{\sigma_i},m_{\sigma_i}}} \otimes \widetilde{M}^c_{ST_{\overline{v}_{\sigma_j},m_{\sigma_j}}})] \\ & +[(DT_{MG_{L_{R_i}}} \otimes DT_{MG_{L_j}})(\widetilde{M}^c_{\overline{MG}_{\overline{v}_{\sigma_i},m_{\sigma_i}}} \otimes \widetilde{M}^c_{MG_{v_{\sigma_j},m_{\sigma_j}}})] \\ & = -[(DT_{M_{L_{R_i}}} \otimes DT_{M_{L_j}})(\widetilde{M}^c_{\overline{M}_{\overline{v}_{\sigma_i},m_{\sigma_i}}} \otimes \widetilde{M}^c_{M_{v_{\sigma_j},m_{\sigma_j}}})] \\ & \longrightarrow \lambda g_{\mu\nu} + G_{\mu\nu} = 8\pi G T_{\mu\nu} \end{aligned}$$

at the conditions that [31]:

a) the sections of the right and left semisheaves " ST ", " MG " and " M ", which are in fact " ST ", " MG " and " M " strings, be viewed as families of geodesics which must be interpreted as the flow lines of a fluid.

b) the directional gradients used in the left hand side of $I^C_{QT\to GR}$ be replaced by covariant derivatives on the right hand side.

Proof.

1. At first, the conditions a) will be precised.

   The sections of the semisheaves " MG " and " M " are assumed to correspond respectively to families of geodesics $\mathcal{P}_{MG}(\lambda_{MG}, n_{MG})$ and $\mathcal{P}_M(\lambda_M, n_M)$, where:

   - λ_{MG} and λ_M are affine parameters telling where we are on a given geodesics,
   - n_{MG} and n_M are selector parameters allowing to distinguish one geodesics from the next [24],

   if these sections have been desingularized in order that the tangent vectors

   $$\vec{u}_{MG} = \frac{\partial \mathcal{P}_{MG}}{\partial \lambda_{MG}} \qquad \vec{u}_M = \frac{\partial \mathcal{P}_M}{\partial \lambda_M}$$

   be parallel on the corresponding geodesics.

   Note that the vectors

   $$\vec{n}_{MG} \equiv \frac{\partial \mathcal{P}}{\partial n_{MG}} \qquad \vec{n}_M \equiv \frac{\partial \mathcal{P}}{\partial n_M}$$

   measure the separation between points with the same values λ_{MG} and λ_M on neighbouring geodesics.

2. Then, the terms " ST ", " MG " and " M " will be shown to be in one-to-one correspondence on the left and on the right of $I^c_{QT\to GR}$.

   (a) The term $(T_{ST_{L_{R_i}}} \otimes T_{ST_{L_j}})(\widetilde{M^c_{ST_{\overline{v}_{\sigma_i},m_{\sigma_i}}}} \otimes \widetilde{M^c_{ST_{v_{\sigma_j},m_{\sigma_j}}}})$, describing the internal vacuum structures " ST " at the level of elementary particles and being of expanding space-time nature (which can be seen by the differential nature of the bioperator $(T_{ST_{L_{R_i}}} \times T_{ST_{L_j}})$ acting on H^+_{ST} according to section 3.10), must correspond to the term $\lambda g_{\mu\nu}$ of GR if it is taken into account that:

   - this term $\lambda g_{\mu\nu}$ is not very well shaped, all the information having been smashed in the cosmological constant λ .
   - the internal vacuum energy density $\rho^{(ST)}_{\text{vac}}$, to which $\lambda/8\pi G$ corresponds, must be of expanding space-time nature: for this reason, the space-time differential operator $(T_{ST_{L_{R_i}}} \otimes T_{ST_{L_j}})$ was chosen and not a directional gradient bioperator of the type $(DT_{ST_{L_{R_i}}} \otimes DT_{ST_{L_j}})$ as introduced in proposition 3.11.

(b) The term $(DT_{MG_{L_{R_i}}} \otimes DT_{MG_{L_j}})(\widetilde{M}^c_{\overline{MG}_{\overline{v}_{\sigma_i},m_{\sigma_i}}} \otimes \widetilde{M}^c_{MG_{v_{\sigma_j},m_{\sigma_j}}})$, describing the "middle ground" vacuum structures " MG " of elementary particles and being of contracting space-time nature, is in one-to-one correspondence with the term $G_{\mu\nu}$ of the general relativity equations if:

- the conditions (a) and (b) of this proposition are taken into account.
- it is noted that each right section $\widetilde{M}^c_{\overline{MG}_{\overline{v}_{\sigma_i},m_{\sigma_i}}}$ has been projected onto each left section $\widetilde{M}^c_{MG_{\overline{v}_{\sigma_j},m_{\sigma_j}}}$) in such a way that they are confounded: they can then be rewritten according to:

$$(\widetilde{M}^c_{\overline{MG}_{\overline{v}_{\sigma_i},m_{\sigma_i}}} \otimes \widetilde{M}^c_{MG_{v_{\sigma_j},m_{\sigma_j}}}) \longrightarrow \widetilde{M}^c_{MG_{v_{\sigma_{i-j}},m_{\sigma_{i-j}}}} .$$

So, to the set $\{\widetilde{M}^c_{MG_{v_{\sigma_{i-j}},m_{\sigma_{i-j}}}}\}$ of bisections of " MG " will correspond a family $\mathcal{P}_{MG_{L_{R-L}}}(\lambda_{MG}, n_{MG})$ of products of left geodesics, localized in the upper half space, by projected symmetric right geodesics localised in the lower half space, in such a way that to the term $(DT_{MG_{L_{R_i}}} \otimes DT_{MG_{L_j}})(\widetilde{M}^c_{MG_{v_{\sigma_{i-j}},m_{\sigma_{i-j}}}})$ will correspond the term $\vec{\nabla}_{\vec{u}_{MG}} \cdot \vec{\nabla}_{\vec{u}_{MG}} \cdot \vec{n}_{MG}$.

$\vec{\nabla}_{\vec{u}_{MG}} \vec{n}_{MG}$ is the covariant derivative of the vector field $\vec{n}_{MG}$ along a product, right by left, $\mathcal{P}_{MG_{L_{R-L}}}(\lambda_{MG})$ of symmetric geodesics with tangent vector

$$\vec{u}_{MG} = \frac{\partial \mathcal{P}_{MG}}{\partial \lambda_{MG}} ,$$

and $\vec{\nabla}_{\vec{u}_{MG}} \vec{\nabla}_{\vec{u}_{MG}} \vec{n}_{MG}$ is the corresponding relative acceleration.

As the relative acceleration of geodesics allows to define the Riemann curvature tensor [24] by:

$$\vec{\nabla}_{\vec{u}_{MG}} \vec{\nabla}_{\vec{u}_{MG}} \vec{n}_{MG} + \text{Riemann}(\dots \vec{u}_{MG}, \vec{n}_{MG}, \vec{u}_{MG}) = 0$$

which leads to the components of the tensor of Riemann $R^{\lambda}_{\mu\lambda\nu}$ in a coordinate basis.

Due to the antisymmetric property of $R^{\lambda}_{\mu\lambda\nu}$ [39], there are only two tensors which can be generated by contraction from $R^{\lambda}_{\mu\lambda\nu}$: it is the Ricci tensor $R_{\mu\nu} \equiv R^{\lambda}_{\mu\lambda\nu}$ and the curvature scalar $R = R^{\mu}_{\mu}$ [7].

So, the only tensor which can be formed from $R_{\mu\nu}$ and R is the tensor $G_{\mu\nu} = R_{\mu\nu} - \frac{1}{2} g_{\mu\nu} R$ which is thus in one-to-one correspondence with the " MG " term on the left hand side of $I^c_{QT\to GR}$.

(c) Finally, the term $(DT_{M_{L_{R_i}}} \otimes DT_{M_{L_j}})(\widetilde{M}^c_{\overline{M}_{\overline{v}_{\sigma_i},m_{\sigma_i}}} \otimes \widetilde{M}^c_{M_{v_{\sigma_j},m_{\sigma_j}}})$, describing the "mass" structures of elementary particles and being also of contracting space-time nature, is in one-to-one correspondence with the tensor $T_{\mu\nu}$ of the equations of general relativity.

Indeed, if it is taken into account that the "mass" term " M " and the "middle ground" term " MG " on the left of $I^c_{QT\to GR}$ have the same structure, it is immediate to associate with $(DT_{M_{L_{R_i}}} \otimes DT_{M_{L_j}})(\widetilde{M^c_{\overline{M}_{v\sigma_{i-j},m\sigma_{i-j}}}})$ the "mass" curvature tensor of Riemann $R^{\lambda}_{\mu\lambda\nu}$ from which the "mass" tensor

$$G^{(M)}_{\mu\nu} = R^{(M)}_{\mu\nu} - \frac{1}{2} g^{(M)}_{\mu\nu} R^{(M)}$$

can be formed.

Considering the Einstein equations

$$G_{\mu\nu} = 8\pi G T_{\mu\nu} ,$$

it is evident that:

(1) the tensor $G^{(M)}_{\mu\nu}$ is equal to the "mass" energy-momentum tensor $8\pi G T_{\mu\nu}$.

(2) to the mass term on the left hand side of $I^c_{QT\to GR}$ corresponds the tensor $T_{\mu\nu}$. ■

4.9. Proposition

In the context of the new interpretation of the equations of general relativity describing the dynamics of the generation of the mass shells of the elementary particles from their vacuum structures, we have that:

1. *the cosmological constant will likely have a high value and will be noted λ_{ST} .*
2. *the internal vacuum structure " ST " itself (without the generated " MG " and " M " structure) at the elementary particle level is probably responsible for the dark energy.*

Proof.

1. The cosmological constant λ_{ST} , referring now to the internal vacuum structures of elementary particles, is now directly related to the Planck scale; consequently, λ_{ST} will have a high value.
2. The dark energy, being of expanding space-time nature, will probably correspond to sets of elementary particles endowed only with their internal vacuum structures " ST ". ■

4.10. The Algebraic Quantum Theory Is a Quantum Gravity Theory

The algebraic quantum theory [33], recalled at the beginning of this chapter for a set of interacting bisemiparticles, is a quantum gravity theory.

Indeed, the gravity is no more introduced ontologically as resulting from the curvature of space-time but from the diagonal interactions between right and left semisheaves " ST ", " MG " and " M " belonging to different bisemiparticles as it was developed in [33]. We refer thus to the preprint "Algebraic quantum theory" [33] for a description of quantum gravity as resulting from the diagonal interactions between pairs of semiobjects.

References

[1] Abbott, L.: Le mystère de la constante cosmologique. *Pour la Science*, 129 (1988), 48–56.

[2] Alvares, E.: Quantum gravity: an introduction to some recent results. *Rev. Mod. Phys.*, 61 (1989), 561–604.

[3] Ashtekar, A.: New perspectives in canonical gravity. Lect. on non-perturbative canonical gravity, *Adv. Ser. in astrophysics and cosmology*, Vol. 6, World Scientif. Singapore (1994).

[4] Atiyah, M.: Topological quantum field theory. *Publ. Math. IHES*, 68 (1989), 175.

[5] Baez, J.: Spin foam models. *Class quantum gravity*, 15 (1998), 1827–1858.

[6] Banks, T.: TCP, quantum gravity, the cosmological constant and all that. *Nucl. Phys.*, B249 (1985), 332–360.

[7] Darboux, G.: *Leçons sur la théorie générale des surfaces*. Gauthiers Villars (1887), Vol III, Livre V, ch. 4.

[8] De Witt, B.: Quantum theory of gravity, I – The canonical theory. *Phys. Rev.*, 160 (1967), 1113–1148.

[9] Dÿkgraaf, R. Vafa, C.: *A perturbative window into non perturbative physics*. hep-th/0208048.

[10] Eddington, A.: On the value of the cosmological constant. *Phys. Roy. Soc.*, 133 (1931), 605–615.

[11] Einstein, A.: Zür Elektrodynamik Beweg ter Körper. *Annal Phys.*, 17 (1905).

[12] Einstein, A.: *The principle of relativity*. Dover Publ. (1923).

[13] Einstein, A.: Sitzungber. *K. Preuss. Akad. Wiss.*, 142 (1917).

[14] Eguchi, T., Gilkey, P., Hanson, A.: Gravitation, gauge theories and differential geometry. *Phys. Rep.*, 66 (1980), 213–393.

[15] Fulton, W. McPherson, R.: A compactification of configuration spaces. *Annals of Math.*, 139 (1994), 183–225.

[16] Friedmann, A.: *Z. Phys.* 21 (1924), 326– .

[17] Gliner, E.: Quantum theory of fields and origin of gravity. *Int. J. Theor. Phys.*, 25 (1986), 477–492.

[18] Gross, D.: Is quantum gravity unpredictable? *Nucl. Phys.*, B236 (1984), 349–367.

[19] Green, M.B., Schwarz, J. Witten, E.: *Superstring theory*. Cambridge Univ. Press (1987).

[20] 't Hooft, G.: *How does God play dice?* (pre)determinism at the Planck scale, hep-th/0104219

[21] 't Hooft, G.: *Dimensional reduction in quantum gravity*, gr-qc/931006.

[22] Hacyan, S., Sarminento, A., Cocko, G. Soto, F.: Zero-point field in curved spaces. *Phys. Rev.*, D32 (1985), 914–919.

[23] Lemaître, G.: *Proc. Natl. Acad. Sci.* (USA), 20 (1934).

[24] Misner, C., Thorne, K., Wheeler, J.A.: *Gravitation*. Freeman & Comp. (1972).

[25] Van Nieuwenhuizen: An introduction to covariant quantization and gravitation. *Proc. of the 1-th M. Grossmann Meeting on Gen. Rel.*, R. Ruffini – North Holland (1977).

[26] Pagels, H.: Gauge theory of gravity. *Phys. Rev.* D27, (1983), 2299–2304.

[27] Parker, L.: Quantized fields and particle creation in expanding universes. *Phys. Rev.* 183, (1969), 1057–1068.

[28] Peebles, P.: The cosmological constant of dark energy. *Rev. Mod. Phys.* 75, (2003), 559–606.

[29] Pierre, C.: *Brane and string field structures of elementary particles*. Arxiv Phys/0607249 (2006).

[30] Pierre, C.: *n-dimensional global correspondences of Langlands*. Arxiv Math RT/0510348 v. 3 (2005).

[31] Pierre, C.: The dark matter and the dark energy proceeding from elementary particles. *Int. J. Pure and Applied Phys.*, 6, (2010), 383–412.

[32] Pierre, C.: *n-dimensional global correspondences of Langlands over singular schemes*. Arxiv Math RT/0606342 (2006).

[33] Pierre, C.: *Algebraic quantum theory*. ArXiv:math-ph/0404024 (2004).

[34] Polchinski, J.: *String theory*. Vol. 1 and 2. Cambridge Univ. Press, (1998).

[35] Rovelli, C., Smolin, L.: The physical Hamiltonian in non-perturbative quantum gravity. *Phys. Rev. Lett.*, 72 (1992), 446– .

[36] Sakharov, A.: Vacuum quantum fluctuations in curved space-time and the theory of gravitation. *Soviet. Phys. Dokl.*, 12 (1968), 1040–1041.

[37] Smolin, L.: *How far are we from the quantum theory of gravity?* hep-th/0303185.

[38] Wald, R.: Quantum gravity and time reversibility. *Phys. Rev.*, D21 (1980), 2712–2755.

[39] Weinberg, S.: *Gravitation and cosmology principles and applications of the general theory of relativity*. J. Wiley (1972).

[40] Weinberg, S.: The cosmological constant problem. *Rev. Mod. Phys.*, 61 (1989), 1–23.

[41] Wheeler, J.A.: Geometrodynamics steering principles reveals the determinacy of inertia. *Intern. J. Mod. Phys.*, 3 (1988), 2207–2247.

[42] Witten, E.: *Quantum gravity in the Sitter space*. hep-th/0106109.

[43] Witten, E.: *Singularities in string theory*. ICM (2002), Vol. 1, 495–504.

[44] Zel'dovich, Y.: The cosmological constant and the theory of elementary particles. *Sov. Phys. usp.*, (1968), 381–393.

[45] Zel'dovich, Y.: The creation of particles and antiparticles in electric and gravitational fields. In: *J.A. Wheeler - Magic Without Magic*, Ed. Klauder-Freeman (1972), 277–288.

In: Einstein and Others: Unification
Editor: Valeriy V. Dvoeglazov
ISBN: 978-1-63463-276-8

Chapter 6

EXTRACTING ENERGY FROM AN EXTERNAL MAGNETIC FIELD

Waldyr Alves Rodrigues Jr. * ***and Edmundo Capelas de Oliveira*** †
Institute of Mathematics, Statistics and Scientific Computation
IMECC-UNICAMP, Brazil

Abstract

In this chapter we present the theory of a device that can extract energy from an external magnetic field. The device is a cylindrical magnetic insulator which, once it is set in rotation, makes electromagnetic angular momentum to be stored in the electromagnetic field in a direction contrary the direction of the device's mechanical angular momentum. As a consequence of total angular momentum conservation the device's angular velocity increases (when $\varepsilon\mu > 1$ in natural units) and all conservation laws are rigorously satisfied. The voltage generated by such device is found by solving explicitly Maxwell equations for a rotating magnetic insulator in an external field, a subject that has provoked lots of polemics in the literature and which we hope to clarify here by means ofa a pedagogical presentation.

PACS: 41.20.-q

Keywords: Maxwell theory, rotating frames, differential forms, electromagnetic generators

1. Introduction

In this chapter we introduce the theory of a device that can, in principle, extract energy from an external magnetic field. The idea of such a device arose from the theoretical analysis of the 1913 Wilson & Wilson [21] experiment (WWE)[1] described in Section 5. As the correct theoretical explanation of the experiment gave rise to controversies[2] in the literature

*E-mail addresses: walrod@ime.unicamp.br; walrod@mpc.com.br
†E-mail address: capelas@ime.unicamp.br

[1]The experiment was repeated by Hertzberg and collaborators [5] in 2001.

[2]A sample may be found in [11, 13, 17, 18, 20] and in the numerous references to related issues in those papers.

due to difficulties in properly applying Maxwell theory in rotating frames[3] we decided to give a very pedagogical presentation of the problem of expressing Maxwell equations in a coordinate invariant way using the theory of differential forms. Then, using that formalism and the correct jump conditions for the electromagnetic field variables at the boundaries separating two different media we obtain in a very clear way the electromagnetic field measured in the WWE.[4] We think that our approach will clear up what is wrong with some other attempts to find a correct description of the WWE in the sample of papers quoted. The description of the machine for extracting energy from an external magnetic field is given in section 6, where it is observed that the machine, once set in motion by an external device, makes electromagnetic angular momentum to be stored in the electric plus magnetic fields in a direction opposite to the direction of the device's mechanical angular momentum (when $\mu\varepsilon > 1$) and thus, due to angular momentum conservation, the machine has its mechanical angular momentum increased. Of course, all conservation laws are in operation in our device and the energy generated comes from the external electromagnetic field.

2. Maxwell Equations

The formulation of Maxwell equations in intrinsic form requires as preliminary mathematical structures:

(i) a 4-dimensional orientable[5] manifold M, the bundle of non homogeneous differential forms $\bigwedge T^*M = \bigoplus_{r=0}^{r=4} \bigwedge^r T^*M$, where $\bigwedge^r T^*M$ is the bundle of r-forms and

(ii) the differential operator $d : \bigwedge^r T^*M \to \bigwedge^{r+1} T^*M$, $d^2 = 0$.

Indeed, Maxwell equations deal with a field $F \in \sec \bigwedge^2 T^*M$ which is exact, i.e., $F = dA$, where $A \in \sec \bigwedge^1 T^*M$, and a current $\mathbf{J} \in \sec \bigwedge^3 T^*M$ which is also exact, i.e., $\mathbf{J} = dH$, where $H \in \sec \bigwedge^2 T^*M$. The set of equations

$$dF = 0, \quad dH = -\mathbf{J}, \tag{1}$$

is known as Maxwell equations and of course, we have

$$d\mathbf{J} = 0, \tag{2}$$

i.e., the current is conserved.

Maxwell equations are *invariant under diffeomorphisms* as it happens with all theories formulated with differential forms. This means that if $h : M \to M$ is a diffeomorphism, then, denoting the pullback mapping by h^*, since the differential d commutes with the

[3]See, e.g., the most quoted Schiff's paper [19] in which it is (wrongly) claimed that one needs to use General Relativity to explain electrodynamic phenomena in rotating frames. Indeed, there is in each generation a pervasive difficulty to understand even simple electromagnetic phenomena whenever moving boundaries are involved as, e.g., in the application of Faraday's law of induction to a homopolar generator. A thoughtful discussion of the application of Faraday's law in its many equivalent formulations can be found in [16].

[4]Applications of differential forms formalism to the WWE already appeared in [1, 6]. Our approach details each step in the derivation and hopefully is a little bit more pedagogical.

[5]By orientable we mean that there exists in M a global 4-form $\tau_g \in \sec \bigwedge^4 T^*M$. If the manifold is not orientable then it is necessary to use besides the concepts of *pair* form fields also the concepts of *impair* form fields. Details may be found in [6]. See also a thoughtful discussion in [14].

pullback, i.e., $dh^* = h^*d$, we have that the fields $F' = h^*F, H' = h^*H$ and $\mathbf{J}' = h^*\mathbf{J}$ satisfy

$$dF' = 0, \quad dH' = -\mathbf{J}'. \tag{3}$$

Notice that we have not made any assumptions concerning the topology of manifold M.

We now take *arbitrary coordinates* $\langle x^\mu \rangle$ covering $U \subset M$. Denoting by[6] $\langle e_\mu = \boldsymbol{\partial}/\boldsymbol{\partial} x^\mu = \boldsymbol{\partial}_\mu \rangle$ the corresponding coordinate basis for TU and by $\langle \vartheta^\mu = dx^\mu \rangle$ the basis for T^*U dual to the basis $\langle e_\mu \rangle$ we can write

$$\begin{aligned} F &= \frac{1}{2} F_{\mu\nu} \vartheta^\mu \wedge \vartheta^\nu, \\ H &= \frac{1}{2} H_{\mu\nu} \vartheta^\mu \wedge \vartheta^\nu, \\ \mathbf{J} &= \frac{1}{3!} J_{\mu\nu\rho} \vartheta^\mu \wedge \vartheta^\nu \wedge \vartheta^\nu. \end{aligned} \tag{4}$$

Maxwell equations are supposed to describe the behavior of electromagnetic fields in vacuum or in material media. But whereas, according to Feynman, field F is fundamental (it defines the Lorentz force acting on charged probe particles moving in the field), field H is phenomenological[7]. In fact, suppose the following Fourier representation for F and H;

$$\begin{aligned} F_{\mu\nu}(x) &= \textstyle\int d^4k \breve{F}_{\mu\nu}(k) e^{-i(k_0x^0 + k_1x^1 + k_2x^2 + k_3x^3)} := \int d^4k \mathfrak{F}_{\mu\nu}(k, x), \\ H_{\mu\nu}(x) &= \textstyle\int d^4k \ \breve{H}_{\mu\nu}(k) e^{-i(k_0x^0 + k_1x^1 + k_2x^2 + k_3x^3)} := \int d^4k \mathfrak{H}_{\mu\nu}(k, x). \end{aligned} \tag{5}$$

Defining the 2-form valued function $\breve{\boldsymbol{\varkappa}} : \mathbb{R} \times \sec \bigwedge^2 T^*M \to \sec \bigwedge^2 T^*M$, the frequency constituent equations of the medium, we have in general that

$$\mathfrak{H}(k_0, x) = \breve{\boldsymbol{\varkappa}}(\mathfrak{F}(k_0, x)). \tag{6}$$

In some media, $\breve{\boldsymbol{\varkappa}}$ may be a nonlinear function of $\mathfrak{F}$ (see, e.g., [2]) but in what follows we consider only non dispersive media. This means that $\breve{\boldsymbol{\varkappa}}$ is independent of k_0, so that we can define a linear constituent function (an extensor field)

$$\begin{aligned} \boldsymbol{\varkappa} : \sec \textstyle\bigwedge^2 T^*M &\to \sec \textstyle\bigwedge^2 T^*M, \\ H &= \boldsymbol{\varkappa}(F). \end{aligned} \tag{7}$$

In this case we have for the components $H_{\mu\nu}$ of H:

$$H_{\mu\nu} = \frac{1}{2} \boldsymbol{\varkappa}_{\mu\nu \cdot\cdot}^{\cdot\cdot\alpha\beta} F_{\alpha\beta}. \tag{8}$$

Of course, we have the obvious symmetries for the components of the constituent extensor[8] $\boldsymbol{\varkappa}$,

$$\boldsymbol{\varkappa}_{\mu\nu \cdot\cdot}^{\cdot\cdot\alpha\beta} = -\boldsymbol{\varkappa}_{\nu\mu \cdot\cdot}^{\cdot\cdot\alpha\beta}, \quad \boldsymbol{\varkappa}_{\mu\nu \cdot\cdot}^{\cdot\cdot\alpha\beta} = -\boldsymbol{\varkappa}_{\mu\nu \cdot\cdot}^{\cdot\cdot\beta\alpha}, \quad \boldsymbol{\varkappa}_{\mu\nu \cdot\cdot}^{\cdot\cdot\alpha\beta} = \boldsymbol{\varkappa}_{\nu\mu \cdot\cdot}^{\cdot\cdot\beta\alpha}. \tag{9}$$

[6]Remark that coordinate vector fields are denoted by a bold symbol $\boldsymbol{\partial}$, e.g., $\boldsymbol{\partial}/\boldsymbol{\partial} x^\mu = \boldsymbol{\partial}_\mu$, while we use the symbol $\partial/\partial x^\mu := \partial_\mu$ to denote the usual partial derivatives. This means the following: Let $f : M \to \mathbb{R}$ be a differentiable function and let (U, ψ) be a chart of the atlas of M such that for $e \in U$, $\psi(e) = (x^0, x^1, x^2, x^3)$. Then, $\frac{\boldsymbol{\partial}}{\boldsymbol{\partial} x^\mu} f = \frac{\partial}{\partial x^\mu} f \circ \psi$.

[7]At least at the classical level. This is so because the calculation of H needs in the general case a sophisticated use of quantum theory.

[8]We recall that these are the same symmetries of the components of the Riemann tensor of a metric compatible connection.

2.1. Enter $\langle M, \boldsymbol{g}, \boldsymbol{\nabla}, \boldsymbol{\tau_g}, \uparrow\rangle$

We cannot proceed with the formulation of Maxwell theory without introducing additional mathematical objects in manifold M. If you are interested in knowing the details, please consult [6]. From now on we suppose that electromagnetic fields are to be described in a Lorentzian spacetime structure $\langle M, \boldsymbol{g}, \boldsymbol{\nabla}, \boldsymbol{\tau_g}, \uparrow\rangle$, where $\langle M, \boldsymbol{g}\rangle$ is a Lorentzian manifold, $\boldsymbol{\nabla}$ is the Levi-Civita connection of $\boldsymbol{g}$, $\boldsymbol{\tau_g} = \sqrt{|\det \boldsymbol{g}|}\vartheta^0 \wedge \vartheta^1 \wedge \vartheta^2 \wedge \vartheta^3$ is the metrical volume element and $\uparrow$ denotes that the structure $\langle M, \boldsymbol{g}\rangle$ is time orientable.

Remark 1. *As it is well known a spacetime structure* $(M, \boldsymbol{g}, \boldsymbol{\nabla}, \boldsymbol{\tau_g}, \uparrow\rangle$ *such that the Riemann curvature tensor of* $\boldsymbol{\nabla}$ *is non null represents the gravitational field generated by an energy-momentum tensor* T *in Einstein's General Relativity* (GR). *It is a simple exercise to find an effective spacetime structure, say* $\langle M, \mathrm{g}, \boldsymbol{\nabla}, \boldsymbol{\tau}_{\mathrm{g}}, \uparrow\rangle$, *that can be used to describe some material media. However, such structure has nothing to do with GR.*

2.1.1. Reference Frames

In what follows we shall need the concepts of reference frames, observers and naturally adapted coordinate systems to a reference frame in a general Lorentzian spacetime structure $\langle M, \boldsymbol{g}, \boldsymbol{\nabla}, \boldsymbol{\tau_g}, \uparrow\rangle$.

We define a *reference frame* in $U \subset M$ as a unit timelike vector field $\boldsymbol{Z}$ pointing to the future. We have

$$\boldsymbol{g}(\boldsymbol{Z}, \boldsymbol{Z}) = 1. \tag{10}$$

An *observer* is defined as a timelike curve σ in M (parametrized with proper time) and pointing to the future. We denote by σ_{*s} the tangent vector at $\sigma(s)$ and

$$\boldsymbol{g}(\sigma_{*s}, \sigma_{*s}) = 1. \tag{11}$$

We immediately realize that each of the integral lines of $\boldsymbol{Z}$ is an observer.

Finally we say that a set of coordinates $\langle x^\mu\rangle$ covering $U \subset M$ is a *naturally adapted coordinate system to a reference frame* $\boldsymbol{Z}$ (nacs$|\boldsymbol{Z}$) if

$$\boldsymbol{Z} = \frac{1}{\sqrt{g_{00}}}\frac{\boldsymbol{\partial}}{\boldsymbol{\partial} x^0}\,. \tag{12}$$

A detailed classification of reference frames in a Lorentzian spacetime can be found, e.g., in [15]. For a classification of reference frames in a Riemann-Cartan spacetime, see, e.g., [3].

2.1.2. Formulation of Vacuum Maxwell Equations in $\langle M, \boldsymbol{g}, \boldsymbol{\nabla}, \boldsymbol{\tau_g}, \uparrow\rangle$

Given the structure $\langle M, \boldsymbol{g}, \boldsymbol{\nabla}, \boldsymbol{\tau_g}, \uparrow\rangle$ it is convenient to define

$$G = \underset{\boldsymbol{g}}{\star}^{-1} H \tag{13}$$

where $\underset{\boldsymbol{g}}{\star} : \sec \bigwedge^r T^*M \to \sec \bigwedge^{4-r} T^*M$ is the Hodge star operator and $\underset{\boldsymbol{g}}{\star}^{-1}\underset{\boldsymbol{g}}{\star} = \underset{\boldsymbol{g}}{\star}\underset{\boldsymbol{g}}{\star}^{-1} =$ Id. It is also a standard practice to introduce $J \in \sec \bigwedge^1 T^*M$ such that

$$\mathbf{J} = \underset{\boldsymbol{g}}{\star} J. \tag{14}$$

We define also the (ex)tensor field

$$\boldsymbol{\chi} : \sec \bigwedge^2 T^*M \to \sec \bigwedge^2 T^*M, \qquad G = \boldsymbol{\chi}(F). \tag{15}$$

Writing[9]

$$G = \frac{1}{2} G_{\mu\nu} \vartheta^\mu \wedge \vartheta^\nu = \frac{1}{2} G^{\mu\nu} \vartheta_\mu \wedge \vartheta_\nu \tag{16}$$

we have

$$G_{\mu\nu} = \frac{1}{2} \chi_{\mu\nu\cdot\cdot}^{\cdot\cdot\alpha\beta} F_{\alpha\beta}, \quad G^{\mu\nu} = \frac{1}{2} \chi^{\mu\nu\alpha\beta} F_{\alpha\beta} \tag{17}$$

and, of course, the components of the constituent extensor $\boldsymbol{\chi}$ satisfy

$$\chi^{\mu\nu\alpha\beta} = -\chi^{\nu\mu\alpha\beta}, \quad \chi^{\mu\nu\alpha\beta} = -\chi^{\mu\nu\beta\alpha}, \quad \chi^{\mu\nu\alpha\beta} = \chi^{\nu\mu\beta\alpha}. \tag{18}$$

In vacuum the relation between fields G and F is

$$G = F \tag{19}$$

and we write Maxwell equations as

$$dF = 0, \qquad d \underset{\boldsymbol{g}}{\star} F = - \underset{\boldsymbol{g}}{\star} J. \tag{20}$$

Or better, applying the inverse of Hodge star operator to both members of the non homogenous equation we can write

$$dF = 0, \qquad \underset{\boldsymbol{g}}{\delta} F = -J, \tag{21}$$

where $\underset{\boldsymbol{g}}{\delta}$ is the Hodge codiferential[10]. We recall that[11]

$$\underset{\boldsymbol{g}}{\star} \vartheta^{\mu_1 \cdots \mu_p} = \frac{1}{(n-p)!} \sqrt{|\det \boldsymbol{g}|} g^{\mu_1\nu_1} \ldots g^{\mu_p\nu_p} \epsilon_{\nu_1 \ldots \nu_n} \vartheta^{\nu_{p+1} \ldots \nu_n}. \tag{22}$$

Then, write

$$H = \underset{\boldsymbol{g}}{\star} F = \frac{1}{2} H_{\rho\sigma} \vartheta^\rho \wedge \vartheta^\sigma = \frac{1}{2} H^{\rho\sigma} \vartheta_\rho \wedge \vartheta_\sigma \tag{23}$$

and let us calculate $\underset{\boldsymbol{g}}{\star} F$. We have

$$\begin{aligned} \underset{\boldsymbol{g}}{\star} F &= \frac{1}{2} F_{\mu\nu} \underset{\boldsymbol{g}}{\star} (\vartheta^\mu \wedge \vartheta^\nu) \\ &= \frac{1}{2} \frac{1}{2} F_{\mu\nu} \sqrt{|\det \boldsymbol{g}|} g^{\mu\alpha} g^{\nu\beta} \epsilon_{\alpha\beta\rho\sigma} \vartheta^\rho \wedge \vartheta^\sigma. \end{aligned} \tag{24}$$

---

[9] $\langle \vartheta_\mu \rangle$ is the reciprocal basis of $\langle \vartheta^\mu \rangle$, i.e., $\mathtt{g}(\vartheta^\mu, \vartheta_\nu) = \delta^\mu_\nu$.

[10] We use the conventions in [15], i.e., for $A_p \in \sec \bigwedge^p T^*M$, $\underset{\boldsymbol{g}}{\delta} A_p = (-1)^p \underset{\boldsymbol{g}}{\star}^{-1} d \underset{\boldsymbol{g}}{\star} A_p$.

[11] $\boldsymbol{\vartheta}^{\mu_1 \cdots \mu_p} := \boldsymbol{\vartheta}^{\mu_1 \cdots \mu_p} \wedge \boldsymbol{\vartheta}^{\mu_2} \wedge \cdots \wedge \boldsymbol{\vartheta}^{\mu_p}$.

From Eq.(23) and Eq.(24) we have

$$
\begin{aligned}
H_{\rho\sigma} &= \frac{1}{2}\sqrt{|\det \boldsymbol{g}|} g^{\mu\alpha} g^{\nu\beta} \epsilon_{\alpha\beta\rho\sigma} F_{\mu\nu} \\
&= \frac{1}{2}\sqrt{|\det \boldsymbol{g}|} g^{\mu\alpha} g^{\nu\beta} \epsilon_{\rho\sigma\alpha\beta} F_{\mu\nu} \\
&= \frac{1}{2}\sqrt{|\det \boldsymbol{g}|} \epsilon_{\rho\sigma..}^{\;\;..\mu\nu} F_{\mu\nu}.
\end{aligned} \tag{25}
$$

Taking into account that

$$
H_{\rho\sigma} = {}^{*}G_{\rho\sigma} = \frac{1}{2}\sqrt{|\det \boldsymbol{g}|} \epsilon_{\rho\sigma..}^{\;\;..\mu\nu} G_{\mu\nu} \tag{26}
$$

we get

$$
G^{\rho\sigma} = g^{\rho\mu} g^{\sigma\nu} F_{\mu\nu} = \frac{1}{2}(g^{\rho\mu} g^{\sigma\nu} - g^{\rho\nu} g^{\sigma\mu}) F_{\mu\nu}. \tag{27}
$$

Comparing this equation with Eq.(17) gives for the components $\varkappa^{\rho\sigma\mu\nu}$ of the constitutive (ex)tensor of the vacuum

$$
\chi^{\rho\sigma\mu\nu} = (g^{\rho\mu} g^{\sigma\nu} - g^{\rho\nu} g^{\sigma\mu}). \tag{28}
$$

Remark 2. *Before proceeding let us recall the similarity between the components of* $\chi^{\rho\sigma\mu\nu}$ *and the components* $\boldsymbol{R}^{\rho\sigma\mu\nu}$ *of a Riemann curvature tensor* $\boldsymbol{R}$ *with a constant Riemann curvature scalar* K. *Indeed, we have*

$$
\boldsymbol{R}^{\rho\sigma\mu\nu} = \frac{K}{12}(g^{\rho\mu} g^{\sigma\nu} - g^{\rho\nu} g^{\sigma\mu}) \tag{29}
$$

and thus $K = R = \delta^{\nu}_{\beta} R^{\alpha\beta..}_{..\alpha\nu}$. *Observe that* $\boldsymbol{R}$ *is not the Riemann curvature tensor of the Levi-Civita connection* $\boldsymbol{\nabla}$ *of* $\boldsymbol{g}$, *because, of course, the structure* $(M, \boldsymbol{g}, \boldsymbol{\nabla}, \boldsymbol{\tau}_{\boldsymbol{g}}, \uparrow\rangle$ *in which the electromagnetic field lives does not need to have constant curvature.*

2.2. Maxwell Equations in Components

We now show that when $\boldsymbol{\varkappa}$ has its vacuum value,Maxwell equations $dF = 0$ and $dH = -\underset{\boldsymbol{g}}{\star} J$ can be written in terms of its components as

$$
\partial_\mu F_{\alpha\beta} + \partial_\beta F_{\mu\alpha} + \partial_\alpha F_{\beta\mu} = 0, \tag{30}
$$

$$
\frac{1}{\sqrt{|\det \boldsymbol{g}|}} \partial_\mu (\sqrt{|\det \boldsymbol{g}|} F^{\mu\nu}) = J^\nu. \tag{31}
$$

Indeed, we have

$$
\begin{aligned}
dF &= \frac{1}{2} d(F_{\alpha\beta} \wedge \vartheta^\alpha \wedge \vartheta^\beta) \\
&= \frac{1}{2} \partial_\mu F_{\alpha\beta} \vartheta^\mu \wedge \vartheta^\alpha \wedge \vartheta^\beta \\
&= \frac{1}{3!} (\partial_\mu F_{\alpha\beta} + \partial_\beta F_{\mu\alpha} + \partial_\alpha F_{\beta\mu}) \vartheta^\mu \wedge \vartheta^\alpha \wedge \vartheta^\beta
\end{aligned}
$$

and Eq.(30) follows.

Next we define $J = J_\mu \vartheta^\mu := \rho\vartheta^0 - j_i\vartheta^i$ and get

$$\underset{\boldsymbol{g}}{\star} J = \frac{1}{3!}\sqrt{|\det \boldsymbol{g}|} J_\lambda g^{\lambda\kappa} \epsilon_{\kappa\mu\alpha\beta} \vartheta^\mu \wedge \vartheta^\alpha \wedge \vartheta^\beta. \tag{32}$$

Then equation $dH = -\underset{\boldsymbol{g}}{\star} J$ can be written

$$\frac{1}{3!}(\partial_\mu H_{\alpha\beta} + \partial_\beta H_{\mu\alpha} + \partial_\alpha H_{\beta\mu})\vartheta^\mu \wedge \vartheta^\alpha \wedge \vartheta^\beta = -\frac{1}{3!}\sqrt{|\det \boldsymbol{g}|} J^\kappa \epsilon_{\kappa\mu\alpha\beta} \vartheta^\mu \wedge \vartheta^\alpha \wedge \vartheta^\beta \tag{33}$$

and

$$(\partial_\mu H_{\alpha\beta} + \partial_\beta H_{\mu\alpha} + \partial_\alpha H_{\beta\mu}) = -\sqrt{|\det \boldsymbol{g}|} J^\kappa \epsilon_{\kappa\mu\alpha\beta}. \tag{34}$$

Multiplying both members of the last equation by $\epsilon^{\lambda\mu\alpha\beta}$ and recalling[12] that

$$\begin{aligned} \epsilon^{\lambda\mu\alpha\beta}\epsilon_{\kappa\mu\alpha\beta} &= 3!\delta^\lambda_\kappa \\ \epsilon^{\lambda\mu\alpha\beta}\epsilon_{\alpha\beta\kappa\iota} &= 2!\delta^{\lambda\mu}_{\kappa\iota} \end{aligned} \tag{35}$$

we have

$$\epsilon^{\lambda\mu\alpha\beta}(\partial_\mu H_{\alpha\beta} + \partial_\beta H_{\mu\alpha} + \partial_\alpha H_{\beta\mu}) = -3!\sqrt{|\det \boldsymbol{g}|} J^\lambda. \tag{36}$$

Now,

$$\begin{aligned} \epsilon^{\lambda\mu\alpha\beta}\partial_\mu H_{\alpha\beta} &= \partial_\mu(\frac{1}{2}\sqrt{|\det \boldsymbol{g}|}\epsilon^{\lambda\mu\alpha\beta}\epsilon_{\alpha\beta\kappa\iota}F^{\kappa\iota}) \\ &= \partial_\mu(\sqrt{|\det \boldsymbol{g}|}\delta^{\lambda\mu}_{\kappa\iota}F^{\kappa\iota}) \\ &= 2\partial_\mu(\sqrt{|\det \boldsymbol{g}|}F^{\lambda\mu}) \\ &= -2\partial_\mu(\sqrt{|\det \boldsymbol{g}|}F^{\mu\lambda}). \end{aligned} \tag{37}$$

Analogous calculations will show that the other two terms in the first member of Eq.(36) are also equal to $-2\partial_\mu(\sqrt{|\det \boldsymbol{g}|}F^{\mu\lambda})$. Then Eq.(36) becomes

$$\frac{1}{\sqrt{|\det \boldsymbol{g}|}}\partial_\mu(\sqrt{|\det \boldsymbol{g}|}F^{\mu\lambda}) = J^\lambda$$

and Eq.(31) is proved.

Remark 3. *There is still another simple proof of* Eq.(31) *which, however, presumes the knowledge of Clifford calculus. We introduce the Dirac operator* $\boldsymbol{\partial} = \vartheta^\mu \boldsymbol{\nabla}_{\partial/\partial x^\mu}$ *and recall that* $dM = \boldsymbol{\partial} \wedge M$ *for any* $M \in \sec \bigwedge T^*M \hookrightarrow \sec \mathcal{C}\ell(M, \mathrm{g})$, *where* $\mathcal{C}\ell(M, \mathrm{g})$ *is the Clifford bundle of differential forms*[13]*. We have*

$$dH = d\underset{\boldsymbol{g}}{\star} G = \boldsymbol{\partial} \wedge \underset{\boldsymbol{g}}{\star} G = -\underset{\boldsymbol{g}}{\star} J$$

[12]See, e.g., p. 111 of [8].

[13]The Clifford product of Clifford fields is denoted by juxtaposition.

Now, we calculate $\boldsymbol{\partial} \wedge \underset{\boldsymbol{g}}{\star} G = \boldsymbol{\partial} \wedge (\tilde{G}\boldsymbol{\tau_g})$ *in orthonormal basis* [14]$\{\mathbf{e_a}\}$ *for* TU *and dual basis* $\{\boldsymbol{\theta}^{\mathbf{a}}\}$ *for* T^*U *with* $U \subset M$. *We have*

$$\begin{aligned}
d\underset{\boldsymbol{g}}{\star} G &= \boldsymbol{\partial} \wedge (\tilde{G}\boldsymbol{\tau_g}) \\
&= -\boldsymbol{\theta}^{\mathbf{a}} \wedge (\boldsymbol{\nabla}_{\mathbf{e_a}}(G\boldsymbol{\tau_g})) \\
&= -\frac{1}{2}(\boldsymbol{\theta}^{\mathbf{a}}\boldsymbol{\nabla}_{\mathbf{e_a}}(G\boldsymbol{\tau_g}) + \boldsymbol{\nabla}_{\mathbf{e_a}}(G\boldsymbol{\tau_g})\boldsymbol{\theta}^{\mathbf{a}}) \\
&= -\frac{1}{2}(\boldsymbol{\theta}^{\mathbf{a}}(\boldsymbol{\nabla}_{\mathbf{e_a}}G)\boldsymbol{\tau_g} + \boldsymbol{\theta}^{\mathbf{a}}G(\boldsymbol{\nabla}_{\mathbf{e_a}}\boldsymbol{\tau_g}) + (\boldsymbol{\nabla}_{\mathbf{e_a}}G)\boldsymbol{\tau_g}\boldsymbol{\theta}^{\mathbf{a}} + G(\boldsymbol{\nabla}_{\mathbf{e_a}}\boldsymbol{\tau_g})\boldsymbol{\theta}^{a}) \\
&= -\frac{1}{2}\left(\boldsymbol{\theta}^{\mathbf{a}}(\boldsymbol{\nabla}_{\mathbf{e_a}}G) - (\boldsymbol{\nabla}_{\mathbf{e_a}}G)\boldsymbol{\theta}^{\mathbf{a}})\boldsymbol{\tau_g} - \frac{1}{2}(-\boldsymbol{\theta}^{\mathbf{a}}G(\boldsymbol{\nabla}_{\mathbf{e_a}}\boldsymbol{\tau_g}) - G(\boldsymbol{\nabla}_{\mathbf{e_a}}\boldsymbol{\tau_g})\boldsymbol{\theta}^{\mathbf{a}}\right).
\end{aligned}$$

But

$$-\frac{1}{2}(\boldsymbol{\theta}^{\mathbf{a}}(\boldsymbol{\nabla}_{\mathbf{e_a}}G) - (\boldsymbol{\nabla}_{\mathbf{e_a}}G)\boldsymbol{\theta}^{\mathbf{a}})\boldsymbol{\tau_g}) = -(\boldsymbol{\partial}\lrcorner G)\boldsymbol{\tau_g}$$

and

$$\boldsymbol{\nabla}_{\mathbf{e_a}}\boldsymbol{\tau_g} = \boldsymbol{\nabla}_{\mathbf{e_a}}(\boldsymbol{\theta}^0 \wedge \boldsymbol{\theta}^1 \wedge \boldsymbol{\theta}^2 \wedge \boldsymbol{\theta}^3) = -\Gamma^{\mathbf{0}\cdot\cdot}_{\cdot\mathbf{a0}}\boldsymbol{\tau_g} - \Gamma^{\mathbf{1}\cdot\cdot}_{\cdot\mathbf{a1}}\boldsymbol{\tau_g} - \Gamma^{\mathbf{2}\cdot\cdot}_{\cdot\mathbf{a2}}\boldsymbol{\tau_g} - \Gamma^{\mathbf{3}\cdot\cdot}_{\cdot\mathbf{a3}}\boldsymbol{\tau_g} = 0,$$

because

$$\Gamma^{\mathbf{0}\cdot\cdot}_{\cdot\mathbf{a0}} = \Gamma^{\mathbf{1}\cdot\cdot}_{\cdot\mathbf{a1}} = \Gamma^{\mathbf{1}\cdot\cdot}_{\cdot\mathbf{a1}} = \Gamma^{\mathbf{1}\cdot\cdot}_{\cdot\mathbf{a1}} = 0.$$

Then,

$$-(\boldsymbol{\partial}\lrcorner G)\boldsymbol{\tau_g} = -J\boldsymbol{\tau_g}$$

or

$$\boldsymbol{\partial}\lrcorner G = J. \tag{38}$$

Now,

$$\begin{aligned}
\boldsymbol{\partial}\lrcorner G &= \frac{1}{2}\vartheta^{\mu}\lrcorner(D_{\mu}G^{\alpha\beta}\vartheta_{\alpha} \wedge \vartheta_{\beta}) = D_{\mu}G^{\mu\beta}\vartheta_{\beta} \\
&= (\partial_{\mu}G^{\mu\beta} + \Gamma^{\mu\cdot\cdot}_{\cdot\mu\kappa}G^{\kappa\beta} + \Gamma^{\beta\cdot\cdot}_{\cdot\mu\kappa}G^{\mu\kappa})\vartheta_{\beta} \\
&= (\partial_{\mu}G^{\mu\beta} + \Gamma^{\mu\cdot\cdot}_{\cdot\mu\kappa}G^{\kappa\beta})\vartheta_{\beta} = (\partial_{\mu}G^{\mu\beta} + \frac{\partial_{\mu}\sqrt{|\det \boldsymbol{g}|}}{\sqrt{|\det \boldsymbol{g}|}}G^{\mu\beta})\vartheta_{\beta} \\
&= \frac{1}{\sqrt{|\det \boldsymbol{g}|}}\partial_{\mu}(\sqrt{|\det \boldsymbol{g}|}G^{\mu\beta})\vartheta_{\beta},
\end{aligned}$$

and recalling that in vacuum $G = F$ *we get*

$$\frac{1}{\sqrt{|\det \boldsymbol{g}|}}\partial_{\mu}(\sqrt{|\det \boldsymbol{g}|}F^{\mu\beta}) = J^{\beta}$$

which is Eq.(31) *again.*

---

[14]Recall that $\boldsymbol{\theta}^{\mathbf{a}}(\boldsymbol{e}_{\mathbf{b}}) = \delta^{\mathbf{a}}_{b}$. Also $\boldsymbol{g}(\boldsymbol{e}_{\mathbf{a}}, \boldsymbol{e}_{\mathbf{b}}) = \eta_{\mathbf{ab}}$ and $\mathrm{g}(\boldsymbol{\theta}^{\mathbf{a}}, \boldsymbol{\theta}^{\mathbf{b}}) = \eta^{\mathbf{ab}} = \boldsymbol{\theta}^{\mathbf{a}} \cdot \boldsymbol{\theta}^{\mathbf{b}} = \boldsymbol{\theta}^{\mathbf{a}} \lrcorner \boldsymbol{\theta}^{\mathbf{b}}$.

3. Maxwell Equations in Minkowski Spacetime

3.1. Vacuum Case

We next suppose that electromagnetic phenomena take place in a non dispersive material medium in Minkowski spacetime, i.e., the structure $M = \langle \mathbb{R}^4, \mathring{\boldsymbol{g}}, \mathring{\boldsymbol{\nabla}}, \boldsymbol{\tau}_{\mathring{\boldsymbol{g}}}, \uparrow \rangle$. Since the Riemann curvature of the Levi-Civita connection $\mathring{\boldsymbol{\nabla}}$ is null (i.e., $\mathring{\boldsymbol{R}}(\mathring{\boldsymbol{\nabla}}) = 0$) there exist *global* coordinates[15] $\langle \mathrm{x}^\mu \rangle$ such that denoting by $\langle \boldsymbol{e}_\mu = \boldsymbol{\partial}/\boldsymbol{\partial}\mathrm{x}^\mu \rangle$ a basis for TM and by $\langle \gamma^\mu = d\mathrm{x}^\mu \rangle$ the basis of T^*M dual to $\langle \boldsymbol{e}_\mu \rangle$ we have

$$\mathring{\boldsymbol{g}} = \eta_{\mu\nu}\gamma^\mu \otimes \gamma^\nu \tag{39}$$

where the matrix with entries $\eta_{\mu\nu}$ is the diagonal matrix $\mathrm{diag}(1, -1, -1, -1)$.

In this case the components of the constitutive (extensor) $\boldsymbol{\chi}$ of the vacuum in the (nacs$|I$) $\{\mathrm{x}^\mu\}$ to the inertial frame $\boldsymbol{L} = \boldsymbol{e}_0 = \boldsymbol{\partial}/\boldsymbol{\partial}\mathrm{x}^0$ are

$$\chi^{\rho\sigma\mu\nu} = (\eta^{\rho\mu}\eta^{\sigma\nu} - \eta^{\rho\nu}\eta^{\sigma\mu}). \tag{40}$$

Thus

$$\mathtt{G}^{\rho\sigma} = \frac{1}{2}(\eta^{\rho\mu}\eta^{\sigma\nu} - \eta^{\rho\nu}\eta^{\sigma\mu})\mathring{F}_{\mu\nu} = \eta^{\rho\mu}\eta^{\sigma\nu}\mathtt{F}_{\mu\nu}. \tag{41}$$

Denoting by $(\mathtt{F}_{\mu\nu})$ and $(\mathtt{G}^{\rho\sigma})$, respectively, the matrices with elements $\mathtt{F}_{\mu\nu}$ and $\mathtt{G}^{\rho\sigma}$ we have

$$(\mathtt{F}_{\mu\nu}) = \begin{pmatrix} 0 & \mathrm{E}_1 & \mathrm{E}_2 & \mathrm{E}_3 \\ -\mathrm{E}_1 & 0 & -\mathrm{B}_3 & \mathrm{B}_2 \\ -\mathrm{E}_2 & \mathrm{B}_3 & 0 & -\mathrm{B}_1 \\ -\mathrm{E}_3 & -\mathrm{B}_2 & \mathrm{B}_1 & 0 \end{pmatrix}, \qquad (\mathtt{G}^{\rho\sigma}) = \begin{pmatrix} 0 & -\mathrm{B}_1 & -\mathrm{B}_2 & -\mathrm{B}_3 \\ \mathrm{B}_1 & 0 & \mathrm{E}_3 & -\mathrm{E}_2 \\ \mathrm{B}_2 & -\mathrm{E}_3 & 0 & \mathrm{E}_1 \\ \mathrm{B}_3 & \mathrm{E}_2 & -\mathrm{E}_1 & 0 \end{pmatrix}.$$

Expressed in term of coordinates in the Einstein-Lorentz Poincaré gauge, Maxwell equations read

$$\frac{\partial \mathtt{F}_{\alpha\beta}}{\partial \mathrm{x}^\mu} + \frac{\partial \mathtt{F}_{\mu\alpha}}{\partial \mathrm{x}^\beta} + \frac{\partial \mathtt{F}_{\beta\mu}}{\partial \mathrm{x}^\alpha} = 0, \tag{42}$$

and

$$\frac{\partial \mathtt{F}^{\mu\nu}}{\partial \mathrm{x}^\mu} = \mathtt{J}^\nu. \tag{43}$$

3.2. The Non Dispersive Homogeneous Isotropic Linear Medium Case

We suppose in what follows that a linear non dispersive homogeneous and isotropic medium (**NDHILM**) is at rest in a given reference frame $\boldsymbol{V}$ in $U \subset M$. This reference frame has an *arbitrary* motion relative to the laboratory frame which is modelled here by the inertial frame $\boldsymbol{L} = \boldsymbol{e}_0 = \boldsymbol{\partial}/\boldsymbol{\partial}\mathrm{x}^0$. Let $\langle x'^\mu \rangle$ be a set of coordinate functions covering U

[15] These coordinates are said to be in Einstein-Lorentz-Poincaré gauge. Note that $\langle \mathrm{x}^\mu \rangle$ is a (nacs$|\boldsymbol{e}_0$) where $\boldsymbol{e}_0 = \boldsymbol{L}$ (the laboratory frame) is an inertial reference system. A reference system is called inertial if and only if $\mathring{\boldsymbol{\nabla}}\boldsymbol{e}_0 = 0$.

and such that $\langle x'^{\mu}\rangle$ are (nacs$|\boldsymbol{V}$). Let $\langle \boldsymbol{\partial}/\boldsymbol{\partial}x'^{\mu}\rangle$ be a basis for TU and $\langle \vartheta'^{\mu} = dx'^{\mu}\rangle$ be the corresponding dual basis for T^*M. Writing in this case

$$\mathring{\boldsymbol{g}} = \mathring{g}'_{\mu\nu}\vartheta'^{\mu}\otimes\vartheta'^{\nu} = \eta_{\mu\nu}\gamma^{\mu}\wedge\gamma^{\nu}, \tag{44}$$

we have

$$\boldsymbol{V} = V'^{\mu}\boldsymbol{\partial}/\boldsymbol{\partial}x'^{\mu} = \frac{1}{\sqrt{\mathring{g}'_{00}}}\boldsymbol{\partial}/\boldsymbol{\partial}x'^{0} = \mathtt{V}^{\mu}\boldsymbol{\partial}/\boldsymbol{\partial}\mathtt{x}^{\mu} = \mathtt{V}^{0}\frac{\boldsymbol{\partial}}{\boldsymbol{\partial}\mathtt{x}^{0}} + \mathtt{V}^{i}\frac{\boldsymbol{\partial}}{\boldsymbol{\partial}\mathtt{x}^{i}}. \tag{45}$$

We now write

$$\begin{aligned} F &= \frac{1}{2}F'_{\rho\sigma}\vartheta'^{\rho}\wedge\vartheta'^{\sigma} = \frac{1}{2}F'^{\rho\sigma}\vartheta'_{\rho}\wedge\vartheta'_{\sigma} \\ &= \frac{1}{2}\mathtt{F}_{\rho\sigma}\gamma^{\rho}\wedge\gamma^{\sigma} = \frac{1}{2}\mathtt{F}^{\rho\sigma}\gamma_{\rho}\wedge\gamma_{\sigma}, \\ G &= \frac{1}{2}G'_{\rho\sigma}\vartheta'^{\rho}\wedge\vartheta'^{\sigma} = \frac{1}{2}G'^{\rho\sigma}\vartheta'_{\rho}\wedge\vartheta'_{\sigma} \\ &= \frac{1}{2}\mathtt{G}_{\rho\sigma}\gamma^{\rho}\wedge\gamma^{\sigma} = \frac{1}{2}\mathtt{G}^{\rho\sigma}\gamma_{\rho}\wedge\gamma_{\sigma}. \end{aligned} \tag{46}$$

We use the following notation for the elements of the matrices $(\mathtt{F}_{\rho\sigma})$, $(F'_{\rho\sigma})$, $(\mathtt{G}_{\rho\sigma})$ and $(G'_{\rho\sigma})$:

$$(\mathtt{F}_{\rho\sigma}) = \begin{pmatrix} 0 & \mathtt{E}_1 & \mathtt{E}_2 & \mathtt{E}_3 \\ -\mathtt{E}_1 & 0 & -\mathtt{B}_3 & \mathtt{B}_2 \\ -\mathtt{E}_2 & \mathtt{B}_3 & 0 & -\mathtt{B}_1 \\ -\mathtt{E}_3 & -\mathtt{B}_2 & \mathtt{B}_1 & 0 \end{pmatrix}, \quad (F'_{\rho\sigma}) = \begin{pmatrix} 0 & E'_1 & E'_2 & E'_3 \\ -E'_1 & 0 & -B'_3 & B'_2 \\ -E'_2 & B'_3 & 0 & -B'_1 \\ -E'_3 & -B'_2 & B'_1 & 0 \end{pmatrix}, \tag{47}$$

$$(\mathtt{G}_{\rho\sigma}) = \begin{pmatrix} 0 & \mathtt{D}_1 & \mathtt{D}_2 & \mathtt{D}_3 \\ -\mathtt{D}_1 & 0 & -\mathtt{H}_3 & \mathtt{H}_2 \\ -\mathtt{D}_2 & \mathtt{H}_3 & 0 & -\mathtt{H}_1 \\ -\mathtt{D}_3 & -\mathtt{H}_2 & \mathtt{H}_1 & 0 \end{pmatrix}, \quad (G'_{\rho\sigma}) = \begin{pmatrix} 0 & D'_1 & D'_2 & D'_3 \\ -D'_1 & 0 & -H'_3 & H'_2 \\ -D'_2 & H'_3 & 0 & -H'_1 \\ -D'_3 & -H'_2 & H'_1 & 0 \end{pmatrix}. \tag{48}$$

Then in the coordinate system $\langle x'^{\mu}\rangle$ (which is a (nacs$|\boldsymbol{V}$)) Maxwell equations are written

$$\frac{\partial F'_{\alpha\beta}}{\partial x'^{\mu}} + \frac{\partial F'_{\mu\alpha}}{\partial x'^{\beta}} + \frac{\partial F'_{\beta\mu}}{\partial x'^{\alpha}} = 0,$$

and

$$\frac{1}{\sqrt{|\det\mathring{\boldsymbol{g}}|}}\frac{\partial}{\partial x'^{\alpha}}\left(\frac{1}{\sqrt{|\det\mathring{\boldsymbol{g}}|}}\mathring{g}'^{\alpha\rho}\mathring{g}'^{\beta\sigma}G'_{\rho\sigma}\right) = J'^{\beta}.$$

3.3. Minkowski Relations

We now want to define, using differential forms, the concepts of electric and magnetic fields and induction fields in a **NDHILM.**

Given an arbitrary reference frame $\boldsymbol{Z}$ in Minkowski spacetime, let $z = \mathring{\boldsymbol{g}}(\boldsymbol{Z},)$ be the physically equivalent 1-form field[1].

[1]We will also call z a reference frame.

Definition 4. *The electric* $\underset{z}{\boldsymbol{E}}$ *and magnetic* $\underset{z}{\boldsymbol{B}}$ 1*-form fields and the* $\underset{z}{\boldsymbol{D}}$ *and* $\underset{z}{\boldsymbol{H}}$ 1*-form fields, according to the observers at rest in* $\boldsymbol{Z}$*, are given by*[2]*:*

$$\underset{z}{\boldsymbol{E}} := \boldsymbol{z} \lrcorner F, \qquad \underset{z}{\boldsymbol{B}} := \boldsymbol{z} \lrcorner \star F, \tag{49}$$

$$\underset{z}{\boldsymbol{D}} := \boldsymbol{z} \lrcorner G, \qquad \underset{z}{\boldsymbol{H}} := \boldsymbol{z} \lrcorner \star G. \tag{50}$$

We immediately have

$$F = \boldsymbol{z} \wedge \underset{z}{\boldsymbol{E}} - \star(\boldsymbol{z} \wedge \underset{z}{\boldsymbol{B}}), \tag{51}$$

$$G = \boldsymbol{z} \wedge \underset{z}{\boldsymbol{D}} - \star(\boldsymbol{z} \wedge \underset{z}{\boldsymbol{H}}). \tag{52}$$

Let $\boldsymbol{V}$ be an arbitrary reference frame in Minkowski spacetime ($\boldsymbol{v} = \mathring{g}(\boldsymbol{V},\,)$) such that the **NDHILM** is at rest in this reference frame.

Definition 5. *The electric* $\underset{v}{\boldsymbol{D}}$ *and the magnetic* $\underset{v}{\boldsymbol{H}}$ 1*-form fields, according to the observers at rest in* $\boldsymbol{V}$*, are related to* $\underset{v}{\boldsymbol{E}}$ *and* $\underset{v}{\boldsymbol{B}}$ *by:*

$$\underset{v}{\boldsymbol{D}} := \varepsilon \underset{v}{\boldsymbol{E}} \qquad \textit{and} \qquad \underset{v}{\boldsymbol{H}} := \frac{1}{\boldsymbol{\mu}} \underset{v}{\boldsymbol{B}}. \tag{53}$$

We immediately have

$$G = \frac{1}{\boldsymbol{\mu}} (\boldsymbol{\varepsilon\mu} - 1) \left[\boldsymbol{v} \wedge (\boldsymbol{v} \lrcorner F)\right] + \frac{1}{\boldsymbol{\mu}} F. \tag{54}$$

which will be called *Minkowski constitutive relation.*

The proof of Eq.(54) is as follows. From Eq.(52) and Definition 5 we have

$$G = \varepsilon \boldsymbol{v} \wedge (\boldsymbol{v} \lrcorner F) - \frac{1}{\mu} \star [\boldsymbol{v} \wedge (\boldsymbol{v} \lrcorner \star F)]. \tag{55}$$

Now using the identities Eq.(137) and Eqs.(138) shown in the Appendix we have

$$\begin{aligned} \star[\boldsymbol{v} \wedge (\boldsymbol{v} \lrcorner \star F)] &= \star[\boldsymbol{v} \wedge \star(\boldsymbol{v} \wedge F)] = \star\star [\boldsymbol{v} \lrcorner (\boldsymbol{v} \wedge F)] = -\boldsymbol{v} \lrcorner (\boldsymbol{v} \wedge F) \\ &= -(\boldsymbol{v} \lrcorner \boldsymbol{v}) F + \boldsymbol{v} \wedge (\boldsymbol{v} \lrcorner F) \end{aligned}$$

and thus

$$\begin{aligned} G &= \varepsilon \boldsymbol{v} \wedge (\boldsymbol{v} \lrcorner F) + \frac{1}{\mu}[F - \boldsymbol{v} \wedge (\boldsymbol{v} \lrcorner F)] \\ &= \frac{1}{\mu} (\varepsilon\mu - 1) \left[\boldsymbol{v} \wedge (\boldsymbol{v} \lrcorner F)\right] + \frac{1}{\mu} F. \end{aligned}$$

[2]The symbol $\star$ denotes the Hodge star operator in Minkowski spacetime.

3.3.1. Coordinate Expression for Minkowski Constitutive Relations

We now want to express Minkowski constitutive relations in arbitrary coordinates $\{x^\mu\}$ covering $U \subset M$. Let $F_{\lambda\mu}$, $G_{\mu\nu}$ and v^μ be the components of F, G and the reference frame $\boldsymbol{V}$ in the arbitrary natural bases $\{\boldsymbol{\partial}_\mu = \boldsymbol{\partial}/\boldsymbol{\partial} x^\mu\}$ and $\{\vartheta^\mu = dx^\mu\}$. Let us calculate $\boldsymbol{v}\lrcorner G$. We have

$$\boldsymbol{v}\lrcorner G = \left(\varepsilon - \frac{1}{\mu}\right)\boldsymbol{v}\lrcorner[\boldsymbol{v} \wedge (\boldsymbol{v}\lrcorner F)] + \frac{1}{\mu}\boldsymbol{v}\lrcorner F.$$

From the identity expressed by Eq.(137) (Appendix) we have

$$\begin{aligned}\boldsymbol{v}\lrcorner[\boldsymbol{v} \wedge (\boldsymbol{v}\lrcorner F)] &= (\boldsymbol{v}\lrcorner\boldsymbol{v})(\boldsymbol{v}\lrcorner F)] - \boldsymbol{v} \wedge (\boldsymbol{v}\lrcorner(\boldsymbol{v}\lrcorner F)) \\ &= \boldsymbol{v}\lrcorner F - \boldsymbol{v} \wedge [(\boldsymbol{v} \wedge \boldsymbol{v})\lrcorner F] \\ &= \boldsymbol{v}\lrcorner F. \end{aligned} \tag{56}$$

Then

$$\boldsymbol{v}\lrcorner G = \boldsymbol{\varepsilon}\boldsymbol{v}\lrcorner F. \tag{57}$$

In components, Eq.(57) is written

$$v^\mu \vartheta_\mu \lrcorner \left(\frac{1}{2} G_{\alpha\beta}\vartheta^\alpha \wedge \vartheta^\beta\right) = \varepsilon v^\mu \vartheta_\mu \lrcorner \left(\frac{1}{2} F_{\alpha\beta}\vartheta^\alpha \wedge \vartheta^\beta\right)$$

and since $\vartheta_\mu \lrcorner \left(\frac{1}{2} G_{\alpha\beta}\vartheta^\alpha \wedge \vartheta^\beta\right) = G_{\mu\nu}\vartheta^\nu$ and $\vartheta_\mu \lrcorner \left(\frac{1}{2} F_{\alpha\beta}\vartheta^\alpha \wedge \vartheta^\beta\right) = F_{\mu\nu}\vartheta^\nu$ we have

$$v^\mu G_{\mu\nu} = \varepsilon v^\mu F_{\mu\nu}. \tag{58}$$

Let us now calculate $\boldsymbol{v}\lrcorner \star G$. First we note that

$$\star G = \left(\boldsymbol{\varepsilon} - \frac{1}{\boldsymbol{\mu}}\right) \star [\boldsymbol{v} \wedge (\boldsymbol{v}\lrcorner F)] + \frac{1}{\boldsymbol{\mu}} \star F.$$

Using the same steps which led to Eq.(57) we first get

$$\boldsymbol{v}\lrcorner \star [\boldsymbol{v} \wedge (\boldsymbol{v}\lrcorner F)] = \star[\boldsymbol{v} \wedge (\boldsymbol{v} \wedge (\boldsymbol{v}\lrcorner F))] = \star(\boldsymbol{v} \wedge \boldsymbol{v}) \wedge (\boldsymbol{v}\lrcorner F) = 0$$

and then

$$\boldsymbol{v}\lrcorner \star G = \frac{1}{\boldsymbol{\mu}}\boldsymbol{v}\lrcorner \star F. \tag{59}$$

Now,

$$\begin{aligned}\star F &= \frac{1}{2}\left(\frac{1}{2}\sqrt{|\det \mathring{\boldsymbol{g}}|} F^{\mu\nu}\varepsilon_{\mu\nu\rho\sigma}\vartheta^\rho \wedge \vartheta^\sigma\right), \\ \star G &= \frac{1}{2}\left(\frac{1}{2}\sqrt{|\det \mathring{\boldsymbol{g}}|} G^{\mu\nu}\varepsilon_{\mu\nu\rho\sigma}\vartheta^\rho \wedge \vartheta^\sigma\right)\end{aligned}$$

and so

$$\begin{aligned}\boldsymbol{v}\lrcorner \star F &= \frac{1}{2}\sqrt{|\det \mathring{\boldsymbol{g}}|} v^\kappa F^{\mu\nu}\varepsilon_{\mu\nu\kappa\sigma}\vartheta^\sigma, \\ \boldsymbol{v}\lrcorner \star G &= \frac{1}{2}\sqrt{|\det \mathring{\boldsymbol{g}}|} v^\kappa G^{\mu\nu}\varepsilon_{\mu\nu\kappa\sigma}\vartheta^\sigma.\end{aligned}$$

Finally, Eq.(59) reads in components

$$\mu v^\kappa G^{\mu\nu}\varepsilon_{\mu\nu\kappa\sigma} = v^\kappa F^{\mu\nu}\varepsilon_{\mu\nu\kappa\sigma} \tag{60}$$

or in equivalent forms

$$\begin{aligned} \mu v_\kappa G_{\mu\nu}\varepsilon^{\mu\nu\kappa\sigma} &= v_\kappa F_{\mu\nu}\varepsilon^{\mu\nu\kappa\sigma}, \\ \mu\varepsilon^{\mu\nu\kappa\sigma}G_{\mu\nu}\mathring{g}_{\kappa\iota}v^\iota &= \varepsilon^{\mu\nu\kappa\sigma}F_{\mu\nu}\mathring{g}_{\kappa\iota}v^\iota. \end{aligned} \tag{61}$$

3.3.2. Polarization, Magnetization, Bound Current and Bound Charge Fields

We define the *polarization 2-form field* Π by

$$\Pi : F - G. \tag{62}$$

Given an arbitrary frame $\boldsymbol{Z}$ we decompose Π as

$$\Pi := \boldsymbol{z} \wedge \underset{\boldsymbol{z}}{\boldsymbol{P}} - \star(\boldsymbol{z} \wedge \underset{\boldsymbol{z}}{\boldsymbol{M}}), \tag{63}$$

where $\underset{\boldsymbol{z}}{\boldsymbol{P}}$ and $\underset{\boldsymbol{z}}{\boldsymbol{M}}$ are called respectively the polarization 1-form field and magnetization 1-form field in $\boldsymbol{Z}$. Moreover, from Eq.(53) and Eq.(54) we get

$$\underset{\boldsymbol{z}}{\boldsymbol{P}} = \underset{\boldsymbol{z}}{\boldsymbol{E}} - \underset{\boldsymbol{z}}{\boldsymbol{D}} \text{ and } \underset{\boldsymbol{z}}{\boldsymbol{M}} = \underset{\boldsymbol{z}}{\boldsymbol{B}} - \underset{\boldsymbol{z}}{\boldsymbol{H}}.$$

From the non homogeneous Maxwell equation $d \star G = -\mathbf{J}$ we have

$$d \star F = -\mathbf{J} + d \star \Pi. \tag{64}$$

The field

$$\mathcal{J} := -d \star \Pi \tag{65}$$

is called the *bound current* 3-form. Given an arbitrary frame $\boldsymbol{Z}$ we decompose $\mathcal{J}$ as

$$\mathcal{J} := \boldsymbol{z} \wedge \underset{\boldsymbol{z}}{\mathcal{J}} + \underset{\boldsymbol{z}}{\rho}, \tag{66}$$

where $\underset{\boldsymbol{z}}{\mathcal{J}}$ is called the *bound current* 2-form field and $\underset{\boldsymbol{z}}{\rho}$ is called the *bound charge* 3-form field.

In the coordinate systems $\langle \mathrm{x}^\mu \rangle$ and $\langle x'^\mu \rangle$ respectively adapted to the inertial laboratory frame $\boldsymbol{L} = \boldsymbol{e}_0$ and to an arbitrary frame $\boldsymbol{V}$ we write

$$\begin{aligned} \Pi &= \frac{1}{2}\Pi_{\rho\sigma}\gamma^\rho \wedge \gamma^\sigma = \frac{1}{2}\Pi^{\rho\sigma}\gamma_\rho \wedge \gamma_\sigma \\ &= \frac{1}{2}\Pi'_{\rho\sigma}\vartheta'^\rho \wedge \vartheta'^\sigma = \frac{1}{2}\Pi'^{\rho\sigma}\vartheta'_\rho \wedge \vartheta'_\sigma, \end{aligned} \tag{67}$$

where the entries of matrices $(\Pi_{\rho\sigma})$ and $(\Pi'_{\rho\sigma})$ are denoted as usual $(\vec{P}, \vec{M})$ and $(\vec{P}', \vec{M}')$ and are easily calculated from Eq.(63) with $\boldsymbol{Z} = \boldsymbol{L}$ and $\boldsymbol{Z} = \boldsymbol{V}$, respectively.

3.4. Constitutive Relations in a Uniformly Rotating Frame $\boldsymbol{V}$

We introduce, in addtion to coordinates $\{\mathrm{x}^\mu\}$, a set of *cylindrical coordinates* $\langle x^\mu \rangle$ in $\mathbb{R}^4 - \{0\}$ which are also (nacs$|\boldsymbol{L} = \boldsymbol{e}_0$). As usual we write

$$\begin{aligned} \mathrm{x}^0 &= x^0, \\ \mathrm{x}^1 &= x^1 \cos x^2, \\ \mathrm{x}^2 &= x^1 \sin x^2, \\ \mathrm{x}^3 &= x^3. \end{aligned} \tag{68}$$

We will simplify the notation, when convenient, by writing $\{\mathrm{x}^0, \mathrm{x}^1, \mathrm{x}^2, \mathrm{x}^3\} := \{t, \mathrm{x}, \mathrm{y}, \mathrm{z}\}$ and $\{x^0, x^1, x^2, x^3\} := \{t, r, \phi, z\}$.

Next we introduce a particular *rotating* reference frame $\boldsymbol{V} \in \sec TM$ in Minkowski spacetime:

$$\boldsymbol{V} = \frac{1}{\sqrt{1-\mathrm{v}^2}} \frac{\boldsymbol{\partial}}{\boldsymbol{\partial} t} + \frac{\omega}{\sqrt{1-\mathrm{v}^2}} \left(-\mathrm{y} \frac{\boldsymbol{\partial}}{\boldsymbol{\partial} \mathrm{x}} + \mathrm{x} \frac{\boldsymbol{\partial}}{\boldsymbol{\partial} \mathrm{y}} \right) \tag{69}$$

where

$$\mathrm{v} := \sqrt{\mathrm{v}_\mathrm{x}^2 + \mathrm{v}_\mathrm{y}^2} = \omega r, \tag{70}$$

with ω the (classical) angular velocity of $\boldsymbol{V}$ relative to the inertial laboratory frame $\boldsymbol{L} = \boldsymbol{e}_0$.

Since

$$\frac{\boldsymbol{\partial}}{\boldsymbol{\partial} \phi} = -y \frac{\boldsymbol{\partial}}{\boldsymbol{\partial} x} + x \frac{\boldsymbol{\partial}}{\boldsymbol{\partial} y} \tag{71}$$

we can also write, taking into account that $\gamma = 1/\sqrt{1-\mathrm{v}^2}$,

$$\begin{aligned} \boldsymbol{V} &= \frac{1}{\sqrt{1-\mathrm{v}^2}} \frac{\boldsymbol{\partial}}{\boldsymbol{\partial} t} + \frac{\omega}{\sqrt{1-\mathrm{v}^2}} \frac{\boldsymbol{\partial}}{\boldsymbol{\partial} \phi} \\ &= \gamma \left(\boldsymbol{\partial}_t + \mathrm{v} \frac{1}{r} \boldsymbol{\partial}_\phi \right) \end{aligned} \tag{72}$$

and

$$\boldsymbol{v} := \mathring{\boldsymbol{g}}(\boldsymbol{V}, \ ) = \gamma \left(dt + \mathrm{v} r d\phi \right). \tag{73}$$

Remark 6. *It is important to recall that although coordinates $\{t, r, \phi, z\}$ cover $\mathbb{R}^4 - \{0\}$ the reference frame $\boldsymbol{V}$ if realized by a physical system can only have material support for $r < 1/\omega$.*

We next introduce the vector field

$$\begin{aligned} \mathbf{v} &= -\omega \mathrm{y} \boldsymbol{\partial}_\mathrm{x} + \omega \mathrm{x} \boldsymbol{\partial}_\mathrm{y} \\ &= \mathrm{v} \frac{1}{r} \boldsymbol{\partial}_\phi = \mathrm{v} \boldsymbol{e}_\phi \end{aligned} \tag{74}$$

It is quite obvious that the vector field $\mathbf{v}$ represents the classical 3-velocity of a (material) point whose 3-dimensional trajectory in the spacelike section $\mathbb{R}^3$ determined by $\boldsymbol{L} = \boldsymbol{e}_0$ has parametric equations $(r \cos \omega t, r \sin \omega t, z_0)$, where z_0 is an arbitrary real constant.

Observe that in engineering notation the vector field $\mathbf{v}$ is usually denoted by $\vec{v} = \mathrm{v} \hat{\mathbf{e}}_\phi$. We will use engineering notation when convenient in some of the formulas that follow, for pedagogical reasons.

A (nacs$|\boldsymbol{V}$) Before we proceed we introduce explicitly the transformation law between coordinates $\langle \mathrm{x}^{\mu}\rangle$, which are (nacs$|\boldsymbol{e}_0$) and $\langle x'^{\mu}\rangle$, which are (nacs$|\boldsymbol{V}$). They are relatec by:

$$
\begin{aligned}
\mathrm{x}^0 &= x'^0, \\
\mathrm{x}^1 &= x'^1 \cos\omega\mathrm{x}^0 - x'^2 \sin\omega\mathrm{x}^0, \\
\mathrm{x}^2 &= x'^1 \sin\omega\mathrm{x}^0 + x'^2 \cos\omega\mathrm{x}^0. \\
\mathrm{x}^3 &= x'^3.
\end{aligned} \tag{75}
$$

To simplify the notation we will write $\{x'^0, x'^1, x'^2, x'^3) := (t', x', y', z')$.

We also define cylindrical coordinates $\langle \hat{x}'^{\mu}\rangle$ naturally adapted to $\boldsymbol{V}$ by

$$
\begin{aligned}
x'^0 &= \hat{x}'^0, \\
x'^1 &= \hat{x}'^1 \cos\hat{x}'^2, \\
x'^2 &= \hat{x}'^1 \sin\hat{x}'^2, \\
x'^3 &= \hat{x}'^3.
\end{aligned} \tag{76}
$$

We again simplefy the notation by writing $\{\hat{x}'^0, \hat{x}'^1, \hat{x}'^2, \hat{x}'^3) := (t', r', \phi', z')$.

Now, it is trivial to see that

$$
t' = t, \quad r' = r, \quad \phi' = \phi - \omega t, \quad z' = \mathrm{z}. \tag{77}
$$

This has as consequence the obvious relations

$$
dt' = dt, \quad dr' = dr, \quad d\phi' = d\phi - \omega dt, \quad dz' = d\mathrm{z} \tag{78}
$$

and the not so obvious ones

$$
\frac{\partial}{\partial t'} = \frac{\partial}{\partial t} + \omega\frac{\partial}{\partial \phi}, \quad \frac{\partial}{\partial r'} = \frac{\partial}{\partial r}, \quad \frac{\partial}{\partial \phi'} = \frac{\partial}{\partial \phi}, \quad \frac{\partial}{\partial z'} = \frac{\partial}{\partial \mathrm{z}}. \tag{79}
$$

We then see that

$$
\boldsymbol{V} = \frac{1}{\sqrt{1-\mathrm{v}^2}}\frac{\partial}{\partial t'} = \frac{1}{\sqrt{\mathring{g}_{00}}}\frac{\partial}{\partial t'}, \tag{80}
$$

which shows that $\langle \hat{x}'^{\mu}\rangle$ is indeed a (nacs$|\boldsymbol{Z}$).

We also notice that since $r' = r$ and $\frac{\partial}{\partial\phi'} = \frac{\partial}{\partial\phi}$ we can write Eq.(74) as[3]

$$
\begin{aligned}
\mathbf{v} &= \omega\mathrm{v}\frac{1}{r}\frac{\partial}{\partial\phi} = \omega\mathrm{v}\frac{1}{r'}\frac{\partial}{\partial\phi'} = -\omega y'\frac{\partial}{\partial x'} + \omega x'\frac{\partial}{\partial y'} \\
&= v_{x'}\frac{\partial}{\partial x'} + v_{y'}\frac{\partial}{\partial y'} \\
&= v_1'\frac{\partial}{\partial x'^1} + v_2'\frac{\partial}{\partial x'^2}.
\end{aligned} \tag{81}
$$

[3]Paying due attention to the notation used.

We write the metric $\mathring{\boldsymbol{g}}$ in terms of coordinates $\langle x^{\mu}\rangle$ and $\langle \hat{x}^{\mu}\rangle$ as

$$\begin{aligned}\mathring{\boldsymbol{g}} &= \mathring{g}'_{\mu\nu}dx'^{\mu}\otimes dx'^{\nu}\\ &= (1-\omega^2r'^2)dt'\otimes dt' + 2\omega y' dx'\otimes dt' - 2\omega x' dy\otimes dt' - dx'\otimes dx' - dy'\otimes dy'\\ &- dz'\otimes dz', \end{aligned} \tag{82}$$

or

$$\begin{aligned}\mathring{\boldsymbol{g}} &= \hat{g}_{\mu\nu}d\hat{x}'^{\mu}\otimes d\hat{x}'^{\nu}\\ &= (1-\mathrm{v}^2)dt\otimes dt - 2\omega d\phi'\otimes dt - dr'\otimes dr' - r'^2 d\phi'\otimes d\phi' - dz'\otimes dz'.\end{aligned} \tag{83}$$

We immediately see from Eq.(82) that

$$\mathring{g}'_{00} = (1-\mathrm{v}^2) = 1/\gamma^2,\ \ \mathring{g}'_{10} = -v'_1 = \omega y',\ \ \mathring{g}'_{20} = -v'_2 = \omega x',\ \ \mathring{g}'_{30} = 0. \tag{84}$$

After this (long) digression we return to Eq.(58) and Eq.(61) which, in the (nacs$|\boldsymbol{V}$) coordinate system $\langle x'^{\mu}\rangle$ can be immediately written, in engineering notation —the notation of vector calculus—, as:

$$\vec{D}' = \boldsymbol{\varepsilon}\vec{E}', \tag{85}$$

$$\boldsymbol{\mu}[\frac{1}{\gamma^2}\vec{H}' - \vec{v}\times\vec{D}'] = [\frac{1}{\gamma^2}\vec{B}' - \vec{v}\times\vec{E}']. \tag{86}$$

We finally get

$$\vec{D}' = \boldsymbol{\varepsilon}\vec{E}', \tag{87}$$

$$\vec{H}' = \frac{1}{\boldsymbol{\mu}}[\vec{B}' + \gamma^2(\boldsymbol{\varepsilon}\boldsymbol{\mu}-1)\vec{v}\times\vec{E}']. \tag{88}$$

In engineering notation the *polarization* and *magnetization* vector fields are, respectively,

$$\vec{P}' = (\boldsymbol{\varepsilon}-1)\vec{E}', \tag{89}$$

$$\vec{M}' = (1-\frac{1}{\boldsymbol{\mu}})\vec{B}' + \frac{1}{\boldsymbol{\mu}}\gamma^2(1-\boldsymbol{\varepsilon}\boldsymbol{\mu})\vec{v}\times\vec{E}'. \tag{90}$$

To obtain the expression of those fields in the laboratory (in coordinates $\langle \mathrm{x}^{\mu}\rangle$ naturally adapted to $\boldsymbol{l} = \boldsymbol{e}_0 = \boldsymbol{\partial}/\boldsymbol{\partial}\mathrm{x}^0 = \boldsymbol{\partial}/\boldsymbol{\partial}t$) it is only necessary to recall that

$$F'_{\mu\nu} = \frac{\partial \mathrm{x}^{\alpha}}{\partial x'^{\mu}}\frac{\partial \mathrm{x}^{\beta}}{\partial x'^{\nu}}\mathrm{F}_{\alpha\beta},\ \ H'_{\mu\nu} = \frac{\partial \mathrm{x}^{\alpha}}{\partial x'^{\mu}}\frac{\partial \mathrm{x}^{\beta}}{\partial x'^{\nu}}\mathrm{H}_{\alpha\beta},\ \ \Pi'_{\mu\nu} = \frac{\partial \mathrm{x}^{\alpha}}{\partial x'^{\mu}}\frac{\partial \mathrm{x}^{\beta}}{\partial x'^{\nu}}\Pi_{\alpha\beta}. \tag{91}$$

Now, from Eq.(75) we have

$$\left(\frac{\partial \mathrm{x}^{\alpha}}{\partial x'^{\mu}}\right) = \begin{pmatrix} 1 & 0 & 0 & 0\\ -\omega \mathrm{y} & \cos\omega t' & -\sin\omega t' & 0\\ \omega \mathrm{x} & \sin\omega t' & \cos\omega t' & 0\\ 0 & 0 & 0 & 1\end{pmatrix}. \tag{92}$$

Thus we get:

$$\begin{aligned}
E'_x &= \mathrm{E_x}\cos\omega t + \mathrm{E_y}\sin\omega t + \omega x'\mathrm{B_z},\\
E'_y &= -\mathrm{E_x}\sin\omega t + \mathrm{E_y}\cos\omega t + \omega y'\mathrm{B_z},\\
E'_z &= \mathrm{E_z} - \omega \mathrm{y}\mathrm{B_y} - \omega \mathrm{x}\mathrm{B_x},\\
B'_x &= \mathrm{B_x}\cos\omega t + \mathrm{B_y}\sin\omega t,\\
B'_y &= -\mathrm{B_x}\sin\omega t + \mathrm{B_y}\cos\omega t,\\
B'_z &= \mathrm{B_z}. \qquad (93)
\end{aligned}$$

We recall also that writing the position vector of a point $p = (\mathrm{x}, \mathrm{y}, \mathrm{z})$ at time t in the laboratory frame, in engineering notation, as $\tilde{\mathrm{x}} = \mathrm{x}\hat{\mathrm{e}}_\mathrm{x} + \mathrm{y}\hat{\mathrm{e}}_\mathrm{y} + \mathrm{z}\hat{\mathrm{e}}_\mathrm{z}$ and the (3-dimensional) angular velocity *field* of frame $\boldsymbol{V}$ as $\vec{\omega} = \Omega\hat{\mathrm{e}}_\mathrm{z}$ we have

$$\vec{v} = \vec{\Omega} \times \tilde{\mathrm{x}}. \qquad (94)$$

We finally write the relation between the charge and current densities as observed in the laboratory and in the rotating frame. From $J = J'^{\mu}\vartheta'_{\mu} = J^{\mu}\gamma_{\mu}$, with $(J^0, J^1, J^2, J^3) = (\rho, j_x, j_y, j_z)$ and $(J'^0, J'^1, J'^2, J'^3) = (\rho', j'_x, j'_y, j'_z)$ we have

$$J'^{\mu} = \frac{\partial x'^{\mu}}{\partial x^{\alpha}} J^{\alpha}, \qquad (95)$$

which gives

$$\rho' = \rho, \qquad (96)$$
$$j'_x = \omega y'\rho + \mathrm{j_x}\cos\omega t + \mathrm{j_y}\sin\omega t, \qquad (97)$$
$$j'_y = -\omega x'\rho - \mathrm{j_x}\sin\omega t + \mathrm{j_y}\cos\omega t, \qquad (98)$$
$$j'_z = \mathrm{j_z}. \qquad (99)$$

4. Jump Conditions for Fields F and G at the Boundary of a Moving NDHILM

The interface between a *moving* **NDHILM** described by the velocity field $\boldsymbol{V}$ and the vacuum defines a hypersurface $\Xi = 0$ in Minkowski spacetime. The jump conditions can be deduced from Maxwell equations and are well known. A very simple deduction of the discontinuities of fields $\vec{E}$, $\vec{B}$, $\vec{D}$ and $\vec{H}$ is given in [10]. Here we write the jump conditions in differential form for the case in which the free current $\star J = 0$. Denoting by $[F]$, $[G]$ and $[\star G]$ the respective discontinuities of F, G and $\star G$ at the boundary between medium and vacuum we have:

$$[F] \wedge d\Xi|_{\Xi=0} = 0, \qquad (100)$$
$$[G] \llcorner d\Xi|_{\Xi=0} = 0. \qquad (101)$$

Observe that $[G]\llcorner d\Xi|_{\Xi=0} = 0$ implies, of course,

$$\star[[G]\llcorner d\Xi|_{\Xi=0}] = -\star[[G]\lrcorner d\Xi|_{\Xi=0}] = [\star G] \wedge d\Xi|_{\Xi=0} = 0.$$

Besides, $[F]\wedge d\Xi|_{\Xi=0} = d\Xi\wedge[F]|_{\Xi=0} = \star(d\Xi\lrcorner\star F)|_{\Xi=0} = 0$ so that we can also write the jump conditions as:

$$[\star F]\llcorner d\Xi|_{\Xi=0} = 0, \tag{102}$$

$$[\star G]\wedge d\Xi|_{\Xi=0} = 0. \tag{103}$$

Now, in the (nacs$|\boldsymbol{L}=\boldsymbol{e}_0$) $\langle \mathrm{x}^\mu\rangle$ we have

$$n := d\Xi = \frac{\partial\Xi}{\partial \mathrm{x}^\mu}\boldsymbol{\vartheta}^\mu = n_\mu\boldsymbol{\vartheta}^\mu = n^\mu\boldsymbol{\vartheta}_\mu. \tag{104}$$

Define[4] $\mathbf{n} := -\frac{\partial\Xi}{\partial \mathrm{x}^i}\frac{\boldsymbol{\partial}}{\boldsymbol{\partial}\mathbf{x}^i}$ the spatial vector field which is normal to the moving boundary $\Xi(\mathtt{t},\mathtt{x},\mathtt{y},\mathtt{z})=0$. Now, each spatial point of the moving boundary at time $\mathtt{t}$ has Newtonian velocity $\mathbf{v}$. Observe that during a time interval $\triangle\mathrm{x}^0 = \triangle\mathtt{t}$ a point at the boundary with arbitrary coordinates $(\mathtt{x},\mathtt{y},\mathtt{z})$ will arrive at the point $(\mathtt{x}+v_\mathtt{x}\triangle\mathtt{t},\mathtt{y}+v_\mathtt{y}\triangle\mathtt{t},\mathtt{z}+v_\mathtt{z}\triangle\mathtt{t})$. The hypersurface Ξ at $(\mathtt{t}+\triangle\mathtt{t},\mathtt{x}+v_\mathtt{x}\triangle\mathtt{t},\mathtt{y}+v_\mathtt{y}\triangle\mathtt{t},\mathtt{z}+v_\mathtt{z}\triangle\mathtt{t})$ will satisfy

$$\begin{aligned}
&\Xi(\mathtt{t}+\triangle\mathtt{t},\mathtt{x}+v_\mathtt{x}\triangle\mathtt{t},\mathtt{y}+v_\mathtt{y}\triangle\mathtt{t},\mathtt{z}+v_\mathtt{z}\triangle\mathtt{t})\\
&=\Xi(\mathtt{t},\mathtt{x},\mathtt{y},\mathtt{z}) + \left.\frac{\partial\Xi}{\partial\mathrm{x}^\mu}\right|_{(\mathtt{t},\mathtt{x},\mathtt{y},\mathtt{z})}\triangle\mathrm{x}^\mu = 0\\
&=\triangle\mathtt{t}\left.\frac{\partial\Xi}{\partial\mathtt{t}}\right|_{(\mathtt{t},\mathtt{x},\mathtt{y},\mathtt{z})} + v_\mathtt{x}\triangle\mathtt{t}\left.\frac{\partial\Xi}{\partial\mathtt{x}}\right|_{(\mathtt{t},\mathtt{x},\mathtt{y},\mathtt{z})} + v_\mathtt{y}\triangle\mathtt{t}\left.\frac{\partial\Xi}{\partial\mathtt{y}}\right|_{(\mathtt{t},\mathtt{x},\mathtt{y},\mathtt{z})} + v_\mathtt{z}\triangle\mathtt{t}\left.\frac{\partial\Xi}{\partial\mathtt{z}}\right|_{(\mathtt{t},\mathtt{x},\mathtt{y},\mathtt{z})} = 0.
\end{aligned} \tag{105}$$

Thus we get, in engineering notation, taking into account that $\vec{n} = -\nabla\Xi$, that Eq.(105) implies

$$\frac{\partial\Xi}{\partial\mathtt{t}} - \vec{n}\bullet\vec{v} = 0. \tag{106}$$

We can then write the jump conditions in the usual engineering notation (the subscript $\boldsymbol{l}$ means that the fields are measured in the laboratory) as

$$\vec{n}\bullet[\underset{\boldsymbol{l}}{\vec{B}}] = 0, \quad (\vec{n}\bullet\vec{v})[\underset{\boldsymbol{l}}{\vec{B}}] + \vec{n}\times[\underset{\boldsymbol{l}}{\vec{E}}] = 0, \tag{107}$$

$$\vec{n}\bullet[\underset{\boldsymbol{l}}{\vec{D}}] = 0, \quad (\vec{n}\bullet\vec{v})[\underset{\boldsymbol{l}}{\vec{D}}] + \vec{n}\times[\underset{\boldsymbol{l}}{\vec{H}}] = 0. \tag{108}$$

Remark 7. *For the sake of completeness, before we proceed, we show that* Eq.(101) *implies the formulas in* Eq.(108).

We calculate $[G]\llcorner d\Xi|_{\Xi=0} = [G]\llcorner n|_{\Xi=0} = n\lrcorner[G]|_{\Xi=0} = 0$. *Now, from* Eq.(52) *we have*

$$n\lrcorner G = n\lrcorner(\boldsymbol{l}\wedge\underset{\boldsymbol{l}}{\boldsymbol{D}}) - n\lrcorner[\star(\boldsymbol{i}\wedge\underset{\boldsymbol{l}}{\boldsymbol{H}})] \tag{109}$$

and since $\boldsymbol{l}\lrcorner\boldsymbol{D} = \gamma^0\lrcorner(D_i\gamma^i) = 0$ *we get*

$$n\lrcorner(\boldsymbol{l}\wedge\underset{\boldsymbol{l}}{\boldsymbol{D}}) = (n\lrcorner\boldsymbol{l})\underset{\boldsymbol{l}}{\boldsymbol{D}} - n\wedge(\boldsymbol{l}\lrcorner\underset{\boldsymbol{l}}{\boldsymbol{D}}) = n_0\underset{\boldsymbol{l}}{\boldsymbol{D}} \tag{110}$$

[4]The minus sign is necessary due to the signature of the metric.

which in engineering notation reads

$$(\vec{n} \bullet \vec{v})\vec{D}. \tag{111}$$

In addition,

$$\begin{aligned} n\lrcorner[\star(\boldsymbol{l} \wedge \boldsymbol{H})] &= \star(n \wedge \boldsymbol{l} \wedge \boldsymbol{H}) \\ &= \star(n_i H_j \boldsymbol{\gamma}^i \wedge \boldsymbol{\gamma}^0 \wedge \boldsymbol{\gamma}^j) \\ &= -n_i H_j \star (\boldsymbol{\gamma}^0 \wedge \boldsymbol{\gamma}^i \wedge \boldsymbol{\gamma}^j) \\ &= -n_i H_j \eta^{0k} \eta^{im} \eta^{jn} \varepsilon_{kmnl} \boldsymbol{\gamma}^j \\ &= -n_i H_j \eta^{im} \eta^{jn} \varepsilon_{0mnl} \boldsymbol{\gamma}^j \\ &= -n^m H^n \varepsilon_{0mnl} \boldsymbol{\gamma}^j, \end{aligned} \tag{112}$$

which is equivalent, in engineering notation, to

$$-\vec{n} \times \vec{H}. \tag{113}$$

Using Eq.(111) *and* Eq.(113) *we can write the equation* $n\lrcorner[G]|_{\Xi=0} = 0$ *in vector calculus notation as*

$$(\vec{n} \bullet \vec{v})[\underset{\boldsymbol{l}}{\vec{D}}] + \vec{n} \times [\underset{\boldsymbol{l}}{\vec{H}}] = 0.$$

To show that $n\lrcorner[G]|_{\Xi=0} = 0$ *implies also* $\vec{n} \bullet [\underset{\boldsymbol{l}}{\vec{D}}] = 0$*, it is enough to observe that*

$$\begin{aligned} \boldsymbol{l}\lrcorner(n\lrcorner[G]) &= (\boldsymbol{l} \wedge n)\lrcorner[G] \\ &= n_i(\boldsymbol{\gamma}^0 \wedge \boldsymbol{\gamma}^i)\lrcorner[\frac{1}{2}\mathtt{G}^{\alpha\beta}\gamma_\alpha \wedge \gamma_\beta] \\ &= -n_i(\boldsymbol{\gamma}^0 \wedge \boldsymbol{\gamma}^i) \cdot [\frac{1}{2}\mathtt{G}^{\alpha\beta}\gamma_\alpha \wedge \gamma_\beta] \\ &= -\frac{1}{2} n_i [\mathtt{G}^{\alpha\beta}] \det \begin{pmatrix} \gamma^0 \cdot \gamma_\alpha & \gamma^0 \cdot \gamma_\beta \\ \gamma^i \cdot \gamma_\alpha & \gamma^i \cdot \gamma_\beta \end{pmatrix} \\ &= -\frac{1}{2} n_i [\mathtt{G}^{\alpha\beta}](\delta^0_\alpha \delta^i_\beta - \delta^0_\beta \delta^i_\alpha) \\ &= n_i [\mathtt{G}^{0i}] = 0. \end{aligned} \tag{114}$$

But $\mathtt{G}^{0i} = -\mathtt{D}^i$ *and thus we can write* $\boldsymbol{l}\lrcorner(n\lrcorner[G]) = 0$ *in the notation of vector calculus as* $\vec{n} \bullet [\underset{\boldsymbol{l}}{\vec{D}}] = 0$.

5. Solution of Maxwell Equations for the Wilson & Wilson Experiment

We now show how to find the solution of Maxwell equations

$$dF = 0, \quad d \star G = -\mathbf{J} \tag{115}$$

for the famous Wilson & Wilson experiment of 1913. The experiment is modelled as follows.

A cylindrical magnetic insulator with internal and external radii r_1 and r_2, respectively, with uniform and isotropic electric and magnetic permeabilities ε and $\boldsymbol{\mu}$ is supposed to rotate with constant angular velocity $\boldsymbol{\omega}$ in the z direction in an inertial laboratory $\boldsymbol{L} = \boldsymbol{e}_0$ in which there exists a uniform magnetic field $F_{\mathbf{o}} = \mathrm{B}_{\mathbf{o}} dr \wedge d\phi$ (or, in engineering notation, $\vec{B}_{\mathbf{o}} = \mathrm{B}_{\mathbf{o}} \hat{\boldsymbol{e}}_{\mathbf{z}}$). Figure 1 illustrates the situation just described.

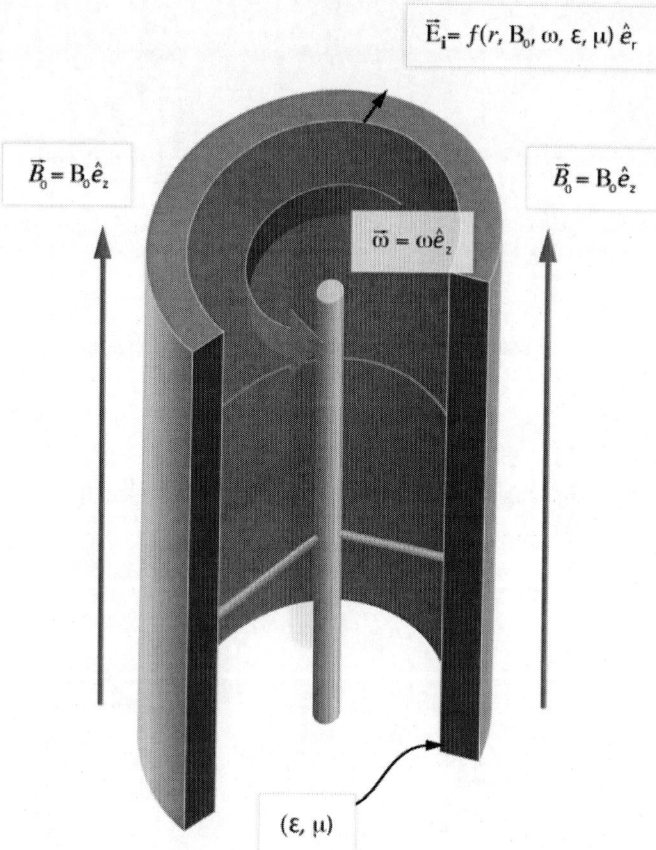

Figure 1. Details of the Wilson & Wilson Experiment. A magnetic insulator rotates with uniform angular velocity $\vec{\omega} = \omega\hat{\boldsymbol{e}}_{\mathbf{z}}$ in an external magnetic field $\vec{B}_{\mathbf{o}} = \mathrm{B}_{\mathbf{o}}\hat{\boldsymbol{e}}_{\mathbf{z}}$. An electric field $\vec{E}_{\mathbf{i}} = f(r, \mathrm{B}_{\mathbf{o}}, \omega, \varepsilon, \boldsymbol{\mu})\hat{\boldsymbol{e}}_r$ is observed.

We begin introducing some useful notations. First we write the Minkowski metric $\mathring{\boldsymbol{g}}$ as

$$\mathring{\boldsymbol{g}} = \boldsymbol{\eta}_{\mu\nu}\boldsymbol{\theta}^{\mu} \otimes \boldsymbol{\theta}^{\nu}, \tag{116}$$

with

$$\boldsymbol{\theta}^0 = dt, \quad \boldsymbol{\theta}^1 = dr, \quad \boldsymbol{\theta}^2 = rd\phi, \quad \boldsymbol{\theta}^3 = dz. \tag{117}$$

In our problem the boundary of the moving dieletric has equations $\Xi_1 = r - r_1 = 0$ and $\Xi_2 = r - r_2 = 0$ in the coordinates naturally adapted to the laboratory ($\boldsymbol{L}$) frame[5], so the normal to this surface is $d\Xi = \boldsymbol{\theta}^1$. In this system we have $\mathbf{J} = 0$. To proceed we write as an *ansatz* the solution for the electromagnetic field inside the material,

$$F_{\mathrm{i}} = \mathcal{E}\boldsymbol{\theta}^0 \wedge \boldsymbol{\theta}^1 + \mathcal{B}\boldsymbol{\theta}^1 \wedge \boldsymbol{\theta}^2, \tag{118}$$

where $\mathcal{E}$ and $\mathcal{B}$ are supposed to be functions of coordinate r only.

We recall from Eq.(73) that the 1-form physically equivalent to the velocity field $\boldsymbol{V}$ is

$$\boldsymbol{v} = \gamma(\boldsymbol{\theta}^0 + \mathrm{v}\boldsymbol{\theta}^2)$$

with $\mathrm{v} = \omega r$ and $\gamma = (1 - \mathrm{v}^2)^{-1/2}$. We now must solve the equation $d \star G = 0$ with the boundary conditions given by Eqs.(102) and (103). To calculate G_{i} we use Eq.(54). First we calculate $\boldsymbol{v}\lrcorner F_{\mathrm{i}}$. We have

$$\begin{aligned} \boldsymbol{v}\lrcorner F_{\mathrm{i}} &= \gamma(\boldsymbol{\theta}^0 + \mathrm{v}\boldsymbol{\theta}^2)\lrcorner(\mathcal{E}\boldsymbol{\theta}^{01} + \mathcal{B}\boldsymbol{\theta}^{12}) \\ &= \gamma(\mathcal{E} + \mathrm{v}\mathcal{B})\boldsymbol{\theta}^1 \end{aligned}$$

and

$$\boldsymbol{v} \wedge (\boldsymbol{v}\lrcorner F_{\mathrm{i}}) = \gamma^2(\mathcal{E} + \mathrm{v}\mathcal{B})\boldsymbol{\theta}^{01} - \gamma^2\mathrm{v}(\mathcal{E} + \mathrm{v}\mathcal{B})\boldsymbol{\theta}^{12}.$$

Then[6]

$$\begin{aligned} G_{\mathrm{i}} &= \frac{\mathcal{E}(\mu\varepsilon - \mathrm{v}^2) + \mathcal{B}\mathrm{v}(\mu\varepsilon - 1)}{\mu(1 - \mathrm{v}^2)}\boldsymbol{\theta}^{01} + \frac{\mathcal{E}\mathrm{v}(1 - \mu\varepsilon) + \mathcal{B}(1 - \mu\varepsilon\mathrm{v}^2)}{\mu(1 - \mathrm{v}^2)}\boldsymbol{\theta}^{12} \\ &:= \mathcal{K}\boldsymbol{\theta}^{01} + \mathcal{L}\boldsymbol{\theta}^{12}. \end{aligned} \tag{119}$$

Since

$$\begin{aligned} \star(\boldsymbol{\theta}^0 \wedge \boldsymbol{\theta}^1) &= (\widetilde{\boldsymbol{\theta}^0 \wedge \boldsymbol{\theta}^1})\lrcorner\boldsymbol{\tau}_{\mathring{g}} = -\boldsymbol{\theta}^2 \wedge \boldsymbol{\theta}^3, \\ \star(\boldsymbol{\theta}^1 \wedge \boldsymbol{\theta}^2) &= (\widetilde{\boldsymbol{\theta}^1 \wedge \boldsymbol{\theta}^2})\lrcorner\boldsymbol{\tau}_{\mathring{g}} = \boldsymbol{\theta}^0 \wedge \boldsymbol{\theta}^3 \end{aligned}$$

we have

$$\star G_{\mathrm{i}} = -\mathcal{K}\boldsymbol{\theta}^{23} + \mathcal{L}\boldsymbol{\theta}^{03}. \tag{120}$$

Then $d \star G = 0$ gives

$$-\frac{d\mathcal{K}}{dr}\boldsymbol{\theta}^{123} - \frac{\mathcal{K}}{r}\boldsymbol{\theta}^{123} - \frac{d\mathcal{L}}{dr}\boldsymbol{\theta}^{013} = 0.$$

So, to solve our problem resumes it is enough to solve the following two trivial ordinary differential equations

$$\frac{d\mathcal{K}}{dr} + \frac{\mathcal{K}}{r} = 0 \text{ and } \frac{d\mathcal{L}}{dr} = 0. \tag{121}$$

[5] $\Xi_1 = r - r_1 = 0$ and $\Xi_2 = r - r_2 = 0$ are also the equations for the boundary in coordinates naturally adapted to the rotating frame $\boldsymbol{V}$. Of course the boundary is at rest in the laboratory frame $\boldsymbol{L}$ and also in the rotating frame $\boldsymbol{V}$, while it may be in motion in an arbitary reference frame $\boldsymbol{Z}$.

[6] There seems to be a misprint sign in the formula appearing in [1].

The solutions are respectively

$$\mathcal{K} = \frac{c_1}{r} \text{ and } \mathcal{L} = c_2 \tag{122}$$

where c_1 and c_2 are integration constants. So we have

$$G_{\mathbf{i}} = \frac{c_1}{r}\boldsymbol{\theta}^{01} + c_2\boldsymbol{\theta}^{12},$$
$$\star G_{\mathbf{i}} = -\frac{c_1}{r}\boldsymbol{\theta}^{23} + c_2\boldsymbol{\theta}^{03}.$$

Now, recalling that outside the magnetic insulator we have $\star G_{\mathbf{o}} = \star F_{\mathbf{o}} = \mathrm{B}_{\mathbf{o}}\boldsymbol{\theta}^{03}$ we get, using the jump condition Eq.(103), that

$$(\star G_{\mathbf{o}} - \star G_{\mathbf{i}}) \wedge \boldsymbol{\theta}^1 = 0, \tag{123}$$

i.e.,

$$-\mathrm{B}_{\mathbf{o}}\boldsymbol{\theta}^{013} + c_2\boldsymbol{\theta}^{013} - \frac{c_1}{r}\boldsymbol{\theta}^{123} = 0 \tag{124}$$

whence it follows that functions $\mathcal{K}$ and $\mathcal{L}$ are given by

$$\mathcal{K} = 0, \quad \mathcal{L} = \mathrm{B}_{\mathbf{o}}. \tag{125}$$

Then from Eq.(119) we have the following system of linear equations for functions $\mathcal{E}$ and $\mathcal{B}$:

$$\begin{cases} \mathcal{E}(\boldsymbol{\mu\varepsilon} - \mathrm{v}^2) + \mathcal{B}\mathrm{v}(\boldsymbol{\mu\varepsilon} - 1) = 0, \\ \frac{\mathcal{E}\mathrm{v}(1-\boldsymbol{\mu\varepsilon}) + \mathcal{B}(1-\boldsymbol{\mu\varepsilon}\mathrm{v}^2)}{\boldsymbol{\mu}(1-\mathrm{v}^2)} = \mathrm{B}_{\mathbf{o}}, \end{cases} \tag{126}$$

whose solution is:

$$\mathcal{E} = \mathrm{B}_{\mathbf{o}}\frac{\omega r(1-\boldsymbol{\mu\varepsilon})}{\boldsymbol{\varepsilon}(1-\omega^2r^2)}, \quad \mathcal{B} = -\mathrm{B}_{\mathbf{o}}\frac{(\omega^2r^2 - \boldsymbol{\mu\varepsilon})}{\boldsymbol{\varepsilon}(1-\omega^2r^2)}. \tag{127}$$

So we finally have

$$F_{\mathbf{i}} = \mathrm{B}_{\mathbf{o}}\frac{\omega r(1-\boldsymbol{\mu\varepsilon})}{\boldsymbol{\varepsilon}(1-\omega^2r^2)}\boldsymbol{\theta}^{01} - \mathrm{B}_{\mathbf{o}}\frac{(\omega^2r^2 - \boldsymbol{\mu\varepsilon})}{\boldsymbol{\varepsilon}(1-\omega^2r^2)}\boldsymbol{\theta}^{12}. \tag{128}$$

The electric and magnetic fields $\underset{l}{\boldsymbol{E}_{\mathbf{i}}}$ and $\underset{l}{\boldsymbol{H}_{\mathbf{i}}}$ inside the material, as determined in the laboratory frame, are[7]

$$\underset{l}{\boldsymbol{E}_{\mathbf{i}}} = \boldsymbol{\theta}^0 \lrcorner F_{\mathbf{i}} = \mathrm{B}_{\mathbf{o}}\frac{\omega r(1-\boldsymbol{\mu\varepsilon})}{\boldsymbol{\varepsilon}(1-\omega^2r^2)}\boldsymbol{\theta}^1, \tag{129}$$

$$\underset{l}{\boldsymbol{H}_{\mathbf{i}}} = \boldsymbol{\theta}^0 \lrcorner \star F_{\mathbf{i}} = -\mathrm{B}_{\mathbf{o}}\frac{(\omega^2r^2 - \boldsymbol{\mu\varepsilon})}{\boldsymbol{\varepsilon}(1-\omega^2r^2)}\boldsymbol{\theta}^3. \tag{130}$$

From Eq.(129) it follows immediately that the potential V between the internal and external surfaces of the material, as shown in Figure 1, when $\omega^2 r^2 << 1$, is given by

$$\begin{aligned} V &= \mathrm{B}_{\mathbf{o}}\omega\frac{1}{\boldsymbol{\varepsilon}}(1-\boldsymbol{\mu\varepsilon})\textstyle\int_{r_1}^{r_2} r dr \\ &= \frac{1}{2}\mathrm{B}_{\mathbf{o}}\omega\frac{1}{\boldsymbol{\varepsilon}}(1-\boldsymbol{\mu\varepsilon})(r_2^2 - r_1^2), \end{aligned} \tag{131}$$

which is the value found by Wilson & Wilson [21].

[7] An expression equivalent to Eq.(128) appeared in [6].

Remark 8. *People are sometimes puzzled when they try to identify the charges and currents responsible for the electric and magnetic fields inside the magnetic insulator. The answer is promptly found if, from the identification of* $\underset{\imath}{\mathbf{P}}$ *and* $\underset{\imath}{\mathbf{M}}$, *we calculate the bound current* $\underset{\imath}{\mathcal{J}}$ *and the bound charge* $\underset{\imath}{\rho}$.

6. Extracting Energy from the Magnetic Field

We now propose a way to extract energy from an external magnetic field using the results obtained above. In order to do that we first recall that when the magnetic insulator in Figure 1, which has momentum of inertia I, is set in rotation by an external motor until it acquires an angular velocity $\vec{\omega}_0 = \omega_0 \hat{\mathbf{e}}_\mathbf{z}$, it also acquires an angular momentum

$$\vec{L}_{0mec} = I\vec{\omega}_0 \tag{132}$$

and, of course, a certain amount of kinetic mechanical energy. We will suppose that when the magnetic insulator reaches angular velocity ω_0 the external motor is turned off. Moreover, we suppose that there are no energy losses due to friction or to Joule effect on electric wires. As a consequence the total energy and angular momentum of the *closed* system (i.e., the rotating dielectric plus the electromagnetic field) must be conserved[8]. We suppose, moreover, that the system does not dissipate energy through radiation. Under these conditions the total angular momentum of the system at any instant of time is

$$\vec{L}_\mathbf{t} = \vec{L}_{mec} + \vec{L}_{elec} = \vec{L}_{0\mathbf{t}} \tag{133}$$

where $\vec{L}_{0\mathbf{t}}$ is the total angular momentum at the moment in which the external driver is turned off, and $\vec{L}_{elec}$ is given by [12] Abraham's formula[9]

$$\vec{L}_{elec} = \int \mathrm{d}x\mathrm{d}y\mathrm{d}z\overrightarrow{\mathrm{x}} \times (\vec{E}_\mathrm{i} \times \vec{H}_\mathrm{i}). \tag{134}$$

Remark 9. *We suppose moreover that the electromagnetic field is non zero only inside a finite region of space of volume* $\mathcal{V}$. *We also make the assumption that the angular velocity of the magnetic insulator is a function of time which can only be determined once we have a specific model for the transformation of the magnetic energy stored in the magnetic field into kinetic energy of the rotating insulator plus energy stored in the electric field. Since this energy transfer cannot be instantaneous, in projecting our machine we can suppose, at a first approximation, that* $\omega = \omega_0 + \alpha t$[10].

---

[8] Of course, the kinetic mechanical energy of the device plus the energy contained in the magnetic field must also be conserved.

[9] At first sight it may seem strange that a magnetic and an electric field arising from different sources might store angular momentum. This is an experimental fact shown for the first time by Graham and Lahoz in 1979 [4, 7].

[10] The hypothesis that there is a transfer of energy from the magnetic field to the kinetic energy field was tested in [7], an experiment in which the capacitor was seen to rotate when the magnetic field was turned off.

We choose our coordinate system so that its origin lies in the mean point of the rotating axis of the dielectric. Under these conditions we have at time t[11]

$$\begin{aligned}\vec{E}_{\mathrm{i}} \times \vec{H}_{\mathrm{i}} &= -\mathrm{B}^2 \frac{\boldsymbol{\mu}(1-\boldsymbol{\mu}\varepsilon)\omega r(\omega^2 r^2 - \boldsymbol{\mu}\varepsilon)}{\varepsilon^2(1-\omega^2 r^2)^2} \hat{\mathbf{e}}_r \times \hat{\mathbf{e}}_{\mathrm{z}} \\ &= f(r, \mathrm{B}, \omega, \boldsymbol{\mu}, \varepsilon)\hat{\mathbf{e}}_\phi.\end{aligned}$$

Then

$$\begin{aligned}\tilde{\mathrm{x}} \times (\vec{E}_{\mathrm{i}} \times \vec{H}_{\mathrm{i}}) &= (\mathrm{z}\hat{\mathbf{e}}_{\mathrm{z}} + r\hat{\mathbf{e}}_r) \times f(r, \mathrm{B}, \omega, \boldsymbol{\mu}, \varepsilon)\hat{\mathbf{e}}_\phi \\ &= \mathrm{z} f(r, \mathrm{B}, \omega, \boldsymbol{\mu}, \varepsilon)\hat{\mathbf{e}}_{\mathrm{z}} \times \hat{\mathbf{e}}_\phi + r f(r, \mathrm{B}, \omega, \boldsymbol{\mu}, \varepsilon)\hat{\mathbf{e}}_r \times \hat{\mathbf{e}}_\phi \\ &= -\mathrm{z} f(r, \mathrm{B}, \omega, \boldsymbol{\mu}, \varepsilon)\hat{\mathbf{e}}_r + r f(r, \mathrm{B}, \omega, \boldsymbol{\mu}, \varepsilon)\hat{\mathbf{e}}_{\mathrm{z}} \\ &= -\mathrm{z} f(r, \mathrm{B}, \omega, \boldsymbol{\mu}, \varepsilon)(\cos\phi\hat{\mathbf{e}}_{\mathrm{x}} + \sin\phi\hat{\mathbf{e}}_{\mathrm{y}}) + r f(r, \mathrm{B}, \omega, \boldsymbol{\mu}, \varepsilon)\hat{\mathbf{e}}_{\mathrm{z}}.\end{aligned}$$

Thus the integral corresponding to the $\mathbf{e}_r$ component vanishes and we have

$$\begin{aligned}\vec{L}_{elec} &= \mathbf{e}_{\mathrm{z}} \int r^2 f(r, \mathrm{B}, \omega, \boldsymbol{\mu}, \varepsilon) dr d\phi dz \\ &= 2\pi Z \mathbf{e}_{\mathrm{z}} \int_{r1}^{r_2} r^2 f(r, \mathrm{B}, \omega, \boldsymbol{\mu}, \varepsilon) dr.\end{aligned} \tag{135}$$

where Z is the height of the cylindrical dielectric. When $\omega^2 r^2 << 1$

$$\begin{aligned}\vec{L}_{elec} &= \hat{\mathbf{e}}_{\mathrm{z}} 2\pi Z \frac{\boldsymbol{\mu}^2}{\varepsilon} \mathrm{B}^2 \omega(1-\boldsymbol{\mu}\varepsilon) \int_{r1}^{r_2} r^3 dr \\ &= -\frac{\pi Z \boldsymbol{\mu}^2}{2\varepsilon} \mathrm{B}^2 \omega(\boldsymbol{\mu}\varepsilon - 1)\left(r_2^4 - r_1^4\right)\hat{\mathbf{e}}_{\mathrm{z}}.\end{aligned} \tag{136}$$

It is a remarkable fact that $\vec{L}_{elec}$ points in the direction opposite to $\vec{L}_{mec}$ if $\boldsymbol{\mu}\varepsilon > 1$, as is the case in the Wilson & Wilson experiment, according to [11]. Since the total angular momentum of our system must be conserved we see that $\left|\vec{L}_{mec}\right|$ must increase if there are no energy losses due to friction and Joule effect in the wires, so that we could use the potential V given by Eq.(131) to power an electric machine. Of course, the energy powering the proposed electric machine can only come from the energy *stocked* in the external magnetic field. It will work while the external magnetic field is present. Of course, extracting energy from the magnetic field will make the value of B decrease, in a way taht can be estimated from Eq.(133), taking into account that the momentum of inertia will be a function of time. With this decrease the potential V will also decrease and at the end the machine will stop working.

Remark 10. *If $\boldsymbol{\mu}\varepsilon < 1$ the electromagnetic angular momentum is in the same direction as the mechanical angular momentum, which must thus decrease after the device is set to rotate. So, the electric field will also decrease, making the electromagnetic momentum to decrease whereas the mechanical angular momentum must then start to increase again. It is not clear at this moment whether the system will oscillate between a minimum and a maximum mechanical angular momentum or will somehow stabilize.*

[11] Of course, at $t = 0$, $\mathrm{B}(0) = \mathrm{B}_0$.

Remark 11. ***Real World Machine.*** *In real world, where friction and Joule effect are always present, extracting energy from the magnetic field requires a machine like the one shown in Figure 2, in which the potential difference V, given by* Eq.(131), *is first used, if necessary, to power up a compensator motor just enough to restore eventual losses due to friction and Joule effect on electric wires. After this is done, V can then be used to power a motor capable of generating useful work.*

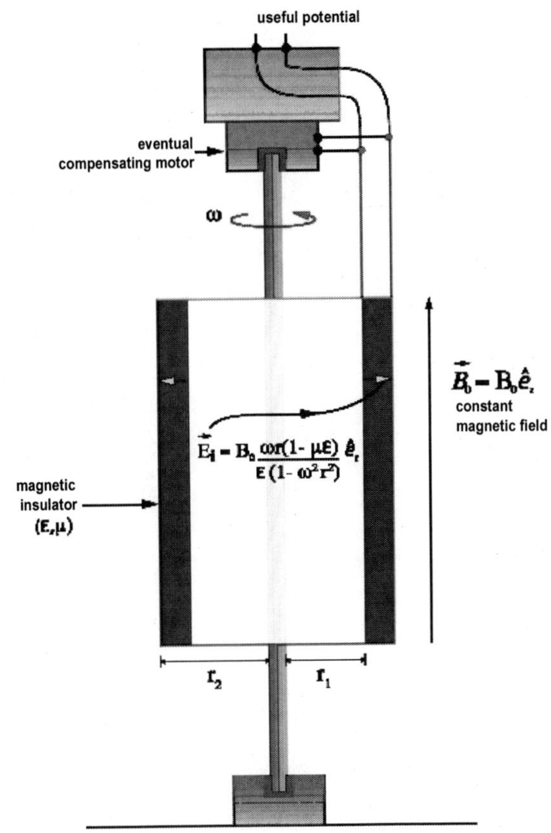

Figure 2. Machine to Extract Energy from an External Magnetic Field.

7. Conclusion

In this chapter we recalled how to correctly solve Maxwell equations to find the electric field in the famous Wilson & Wilson experiment using the theory of differential forms, which allows an intrinsic formulation of the problem. We also presented, in an invariant way, the correct jump conditions for F and G fields when the boundary separating two media is in motion. We give enough details for the chapter to be useful for students and researchers. Moreover, we use the theoretical results to present a surprising result which, for the best of our knowledge, is new: *a machine that can extract energy from an external magnetic field*. Under appropriate conditions ($\mu\varepsilon$ >1) the machine, once set in motion (of course by an external device which is turned off after the machine acquires angular velocity, say ω_0), will increase the electromagnetic angular momentum stocked in the electromagnetic field in a direction opposite to the original mechanical angular momentum of the device, which thus will begin to increase (theoretically) its angular velocity until $\omega r_2 = 1$, generating a big potential difference, even in a small external magnetic field, that can be use to produce useful work.

Such a machine, if put in orbit around a neutron star, can produce a lot of energy in a simpler way than the hypothetical machine described, e.g., in page 908 of [9], a machine projected to extract energy from a black hole by throwing garbage into it.

A. Some Useful Formulas

In this appendix we present some useful identities of the theory of differential forms which have been used several times in the main text.

For any $a \in \sec \bigwedge^1 T^*M$, $A_r \in \sec \bigwedge^r T^*M$ and $B_r \in \sec \bigwedge^s T^*M$:

$$a \lrcorner (A_r \wedge B_r) = (a \lrcorner A_r) \wedge B_r + \hat{A}_r \wedge (a \lrcorner B_r). \tag{137}$$

$$\begin{aligned}
&A_r \wedge \underset{\boldsymbol{g}}{\star} B_s = B_s \wedge \underset{\boldsymbol{g}}{\star} A_r; \quad r = s\\
&A_r \cdot \underset{\boldsymbol{g}}{\star} B_s = B_s \cdot \underset{\boldsymbol{g}}{\star} A_r; \quad r + s = n\\
&A_r \wedge \underset{\boldsymbol{g}}{\star} B_s = (-1)^{r(s-1)} \underset{\boldsymbol{g}}{\star} (\tilde{A}_r \lrcorner B_s); \quad r \leq s\\
&A_r \lrcorner \underset{\boldsymbol{g}}{\star} B_s = (-1)^{rs} \underset{\boldsymbol{g}}{\star} (\tilde{A}_r \wedge B_s); \quad r + s \leq n\\
&\underset{\boldsymbol{g}}{\star} A_r = \tilde{A}_r \lrcorner \tau_{\boldsymbol{g}}\\
&\underset{\boldsymbol{g}}{\star} \tau_{\boldsymbol{g}} = \operatorname{sgn} \det \boldsymbol{g}; \quad \underset{\boldsymbol{g}}{\star} 1 = \tau_{\boldsymbol{g}}.
\end{aligned} \tag{138}$$

In the formulas above, $\hat{A}_r = (-1)^r A_r$ and $\tilde{A}_r = (-1)^{\frac{r(r-1)}{2}}$ are respectively the grade involution and the reversion operators. Details in [15].

Acknowledgment

The authors are grateful to Flavia Tonelli for the figures.

References

[1] Canovan, C. E. S., and Tucker, R. W., *Maxwell Equations in a Uniformly Rotating Dielectric Medium and the Wilson-Wilson Experiment*. [arXiv:1104.0574v1 [math-phys]]

[2] Jackson, J. D., *Classical Electrodynamics* (third edition), J. Wiley & Sons, New York, 1999.

[3] Giglio, J. F. T., and Rodrigues, W. A. Jr., Locally Inertial Reference Frames in Lorentzian and Riemann-Cartan Spacetimes, *Annalen der Physik* 502, 302-310 (2012).

[4] Graham, G. M , and Lahoz, D. G, Observation of Static Electromagnetic Angular Momentum in Vacuo, *Nature* 285, 154-155 (1980).

[5] Hertzberg, J. B., Bickman, S. R., Hummon, M. T., Krause, D., Jr., Peck, S. K., and Hunter, L .R., Measurement of the Relativisitic Potential Difference Across a Rotating Magnetic Dielectric Cylinder, *Am. J. Phys*. 69, 648-654 (2001).

[6] Hehl, F. W., and Obukhov, Y. N., *Foundations of Classical Electrodynamics*, Birkhäuser, Boston 2003.

[7] Lahoz, D. G. and Graham, G. M ,Observation of Electromagnetic Angular Momentum within Magnetite, *Phys. Rev. Lett*. 42, 1137–1140 (1979).

[8] Lovelock, D. and Rund, H., *Tensors, Differential Forms, and Variational Principles*, John Wiley & Sons, New York, 1975.

[9] Misner, C. W., Thorne, K. S., and Wheeler, J. A., *Gravitation*, W. H. Freeman and Co., San Francisco 1973.

[10] Namias, V., Discontinuity of Electromagnetic Fields, Potentials, and Currents at Fixed and Moving Boundaries, *Am. J. Phys*. 56, 898-904 (1998).

[11] Pellegrini, G. N., and Swift, A. R., Maxwell Equations in a Rotating Medium: Is there a Problem? *Am. J. Phys*. 63, 694-705 (1995).

[12] Ramos, T., Rubilar, G. F., and Obukhov, Y. N., Relativistic Analysis of the Dielectric Einstein Box: Abraham, Minkowski and Total Energy-Momentum Tensors, *Phys. Lett. A*. 375, 1703-1709 (2011).

[13] Ridgely, C. T., Applying Relativistic Electrodynamics to a Rotating Material Medium, *Am. J. Phys*. 66, 114-121 (1998).

[14] da Rocha, R., and Rodrigues, W. A. Jr., Pair and Impair, Even and Odd Form Fields and Electromagnetism, *Annalen der Physik* 19, 6-34 (2010).

[15] Rodrigues, W. A. Jr., and Capelas de Oliveira, E., The Many Faces of Maxwell, Dirac and Einstein Equations. A Clifford Bundle Approach. *Lecture Notes in Physics* 722, Springer, Heidelberg, 2007.[Errata at http://www.ime.unicamp.br/~walrod/svf3a.pdf].

[16] Rodrigues, F. G., On Equivalent Expressions for the Faraday's Law of Induction, *Rev. Bras. Ens. de Fis.* 34, 1309 (2012).[arXiv:1002.2792v2 [physics.class-ph]]

[17] Webster, D. L., Relativity of Moving Circuits and Magnets, *Am J. Phys*. 29, 262- 268 (1960).

[18] Webster, D. L, Schiff's Charges and Currents in Rotating Matter, *Am J. Phys*. 31, 590-597 (1963).

[19] Schiff, L. I. , A Question in General Relativity, *Proc. Nat. Acad. Sci.* 25, 391-395 (1939).

[20] Shiozawa, T., Phenomenological and Electron-Theoretical Study of the Electrodynamics of Rotating Systems, *Proc. of the IEEE* 61, 1694-1702 (1973).

[21] Wilson, M., and Wilson, H.A., On the Electric Effect of Rotating a Magnetic Insulator in a Magnetic Field, *Proc. R. Soc. London Ser. A* 89, 99-106 (1913).

In: Einstein and Others: Unification
Editor: Valeriy V. Dvoeglazov

ISBN: 978-1-63463-276-8

Chapter 7

FIVE-VECTOR OR BIVECTOR?

Alexander Krasulin*
Institute for Nuclear Research of the Russian Academy
of Sciences, Russia†

Abstract

This chapter shows how one can establish a relation between five-dimensional tangent vectors in space-time and spinors that would be invariant under parallel transport and that would reproduce the conventional relation between spinors and four-dimensional tangent vectors. The implications of this relation for choosing the way in which one should generalize the covariant derivative are discussed.

PACS: 05.45-a, 52.35.Mw, 96.50.Fm

Keywords: Five-dimensional tangent vectors, five-vectors, spinors, bivector gauge fields

AMS Subject Classification: 53D, 37C, 65P

1. Introduction

The material presented in this chapter makes essential use of the formalism of five-dimensional tangent vectors on the space-time manifold, as it was presented in refs.[1]–[6]. Within this formalism, one encounters *two* possibilities for generalizing the notion of the covariant derivative. One option is to replace the operator ∇_μ ($\mu = 0, 1, 2, 3$), whose argument is a four-vector, with the operator $\Box_A$ ($A = 0, 1, 2, 3, 5$), whose argument is a five-vector. In the following, this latter operator will be referred to as the *five-vector* covariant derivative. In any regular five-vector basis (where the first four basis five-vectors are orthogonal to the fifth one and correspond to the four basis four-vectors), the operators $\Box_\mu$ coincide with ∇_μ, while $\Box_5$ is a certain local (nondifferential) operator. The operator $\Box$ as a whole can be interpreted as a derivative ∇ of a field in a certain direction calculated by

*E-mail address: krasulin@post.com
†Former affiliation.

using rules of parallel transport with one unusual property: The transports along different directions in space-time are correlated with one another in such a way that the quantity

$$\nabla_{(\mathbf{U}+\mathbf{V})} - \nabla_{\mathbf{U}} - \nabla_{\mathbf{V}}$$

is *not* identically zero for any two four-vectors $\mathbf{U}$ and $\mathbf{V}$, but is proportional to some local operator, which can vary from point to point. For details the reader is referred to ref.[5].

The other possibility of generalizing the ordinary covariant derivative is to replaced ∇_μ with the operator $\mathsf{D}_{AB} = -\mathsf{D}_{BA}$ $(A, B = 0, 1, 2, 3, 5)$, whose argument is a five-vector bivector. In the following, this operator will be referred to as the *bivector* derivative. To explain what the latter is one should recall that within the more rigorous approaches (see e.g., ref.[7]), an ordinary covariant derivative is calculated by (*i*) performing an infinitesimal active *translation* of the differentiated field in the direction specified by the argument of the derivative; (*ii*) subtracting the result from the original field; (*iii*) dividing the difference by the translation parameter s; and (*iv*) taking the limit $s \to 0$. For calculating the bivector derivative one (*i*) makes an infinitesimal active *Poincare* transformation of the differentiated field, which may include both translations *and* rotations; (*ii*) subtracts the result from the original field; (*iii*) divides the difference by the transformation parameter s; and (*iv*) takes the limit $s \to 0$. In the case of four-vector and five-vector fields, the bivector derivative is completely determined by the Riemannian metric of space-time and is more a technical tool that enables one to formulate in a coordinate-independent way the constraints on the five-vector affine connection. However, for the fields whose values are some abstract vectors or tensors whose parallel transport is determined by the connections that physicists call gauge fields, the notion of the bivector derivative enables one to extend the affine geometry associated with such fields and to introduce a new type of connections which in the language of ordinary tangent vectors are described by antisymmetric tensor fields of rank two. For more details the reader is referred to ref.[6].

How one *should* generalize the covariant derivative is a question to be asked of physics, not of mathematics. Here I only wish to present the necessary technicalities that would enable one to understand what *options* one has and, hopefully, to find some *clues* to what the choice should be — five-vector or bivector. Unfortunately, one can find practically no clues by considering scalar, four-vector or four-tensor fields, for the terms in the Lagrangian density containing the field derivatives are typically of the form $(\nabla_\mu\Phi)(\nabla^\mu\Phi)$, where Φ is one of the mentioned fields, and can equally well be replaced with $(\Box_A\Phi)(\Box^A\Phi)$ or with $(D_{AB}\Phi)(D^{AB}\Phi)$. One may hope to find some indications by examining spinor fields, whose mathematics is more complicated than that of boson fields and, hopefully, may give one some guidelines to how the decision in question should be made. To see if this is indeed so, let us now revise the mathematical theory of spinors and see how one can incorporate into it the five-dimensional tangent vectors.

2. 4-Spinors

Algebraically, the 4-spinors commonly used in physics are elements of a complex vector space, S_4, endowed with *two* non-degenerate inner products: (1) an Hermitian inner product $d(\,.\,,.)$ antilinear in its first argument and linear in the other, and (2) an antisymmetric inner product $c(\,.\,,.)$ linear in both of its arguments. The mentioned properties of d can be expressed by the following equations: for any 4-spinors ψ, χ, and ξ and any complex numbers a and b,

$$d(a\psi + b\chi, \xi) = a^* \cdot d(\psi, \xi) + b^* \cdot d(\chi, \xi)$$
$$d(\psi, a\chi + b\xi) = a \cdot d(\psi, \chi) + b \cdot d(\psi, \xi)$$
$$d(\chi, \psi) = d(\psi, \chi)^*,$$

where * denotes ordinary complex conjugation. The properties of c can be expressed in a similar way. To understand the meaning of d, one should recall that any inner product with such properties defines an anti-isomorphism of S_4 onto the space $\widetilde{S}_4$ of corresponding 1-forms (co-spinors): by definition, the image $\overline{\psi}$ in $\widetilde{S}_4$ of $\psi \in S_4$ is such that

$$<\overline{\psi}, \chi>= d(\psi, \chi) \text{ for any } \chi \in S_4.$$

The inner product d is such that $\psi \mapsto \overline{\psi}$ coincides with what physicists call Dirac conjugation. In a similar way, the inner product c defines an isomorphism of S_4 onto $\widetilde{S}_4$: the image $\widehat{\psi}$ of $\psi \in S_4$ relative to this isomorphism is such that

$$<\widehat{\psi}, \chi>= c(\psi, \chi) \text{ for any } \chi \in S_4.$$

It is apparent that the composition of the isomorphism $\psi \mapsto \widehat{\psi}$ and of the anti-isomorphism reverse to $\psi \mapsto \overline{\psi}$ is an anti-isomorphism of S_4 onto itself. The inner product c is such that the latter anti-isomorphism coincides with what physicists call charge conjugation. In other words,

$$\widehat{\psi} = \overline{\psi^c},$$

where ψ^c denotes the charge conjugate of ψ.

Since one is dealing with *two* inner products, in addition to specifying the properties of each of them, one has to specify how they are 'positioned' one to the other. From the mentioned relation of d and c to Dirac and charge conjugations, one can easily derive that in an appropriate 4-spinor basis, these two inner products have the following components:

$$d = \begin{vmatrix} 1 & 0 & 0 & 0 \\ 0 & 1 & 0 & 0 \\ 0 & 0 & -1 & 0 \\ 0 & 0 & 0 & -1 \end{vmatrix} \quad \text{and} \quad c = \begin{vmatrix} 0 & 0 & 0 & 1 \\ 0 & 0 & -1 & 0 \\ 0 & 1 & 0 & 0 \\ -1 & 0 & 0 & 0 \end{vmatrix}. \qquad (1)$$

In the following it will always be assumed that the 4-spinor basis is selected in this particular way. To make the following calculations simpler, it is convenient to present a 4-spinor as a *tensor product* of two two-dimensional complex vectors. Accordingly, each 4×4 matrix associated with 4-spinors can be presented as a tensor product of two 2×2 matrices (or a linear combination of the latter). For example, the matrices in equations (1) can be presented as

$$d = \begin{vmatrix} 1 & 0 \\ 0 & -1 \end{vmatrix} \otimes \begin{vmatrix} 1 & 0 \\ 0 & 1 \end{vmatrix} = \sigma_3 \otimes \mathbf{1} \quad \text{and} \quad c = \begin{vmatrix} 0 & 1 \\ 1 & 0 \end{vmatrix} \otimes \begin{vmatrix} 0 & 1 \\ -1 & 0 \end{vmatrix} = \sigma_1 \otimes i\sigma_2, \quad (2)$$

where $\mathbf{1}$ is the unity 2×2 matrix and σ_k $(k = 1, 2, 3)$ are the standard Pauli matrices.

Let us now determine the symmetry group of S_4. By direct calculation one can find that in the 4-spinor basis specified above, the generators of the transformations in S_4 that preserve the inner products d and c are:

$$\begin{aligned} &\mathbf{1}\otimes\sigma_k\sigma_l \equiv 2M_{kl}\ (k<l),\ \ \sigma_1\otimes\sigma_k \equiv 2M_{0k},\\ &\sigma_2\otimes\sigma_k \equiv 2M_{5k},\ \ i\sigma_3\otimes\mathbf{1} \equiv 2M_{50}. \end{aligned} \tag{3}$$

Supposing that all upper-case Latin indices run 0, 1, 2, 3, and 5 and that $M_{AB} = -M_{BA}$, it is a simple matter to verify that the above generators satisfy the commutation relations for SO(3,2):

$$[M_{KL}, M_{MN}] = \eta_{KM}M_{LN} - \eta_{LM}M_{KN} - \eta_{KN}M_{LM} + \eta_{LN}M_{KM}, \tag{4}$$

where $\eta_{AB} \equiv \mathrm{diag}(+1,-1,-1,-1,+1)$.

Within the conventional approach, the relation between 4-spinors and the space-time manifold is established by postulating that at each point of the latter there is fixed a certain isomorphism between the space of ordinary four-dimensional tangent vectors V_4 and one of the real subspaces of tensors of rank (1,1) over S_4 which is invariant under transformations from a subgroup isomorphic to SO(3,1) of the above mentioned SO(3,2). The components of the images of the basis four-vectors $\mathbf{E}_\alpha$ in the basis in $S_4 \otimes \widetilde{S}_4$ corresponding to the selected 4-spinor basis are denoted as $(\gamma_\alpha)^i{}_j$ and are nothing but Dirac's gamma matrices.

Let us now show that the above mentioned isomorphism of V_4 onto a subspace in $S_4 \otimes \widetilde{S}_4$ can be induced by a 'more fundamental' isomorphism between the space of five-dimensional tangent vectors V_5 and one of the real subspaces in $S_4 \otimes \widetilde{S}_4$ invariant under the transformations from the whole symmetry group of S_4. To this end, let us first find all the invariant subspaces of $S_4 \otimes \widetilde{S}_4$. Apart from the unity subspace (made up by all the tensors proportional to the unity tensor), an evident invariant subspace is the linear envelope of generators (3), which is ten-dimensional and is a Lie algebra. To find the third invariant subspace, let us write out the tensor products of $\mathbf{1}$ and σ_k not included into (3). There are five of them:

$$-i\sigma_2\otimes\mathbf{1} \equiv \Gamma_0,\ \ \sigma_3\otimes\sigma_k \equiv \Gamma_k\ (k=1,2,3),\ \text{ and }\ i\sigma_1\otimes\mathbf{1} \equiv \Gamma_5. \tag{5}$$

It is straightforward to check that Γ_A $(A = 0,1,2,3,5)$ satisfy the following commutation relations:

$$[\Gamma_A, M_{BC}] = -\eta_{AB}\Gamma_C + \eta_{AC}\Gamma_B, \tag{6}$$

which means that the linear envelope of Γ_A is a subspace invariant under the transformations from SO(3,2). Let us denote the real linear envelope of Γ_A as $G^{(5)}$.

It is apparent that Dirac conjugation of 4-spinors induces an anti-isomorphism of $S_4 \otimes \widetilde{S}_4$ onto itself (also referred to as Dirac conjugation). Namely, if a given element of $S_4 \otimes \widetilde{S}_4$ has the components U^i_j, then its Dirac conjugate has the components

$$(\overline{U})^i_j = (d^{li})^*\,(U^k_l)^*\,d_{kj},$$

where d^{ij} denote the components of the matrix inverse to d_{ij} (similar to how $g^{\mu\nu}$ denotes the inverse of $g_{\mu\nu}$). In the matrix form, the latter formula for the components reads as follows: $\overline{U} = d^{-1}\, U^{\dagger} d$, where it has been taken into account that d is Hermitian. It is not difficult to check that

$$\overline{\Gamma}_A = \Gamma_A \quad \text{and} \quad \overline{M}_{AB} = -M_{AB} \tag{7}$$

for all A and B (the latter indices label the elements of $S_4 \otimes \widetilde{S}_4$, not their components). In a similar manner, the isomorphism of S_4 onto $\widetilde{S}_4$ defined by the inner product c induces an isomorphism of $S_4 \otimes \widetilde{S}_4$ onto itself. It is convenient to take that the image of an element with components $U^i_{\ j}$ has the components

$$(\widehat{U})^i_{\ j} = c^{il}\, U^k_{\ l}\, c_{kj},$$

where c^{ij} denote the components of the matrix inverse to c_{ij} . In the matrix form, one has $\widehat{U} = c^{-1}\, U^T c$. It is not difficult to verify that

$$\widehat{\Gamma}_A = \Gamma_A \quad \text{and} \quad \widehat{M}_{AB} = -M_{AB} \tag{8}$$

for all A and B.

In a standard way, the inner product d on S_4 defines an inner product on the space of co-spinors. Since the map $\psi \mapsto \overline{\psi}$ is an anti-isomorphism, any element of $\widetilde{S}_4$ is a Dirac conjugate of some element of S_4, so it is sufficient to define that

$$d(\overline{\psi}, \overline{\chi}) = d(\psi, \chi)^* = d(\chi, \psi)$$

for any ψ and χ. In components, the above inner product equals $(\overline{\psi}_i)^*\, \overline{\chi}_j (d^{ij})^*$, where it has been taken into account that d is Hermitian. It is now a simple matter to define the inner product d for elements of $S_4 \otimes \widetilde{S}_4$. By definition,

$$d(\psi \otimes \overline{\xi}, \phi \otimes \overline{\chi}) = d(\psi, \phi)\, d(\overline{\xi}, \overline{\chi}).$$

So for two arbitrary elements of $S_4 \otimes \widetilde{S}_4$ with components $U^i_{\ j}$ and $V^i_{\ j}$ the considered inner product equals

$$d(U, V) = d_{ik}\, (U^i_{\ j})^* V^k_{\ l}\, (d^{jl})^* = (d^{jl})^*\, (U^i_{\ j})^*\, d_{ik} V^k_{\ l} = (\overline{U})^l_{\ k} V^k_{\ l} = \mathrm{trace}\{\overline{U} V\},$$

where the tensors under the trace sign are regarded as matrices with elements equal to their components.

In a similar way, the inner product c on S_4 induces an inner product on $\widetilde{S}_4$. It is convenient to take that

$$c(\widehat{\psi}, \widehat{\chi}) = c(\psi, \chi) = -\, c(\chi, \psi)$$

for any ψ and χ. This is sufficient to define c for any two co-spinors, since any element of $\widetilde{S}_4$ is an image of some element of S_4 relative to the isomorphism $\psi \mapsto \widehat{\psi}$. In components, the above inner product equals $-\widehat{\psi}_i\, \widehat{\chi}_j\, c^{ij}$, where it has been taken into account that c is antisymmetric. It is now a simple matter to define c for elements of $S_4 \otimes \widetilde{S}_4$. By definition,

$$c(\psi \otimes \widehat{\xi}, \phi \otimes \widehat{\chi}) = c(\psi, \phi)\, c(\widehat{\xi}, \widehat{\chi})\,,$$

so the induced inner product is *symmetric* with respect to its arguments. For two arbitrary elements of $S_4 \otimes \widetilde{S}_4$ with components $U^i_{\ j}$ and $V^i_{\ j}$ the considered inner product equals

$$c(U, V) = -c_{ik}\, U^i_{\ j}\, V^k_{\ l}\, c^{jl} = -c^{jl}\, U^i_{\ j}\, c_{ik} V^k_{\ l} = (c^{lj}\, U^i_{\ j}\, c_{ik}) V^k_{\ l} = (\widehat{U})^l_{\ k} V^k_{\ l} = \mathrm{trace}\{\widehat{U} V\},$$

where the tensors under the trace sign are regarded as matrices with elements equal to their components. According to (7) and (8), for Γ_A one has

$$d(\Gamma_A, \Gamma_B) = c(\Gamma_A, \Gamma_B) = \text{trace}\{\Gamma_A \Gamma_B\},$$

so the inner products induced on $G^{(5)}$ by d and c *are the same.*

As for any other type of tensors of rank (1,1), for the elements of $S_4 \otimes \widetilde{S}_4$ one can define *multiplication* as

$$(AB)^i_{\ j} = A^i_{\ k} B^k_{\ j}.$$

By direct calculation, one can verify that for the Γ_A given by equation (5), one has

$$\Gamma_A \Gamma_B + \Gamma_B \Gamma_A = -2\eta_{AB} \cdot \mathbb{1}, \tag{9}$$

where $\mathbb{1}$ is the unity tensor. Furthermore, it is easy to prove that $\Gamma_0 \Gamma_1 \Gamma_2 \Gamma_3 \Gamma_5 = -\mathbb{1}$ and that for all $A \neq B$,

$$M_{AB} = (1/4) \cdot (\Gamma_A \Gamma_B - \Gamma_B \Gamma_A), \tag{10}$$

so $S_4 \otimes \widetilde{S}_4$ is a Clifford algebra whose constituents are any four elements out of the quintuple Γ_A.

It would seem that a relation between 4-spinors and the space-time manifold can now be established by postulating that at each point of the latter there is fixed a certain isomorphism from the vector space V_5 of five-dimensional tangent vectors onto the subspace $G^{(5)}$. To each standard orthonormal basis in V_5 one can put into correspondence a basis in S_4 by requiring that in this latter basis the components of the inner products d and c and the components $(\Gamma_A)^i_j$ of the images of the basis five-vectors $\mathbf{e}_A$ in the corresponding basis in $S_4 \otimes \widetilde{S}_4$ be given by certain numerical matrices, for example, by those appearing in equations (1) or (2) and in equation (5).

The described isomorphism of V_5 onto $G^{(5)}$ can be used to establish an isomorphism between V_4 and a subspace in $S_4 \otimes \widetilde{S}_4$ made up by the elements $i\,\Gamma_\alpha \Gamma_5$ ($\alpha = 0, 1, 2, 3$) and their linear combinations with real coefficients. Actually, what one has here is an isomorphism of V_4 onto one of the maximal vector spaces of simple bivectors over V_5 with $\mathbf{E}_\alpha \mapsto \mathbf{e}_\alpha \wedge \mathbf{e}_5$ (see ref.[1]). If the components of d and c and $(\Gamma_A)^i_j$ are given by equations (2) and (5), then the components of the images of the basis four-vectors $\mathbf{E}_\alpha$ in this basis in $S_4 \otimes \widetilde{S}_4$ are

$$(\gamma_0)^i_j = \sigma_3 \otimes \mathbf{1} = \begin{vmatrix} \mathbf{1} & \mathbf{0} \\ \mathbf{0} & -\mathbf{1} \end{vmatrix} \quad \text{and} \quad (\gamma_k)^i_j = i\sigma_2 \otimes \sigma_k = \begin{vmatrix} \mathbf{0} & \sigma_k \\ -\sigma_k & \mathbf{0} \end{vmatrix}$$

and coincide with the conventional gamma matrices in the Dirac basis.

3. 8-Spinors

Unfortunately, there is one substantial fly in the ointment: Under the assumptions traditionally made about 4-spinors, the isomorphism between V_5 and $G^{(5)}$ described above is *not preserved* by parallel transport. To see what is exactly the problem, let us consider flat space-time with zero torsion. Within the traditional approach, it is assumed that the

4-spinor basis corresponding to a Lorentz four-vector basis (in the sense explained above) is *self-parallel*, meaning that all the 4-spinor connection coefficients for it are identically zero. Since the basis in S_4 is defined in such a way that the matrices $(\gamma_\alpha)^i_j$ are the same at every point in space-time, an element of $S_4 \otimes \widetilde{S}_4$ with such components is transported into an element of $S_4 \otimes \widetilde{S}_4$ with the same components. This means that the image of a basis four-vector $\mathbf{E}_\alpha$ at one point is transported into the image of $\mathbf{E}_\alpha$ at any other point, which is in agreement with the fact that a Lorentz four-vector basis is self-parallel. Unfortunately, by the same token, the image of a basis *five*-vector $\mathbf{e}_A$ at one point will be transported into the image of $\mathbf{e}_A$ at another point, whereas the basis five-vector $\mathbf{e}_\alpha$ itself is *not* necessarily transported into $\mathbf{e}_\alpha$, but may acquire an additional component proportional to $\mathbf{e}_5$. For instance, for a standard orthonormal basis associated with a Lorentz frame one has [1]

$$\nabla_\mu \mathbf{e}_\alpha = -\kappa \eta_{\mu\alpha} \mathbf{e}_5 \quad \text{and} \quad \nabla_\mu \mathbf{e}_5 = 0, \tag{11}$$

where κ is a certain constant with dimension of inverse length. We therefore see that to make the relation between V_5 and $S_4 \otimes \widetilde{S}_4$ invariant under parallel transport, one *must* suppose that the 4-spinor basis corresponding to a standard orthonormal five-vector basis associated with a Lorentz frame in flat space-time is *not* self-parallel, meaning that some of the 4-spinor connection coefficients for it are *not* identically zero. Let us define the latter as

$$(\nabla_\mu \psi)^i = \partial_\mu \psi^i + \Theta^i_{\ j\mu} \psi^j.$$

Accordingly, for any $\widetilde{\chi} \in \widetilde{S}_4$, in the dual basis of co-spinors one has

$$(\nabla_\mu \widetilde{\chi})_i = \partial_\mu \widetilde{\chi}_i - \widetilde{\chi}_j \Theta^j_{\ i\mu}.$$

For an element of $S_4 \otimes \widetilde{S}_4$ with components $(\Gamma_A)^i_j$ one will therefore have

$$(\nabla_\mu \Gamma_A)^i_j = \partial_\mu (\Gamma_A)^i_j + \Theta^i_{\ k\mu} (\Gamma_A)^k_j - (\Gamma_A)^i_k \Theta^k_{\ j\mu} = (\Theta_\mu \Gamma_A - \Gamma_A \Theta_\mu)^i_k, \tag{12}$$

where inside the brackets in the right-hand side, Θ_μ and Γ_A are regarded as matrices, and it has been taken into account that $(\Gamma_A)^i_j$ are constants. If one wishes the relation between V_5 and $S_4 \otimes \widetilde{S}_4$ to hold under parallel transport, one should require that

$$\Theta_\mu \Gamma_\alpha - \Gamma_\alpha \Theta_\mu = \nabla_\mu \Gamma_\alpha = -\kappa \eta_{\mu\alpha} \Gamma_5 \tag{13}$$

and

$$\Theta_\mu \Gamma_5 - \Gamma_5 \Theta_\mu = \nabla_\mu \Gamma_5 = 0. \tag{14}$$

Before we move further, it is appropriate to make the following important comment. As it has already been said, within the traditional approach, one establishes a relation between 4-spinors and space-time by supposing that at each point there is fixed an isomorphism between the space of tangent four-vectors V_4 and a certain subspace in $S_4 \otimes \widetilde{S}_4$. Another possibility is to suppose that an isomorphism is fixed between a subspace in $S_4 \otimes \widetilde{S}_4$ and the space of 1-forms, $\widetilde{V}_4$, associated with V_4. In the case of four-vectors, it does not matter which alternative to choose, for there exists an isomorphism between V_4 and $\widetilde{V}_4$, determined by the inner product g, which is invariant under parallel transport, since the latter conserves g.

The situation is different for five-dimensional tangent vectors. Since the nondegenerate inner product h on V_5 is *not* covariantly constant, the isomorphism between V_5 and the space of corresponding 1-forms, $\widetilde{V}_5$, that it defines is *not* preserved by parallel transport. Consequently, an isomorphism between V_5 and a subspace in $S_4 \otimes \widetilde{S}_4$ invariant under parallel transport will *not* induce an isomorphism with similar properties between the same subspace in $S_4 \otimes \widetilde{S}_4$ and $\widetilde{V}_5$, and vice versa. This means that when attempting to establish a relation between spinors and space-time that would involve five-vectors, one does have to specify which of the two spaces—V_5 or $\widetilde{V}_5$—is considered isomorphic to a subspace in $S_4 \otimes \widetilde{S}_4$. Equations (13) and (14) were derived for the case where that space is V_5. If one selects $\widetilde{V}_5$, then instead of equation (12) one will have

$$(\nabla_\mu \Gamma^A)^i_j = \partial_\mu (\Gamma^A)^i_j + \Theta^i_{k\mu}(\Gamma^A)^k_j - (\Gamma^A)^i_k \Theta^k_{j\mu} = (\Theta_\mu \Gamma^A - \Gamma^A \Theta_\mu)^i_k,$$

where $(\Gamma^A)^i_j$ are components of the images of the basis five-vector 1-forms $\widetilde{\mathbf{o}}^A$ dual to the selected five-vector basis $\mathbf{e}_A$, in the corresponding basis in $S_4 \otimes \widetilde{S}_4$. In the following, it will always be supposed that for a standard orthonormal five-vector basis associated with a Lorentz frame, the corresponding basis in S_4 is selected in such a way that $(\Gamma^A)^i_j = \eta^{AB}(\Gamma_B)^i_j$, where $(\Gamma_B)^i_j$ are given by formulae (5). For a basis of five-vector 1-forms associated with a Lorentz frame one has

$$\nabla_\mu \widetilde{\mathbf{o}}^\alpha = 0 \text{ and } \nabla_\mu \widetilde{\mathbf{o}}^5 = \kappa\, \eta_{\mu\beta}\, \widetilde{\mathbf{o}}^\beta.$$

Therefore, if one wishes the isomorphism between $\widetilde{V}_5$ and $S_4 \otimes \widetilde{S}_4$ to hold under parallel transport, instead of relations (13) and (14) one should require that

$$\Theta_\mu \Gamma^\alpha - \Gamma^\alpha \Theta_\mu = \nabla_\mu \Gamma^\alpha = 0 \tag{15}$$

and

$$\Theta_\mu \Gamma^5 - \Gamma^5 \Theta_\mu = \nabla_\mu \Gamma^5 = \kappa \eta_{\mu\beta} \Gamma^\beta. \tag{16}$$

Let us see if the desired operator Θ_μ exists. Let us first consider the case where it is V_5 that is identified with $G^{(5)}$. If one requires that parallel transport conserves the inner products *d* and *c*, then by direct calculation one can verify that the operator satisfying equations (13) is

$$\Theta_\mu = -(\kappa/2)\, \Gamma_\mu \Gamma_5 = \kappa M_{5\mu}. \tag{17}$$

The same result directly follows from commutation relations (6) at $A = \alpha$, $B = 5$, and $C = \mu$. However, from the same relations it follows that the above operator satisfies the equation

$$\Theta_\mu \Gamma_5 - \Gamma_5 \Theta_\mu = \kappa\, \Gamma_\mu$$

rather than equation (14), which was to be expected since $M_{5\mu}$ is a generator of SO(3,2).

Let us now see what happens if it is $\widetilde{V}_5$ that is identified with $G^{(5)}$. It is easy to check that for the $(\Gamma^A)^i_j$ specified above, operator (17) satisfies requirement (16), but not requirements (15). Instead of the latter one has

$$\Theta_\mu \Gamma^\alpha - \Gamma^\alpha \Theta_\mu = -\kappa\, \delta^\alpha_\mu\, \Gamma_5.$$

So in either case the score is one out of two. The latter fact may give one a hint on how to remedy the situation. It is apparent that within the space of complex 4×4 matrices the desired operator Θ_μ does not exist. This means that one has to change either the numbers or the matrices. Let us first present a formal solution to the problem, which at this stage should be regarded merely as a guess. Let us suppose that we have a quantity ε, whose nature is to be specified, that satisfies the equation $\varepsilon \cdot \varepsilon = 0$. And let us construct the operator

$$\Theta_\mu = -(\varepsilon\kappa/2)\,\Gamma_\mu\Gamma_5 = \varepsilon\kappa M_{5\mu}. \tag{18}$$

It is easy to verify that this operator satisfies the relations

$$\Theta_\mu\Gamma_\alpha - \Gamma_\alpha\Theta_\mu = -\kappa\eta_{\mu\alpha}(\varepsilon\Gamma_5) \quad \text{and} \quad \Theta_\mu(\varepsilon\Gamma_5) - (\varepsilon\Gamma_5)\,\Theta_\mu = 0. \tag{19}$$

Consequently, if one supposes that the connection coefficients for the spinor basis corresponding to a standard orthonormal five-vector basis associated with a Lorentz frame (in the sense explained above) are given by formula (18) and that the components of the image of the fifth basis five-vector $\mathbf{e}_5$ in the corresponding spinor basis are $(\varepsilon\Gamma_5)^i_j$ rather than $(\Gamma_5)^i_j$, then the isomorphism between V_5 and the real linear envelope of Γ_α and $\varepsilon\Gamma_5$ will be invariant under parallel transport. In a similar manner, since operator (18) also satisfies the relations

$$\Theta_\mu(\varepsilon\Gamma^\alpha) - (\varepsilon\Gamma^\alpha)\,\Theta_\mu = 0 \quad \text{and} \quad \Theta_\mu\Gamma^5 - \Gamma^5\Theta_\mu = \kappa\eta_{\mu\beta}(\varepsilon\Gamma^\beta), \tag{20}$$

one can make the isomorphism between $\widetilde{V}_5$ and an appropriate subspace of spinor tensors of rank (1,1) invariant under parallel transport by supposing that the components of the images of the first four basis five-vector 1-forms are $(\varepsilon\Gamma^\alpha)^i_j$ rather than $(\Gamma^\alpha)^i_j$, while the image of the fifth basis 1-form still has the components $(\Gamma^5)^i_j$.

Let us now discuss the nature of the quantity ε. One possibility is that the latter is an element of a field of numbers more general than C. This is not a good idea though, since all calculations with spinors should result in complex or real numbers. The other possibility is that ε is an operator, in other words, a matrix, and the simplest variant is for the latter to be a matrix of rank 2. What this means is that we are actually graduating to 8×8 matrices. The latter can be presented either as a tensor product of *three* 2×2 matrices or as a tensor product of a 2×2 matrix and a 4×4 matrix from $S_4 \otimes \widetilde{S}_4$. For example, in the latter notations, the components of the images of the first four basis five-vectors will have the form $\mathbf{1} \otimes \Gamma_\alpha$, where, as before, $\mathbf{1}$ denotes the unity 2×2 matrix; the components of the image of the fifth basis five-vector will have the form $\varepsilon \otimes \Gamma_5$; the operator Θ_μ will be $\varepsilon \otimes \kappa M_{5\mu}$; and so on.

One should now recall a simple theorem stating that for any nonzero 2×2 matrix ε that satisfies the equation $\varepsilon \cdot \varepsilon = 0$, there exists such a unitary transformation that brings ε to the form $\varsigma \cdot \sigma_+$, where $\sigma_+ = \frac{1}{2}(\sigma_1 + i\sigma_2)$ and ς is some positive real number[1]. To be definite, in the following it will always be assumed that the basis in the two-dimensional complex space where ε operates is chosen is such a way that $\varepsilon = \varsigma \cdot \sigma_+$.

Let us denote the space of complex vectors upon which the considered 8×8 matrices act as S_8. The latter is evidently eight-dimensional. The space of corresponding 1-forms

[1]There also exists a unitary transformation that brings any such ε to the form $\varsigma \cdot \sigma_-$, where $\sigma_- = \frac{1}{2}(\sigma_1 - i\sigma_2)$.

(co-spinors) will be denoted as $\widetilde{S}_8$. It is natural to suppose that the inner products d and c defined on S_4 can be extended onto S_8. One can do this by requiring that the resulting inner products be covariantly constant. Let us suppose that the unknown components of d and c for S_8 can be presented as

$$d = x \otimes \sigma_3 \otimes \mathbf{1} \quad \text{and} \quad c = y \otimes \sigma_1 \otimes i\sigma_2,$$

where x and y are some 2×2 matrices. Let us again consider flat space-time with zero torsion. From consideration of homogeneity one may suppose that it is always possible to select a basis in S_8 corresponding to a standard orthonormal five-vector basis associated with a Lorentz frame in such a way that matrices ε, x, and y be constant throughout space-time. By direct calculation with components one can verify that from covariant constancy of d and c follows that x and y should satisfy the relations

$$x \cdot \varepsilon = \varepsilon^{\dagger} \cdot x \quad \text{and} \quad y \cdot \varepsilon = \varepsilon^{\mathrm{T}} \cdot y.$$

Supposing that the basis in S_8 is select in such a way that ε is proportional to σ_+, one finds that x and y should be such that

$$x \cdot \sigma_+ = \sigma_- \cdot x \quad \text{and} \quad y \cdot \sigma_+ = \sigma_- \cdot y,$$

and the only solution of these equations is $x \propto \sigma_1$ and $y \propto \sigma_1$. One can always choose the normalization of d and c in such a way that the proportionality factors between x and σ_1 and y and σ_1 both be unity and have

$$d = \sigma_1 \otimes \sigma_3 \otimes \mathbf{1} \quad \text{and} \quad c = \sigma_1 \otimes \sigma_1 \otimes i\sigma_2. \tag{21}$$

Let us now examine the space $S_8 \otimes \widetilde{S}_8$ in a manner similar to how we have examined the Clifford algebra $S_4 \otimes \widetilde{S}_4$. Let us first determine the transformations under which the above inner products d and c are invariant. To this end, let us use the well-known notation $\sigma_\mu = (\mathbf{1}, \sigma_k)$. One can then present a generator of the transformation in question in the following general form: $X^{\lambda\mu\nu}\sigma_\lambda \otimes \sigma_\mu \otimes \sigma_\nu$. By direct calculation one can obtain the following results for the coefficients $X^{\lambda\mu\nu}$ in the 8-spinor basis specified above:

$X^{\lambda\mu 0}$	$\lambda = 0$	$\lambda = 1$	$\lambda = 2$	$\lambda = 3$
$\mu = 0$	0	0	0	$1/2$
$\mu = 1$	0	0	0	$i/2$
$\mu = 2$	0	0	0	$-i/2$
$\mu = 3$	$i/2$	$i/2$	$-1/2$	0

and

$X^{\lambda\mu k}$	$\lambda = 0$	$\lambda = 1$	$\lambda = 2$	$\lambda = 3$
$\mu = 0$	$i/2$	$i/2$	$-1/2$	0
$\mu = 1$	$1/2$	$1/2$	$i/2$	0
$\mu = 2$	$1/2$	$1/2$	$i/2$	0
$\mu = 3$	0	0	0	$1/2$

,

where all nonzero coefficients have been normalized to have the absolute value $1/2$. From the above tables it follows that there are 36 generators, as there should be, since the dimension of the sympletic Lie algebra of $2n \times 2n$ matrices is $n(2n+1)$. For our purposes it will be sufficient to consider only the subalgebra made up by ten generators of the form $\mathbf{1} \otimes M_{AB}$, where M_{AB} are given by formulae (3). Let us denote this subalgebra as $\mathcal{M}$.

There are 28 triple tensor products of $\mathbf{1}$ and σ_k that are not generators of symmetry transformations. They can be arranged into several sets. For our purposes it will be sufficient to consider only two such sets, each of which makes up a basis of a five-dimensional vector subspace invariant under the transformations from $\mathcal{M}$. These two subspaces are:

G_0: the real linear envelope of the elements $\mathbf{1} \otimes \Gamma_A$, where Γ_A are given by formulae (5);

G_+: the real linear envelope of the elements $\sigma_+ \otimes \Gamma_A$, where Γ_A are given by formulae (5).

It is evident that these two subspaces are made up by elements of the form $\mathbf{1} \otimes U$ and $\sigma_+ \otimes U$, respectively, where U is any element from the subspace $G^{(5)} \subset S_4 \otimes \widetilde{S}_4$ considered earlier. One can therefore establish a natural homomorphism of G_0 onto G_+ by putting $\mathbf{1} \otimes U \mapsto \sigma_+ \otimes U$ for any $U \in G^{(5)}$. In a standard way, one can define the inner products d and c for the elements of $S_8 \otimes \widetilde{S}_8$, similar to how this has been done for $S_4 \otimes \widetilde{S}_4$. One will find that the restrictions of d and c to G_0 result in the same symmetric nondegenerate inner product and that with respect to the latter the elements $\mathbf{1} \otimes \Gamma_A$ are orthonormal. At the same time, one will find that the restrictions of d and c to G_+ are identically zero, so the mentioned correspondence between the elements of G_0 and G_+ is not an isomorphism.

Considering what one has learned in connection with equations (19) and (20), one can describe the relevant part of the structure of $S_8 \otimes \widetilde{S}_8$ in the following way. There exists a five-dimensional real subspace in $S_8 \otimes \widetilde{S}_8$ that has been labeled above as G_0. Its symmetry group is SO(3,2) and the corresponding transformations are produced by the generators from subalgebra $\mathcal{M}$. To construct a subspace that could be identified with V_5, one should first select a four-dimensional subspace in G_0 whose symmetry group would be SO(3,1). The corresponding transformations are produced by the generators from a relevant subalgebra of $\mathcal{M}$. The mentioned four-dimensional subspace is to be put into correspondence with the subspace $\mathcal{Z}$ in V_5 and will be denoted as $S(\mathcal{Z})$. In the following, it will always be assumed that the basis in S_8 is selected in such a way that $S(\mathcal{Z})$ coincides with the real linear envelope of $\mathbf{1} \otimes \Gamma_\alpha$ ($\alpha = 0, 1, 2, 3$).

One's next step is to consider a subspace in G_0 orthogonal to $S(\mathcal{Z})$. It is evident that in the spinor basis specified above, this is the one-dimensional subspace made up by the elements proportional to $\mathbf{1} \otimes \Gamma_5$. One should then consider the image of this subspace in G_+ with respect to the homomorphism mentioned earlier. In the specified spinor basis, this image will apparently consist of all the elements proportional to $\sigma_+ \otimes \Gamma_5$. The latter subspace is to be put into correspondence with the subspace $\mathcal{E}$ in V_5 and will be denoted as $S(\mathcal{E})$.

In agreement with the first equation in (20), the image of the subspace $S(\mathcal{Z})$ in G_+ with respect to the mentioned homomorphism should be put into correspondence with the subspace $\widetilde{\mathcal{Z}}$ in $\widetilde{V}_5$, so it makes sense to denote it as $S(\widetilde{\mathcal{Z}})$. It is evident that the latter coincides with the real linear envelope of the elements $\sigma_+ \otimes \Gamma_\alpha$. Finally, in view of the second equation in (20), the one-dimensional subspace in G_0 made up by the elements proportional to $\mathbf{1} \otimes \Gamma_5$ should be put into correspondence with the subspace $\widetilde{\mathcal{E}}$ in $\widetilde{V}_5$ and will be denoted as $S(\widetilde{\mathcal{E}})$. One therefore has

$$G_0 = S(\mathcal{Z}) \oplus S(\widetilde{\mathcal{E}}) \quad \text{and} \quad G_+ = S(\widetilde{\mathcal{Z}}) \oplus S(\mathcal{E}).$$

To acquire an understanding of these correspondences, let us approach the problem in a slight different way. So far we have been looking for a real five-dimensional vector subspace inside the space of spinor tensors of rank (1,1) that could be identified with V_5. Let us now, so to say, reverse the problem and see if one can construct out of five-vectors a Clifford algebra that could be identified with an appropriate subalgebra inside the space of mentioned spinor tensors. To this end, let us consider a formal vector space $\mathcal{Q}_0$ made up by all the multivectors that one can construct with the five-vectors from the subspace $\mathcal{Z} \subset V_5$. If $\mathbf{e}_A$ is a regular basis in V_5, then the basis in $\mathcal{Q}_0$ will apparently be made up by the following multivectors: four vectors $\mathbf{e}_0$, $\mathbf{e}_1$, $\mathbf{e}_2$, and $\mathbf{e}_3$; six bivectors $\mathbf{e}_0 \wedge \mathbf{e}_1$, $\mathbf{e}_0 \wedge \mathbf{e}_2$, $\mathbf{e}_0 \wedge \mathbf{e}_3$, $\mathbf{e}_1 \wedge \mathbf{e}_2$, $\mathbf{e}_1 \wedge \mathbf{e}_3$, and $\mathbf{e}_2 \wedge \mathbf{e}_3$; four trivectors $\mathbf{e}_0 \wedge \mathbf{e}_1 \wedge \mathbf{e}_2$, $\mathbf{e}_0 \wedge \mathbf{e}_1 \wedge \mathbf{e}_3$, $\mathbf{e}_0 \wedge \mathbf{e}_2 \wedge \mathbf{e}_3$, and $\mathbf{e}_1 \wedge \mathbf{e}_2 \wedge \mathbf{e}_3$; one rank 4 multivector $\mathbf{e}_0 \wedge \mathbf{e}_1 \wedge \mathbf{e}_2 \wedge \mathbf{e}_3$; and one unity scalar *1*, which should be regarded as a multivector of rank 0. An arbitrary element of $\mathcal{Q}_0$ is some linear combination of the mentioned multivectors.[2]

There are two binary operations that one can perform with elements of $\mathcal{Q}_0$: calculation of the inner product h and exterior multiplication $\wedge$. By using these two operations, one can construct a new kind of multiplication for elements of $\mathcal{Q}_0$ that will be denoted with the sign $\circ$. The simplest way to define the latter is by induction:

1. For any scalar f and any multivector $\mathbf{m}_1 \wedge \mathbf{m}_2 \wedge \ldots \wedge \mathbf{m}_k$,

$$f \circ (\mathbf{m}_1 \wedge \mathbf{m}_2 \wedge \ldots \wedge \mathbf{m}_k) = (\mathbf{m}_1 \wedge \mathbf{m}_2 \wedge \ldots \wedge \mathbf{m}_k) \circ f = f \cdot \mathbf{m}_1 \wedge \mathbf{m}_2 \wedge \ldots \wedge \mathbf{m}_k.$$

2. For any two multivectors $\mathbf{m}_1 \wedge \mathbf{m}_2 \wedge \ldots \wedge \mathbf{m}_k$ and $\mathbf{n}_1 \wedge \mathbf{n}_2 \wedge \ldots \wedge \mathbf{n}_s$ with nonzero ranks,

$$\begin{aligned}
&(\mathbf{m}_1 \wedge \mathbf{m}_2 \wedge \ldots \wedge \mathbf{m}_k) \circ (\mathbf{n}_1 \wedge \mathbf{n}_2 \wedge \ldots \wedge \mathbf{n}_s) \\
&\quad = - h(\mathbf{m}_k, \mathbf{n}_1) \cdot (\mathbf{m}_1 \wedge \ldots \wedge \mathbf{m}_{k-1}) \circ (\mathbf{n}_2 \wedge \ldots \wedge \mathbf{n}_s) \\
&\qquad + h(\mathbf{m}_k, \mathbf{n}_2) \cdot (\mathbf{m}_1 \wedge \ldots \wedge \mathbf{m}_{k-1}) \circ (\mathbf{n}_1 \wedge \mathbf{n}_3 \wedge \ldots \wedge \mathbf{n}_s) \\
&\qquad - \ldots \\
&\qquad + (-1)^s\, h(\mathbf{m}_k, \mathbf{n}_s) \cdot (\mathbf{m}_1 \wedge \ldots \wedge \mathbf{m}_{k-1}) \circ (\mathbf{n}_1 \wedge \ldots \wedge \mathbf{n}_{s-1}) \\
&\qquad + (\mathbf{m}_1 \wedge \mathbf{m}_2 \wedge \ldots \wedge \mathbf{m}_{k-1}) \circ (\mathbf{m}_k \wedge \mathbf{n}_1 \wedge \mathbf{n}_2 \wedge \ldots \wedge \mathbf{n}_s)\,.
\end{aligned}$$

It is easy to see that the introduction of multiplication $\circ$ turns $\mathcal{Q}_0$ into a 16-dimensional Clifford algebra. Inside this algebra there exists a five-dimensional subspace H_0, which is the linear envelope of the four vectors $\mathbf{e}_\alpha$ and of the multivector $\mathbf{e}_0 \wedge \mathbf{e}_1 \wedge \mathbf{e}_2 \wedge \mathbf{e}_3$. This subspace is invariant under transformations from an SO(3,2) group whose generators are

[2] To make the following formulae simpler, the definition of a wedge product of k vectors includes the factor $1/k!$. So, for example, $\mathbf{m} \wedge \mathbf{n} = \frac{1}{2}(\mathbf{m} \otimes \mathbf{n} - \mathbf{n} \otimes \mathbf{m})$. Also, one should remember that the signature of the nondegenerate inner product h for five-vectors is assumed to be $(+ - - - +)$.

the six bivectors $\mathbf{e}_\alpha \wedge \mathbf{e}_\beta$ $(\alpha < \beta)$ and four trivectors $\mathbf{e}_\alpha \wedge \mathbf{e}_\beta \wedge \mathbf{e}_\mu$ $(\alpha < \beta < \mu)$. Indeed, introducing for a moment the notations $\Gamma_\alpha = \mathbf{e}_\alpha$, $\Gamma_5 = \mathbf{e}_0 \wedge \mathbf{e}_1 \wedge \mathbf{e}_2 \wedge \mathbf{e}_3$, and

$$\begin{aligned} &M_{\alpha\beta} = -M_{\beta\alpha} = \tfrac{1}{2}\,\mathbf{e}_\alpha \wedge \mathbf{e}_\beta \ (0 \le \alpha < \beta \le 3), \\ &M_{50} = -M_{05} = \tfrac{1}{2}\,\mathbf{e}_1 \wedge \mathbf{e}_2 \wedge \mathbf{e}_3, \ \ M_{51} = -M_{15} = -\tfrac{1}{2}\,\mathbf{e}_0 \wedge \mathbf{e}_2 \wedge \mathbf{e}_3, \\ &M_{52} = -M_{25} = \tfrac{1}{2}\,\mathbf{e}_0 \wedge \mathbf{e}_1 \wedge \mathbf{e}_3, \ \ M_{53} = -M_{35} = -\tfrac{1}{2}\,\mathbf{e}_0 \wedge \mathbf{e}_1 \wedge \mathbf{e}_2, \end{aligned} \tag{22}$$

one can verify that these Γ_A and M_{AB} satisfy the commutation relations (4) and (6), where for any two elements U and V from $\mathcal{Q}_0$ one defines $[U, V] = U \circ V - V \circ U$. Moreover, these Γ_A and M_{AB} also satisfy equations (9) and (10), so it is possible to establish a homomorphism of the complexification of $\mathcal{Q}_0$ onto $S_4 \otimes \widetilde{S}_4$ with the following correspondence:

scalar *1*	unity matrix 1
vector $\mathbf{e}_\alpha$	matrix Γ_α
bivector $\mathbf{e}_\alpha \wedge \mathbf{e}_\beta$	matrix $\Gamma_{[\alpha}\Gamma_{\beta]}$
trivector $\mathbf{e}_\alpha \wedge \mathbf{e}_\beta \wedge \mathbf{e}_\mu$	matrix $\Gamma_{[\alpha}\Gamma_\beta\Gamma_{\mu]}$
multivector $\mathbf{e}_\alpha \wedge \mathbf{e}_\beta \wedge \mathbf{e}_\mu \wedge \mathbf{e}_\nu$	matrix $\Gamma_{[\alpha}\Gamma_\beta\Gamma_\mu\Gamma_{\nu]}$

One can turn this correspondence into an isomorphism by introducing for the elements of $\mathcal{Q}_0$ an Hermitian inner product d, which for the above basis elements is defined as follows:

$$d(1, 1) = 4, \ \ d(\mathbf{e}_\alpha, \mathbf{e}_\beta) = -4\,h(\mathbf{e}_\alpha, \mathbf{e}_\beta),$$

$$d(\mathbf{e}_\alpha \wedge \mathbf{e}_\beta, \mathbf{e}_\sigma \wedge \mathbf{e}_\tau) = -4\,h(\mathbf{e}_\alpha \wedge \mathbf{e}_\beta, \mathbf{e}_\sigma \wedge \mathbf{e}_\tau)$$

$$d(\mathbf{e}_\alpha \wedge \mathbf{e}_\beta \wedge \mathbf{e}_\mu, \mathbf{e}_\sigma \wedge \mathbf{e}_\tau \wedge \mathbf{e}_\rho) = 4\,h(\mathbf{e}_\alpha \wedge \mathbf{e}_\beta \wedge \mathbf{e}_\mu, \mathbf{e}_\sigma \wedge \mathbf{e}_\tau \wedge \mathbf{e}_\rho),$$

$$d(\mathbf{e}_\alpha \wedge \mathbf{e}_\beta \wedge \mathbf{e}_\mu \wedge \mathbf{e}_\nu, \mathbf{e}_\sigma \wedge \mathbf{e}_\tau \wedge \mathbf{e}_\rho \wedge \mathbf{e}_\omega) = 4h(\mathbf{e}_\alpha \wedge \mathbf{e}_\beta \wedge \mathbf{e}_\mu \wedge \mathbf{e}_\nu, \mathbf{e}_\sigma \wedge \mathbf{e}_\tau \wedge \mathbf{e}_\rho \wedge \mathbf{e}_\omega),$$

where the inner product h of two multivectors of same rank is defined in the standard way, and the inner product d of any two multivectors with different ranks is taken to be identically zero.

Let us now see what happens to the elements of $\mathcal{Q}_0$ under parallel transport. Let us once again consider flat space-time with zero torsion. In view of equations (11), all the elements of $\mathcal{Q}_0$ except the scalars will be transported into multivectors not entirely belonging to $\mathcal{Q}_0$. It is apparent that the additional components they acquire in the process of transport are wedge products that involve the fifth basis element $\mathbf{e}_5$. There are 16 basis wedge products like that and it is evident that all of them can be obtained by multiplying the 16 basis elements of $\mathcal{Q}_0$ exteriorly by the vector $\mathbf{e}_5$. These wedge products and their linear combinations make up a vector subspace $\mathcal{Q}_+$, and it is obvious that there exists a natural one-to-one correspondence between the elements of $\mathcal{Q}_0$ and $\mathcal{Q}_+$ given by the rule $U \leftrightarrow U \wedge \mathbf{e}_5$ for any $U \in \mathcal{Q}_0$, which is not very useful though.

For the elements of $\mathcal{Q}_+$ one can define multiplication $\circ$ in a manner similar to how it has been done for $\mathcal{Q}_0$ and verify that $\mathcal{Q}_+ \circ \mathcal{Q}_0 \subset \mathcal{Q}_+$. Likewise, one can define on $\mathcal{Q}_+$ the inner product d as it has been done above. One should observe that the elements of $\mathcal{Q}_+$ are parallel transported into themselves, which means that any homomorphism between $\mathcal{Q}_0$ and $\mathcal{Q}_+$ is a priori not preserved by parallel transport.

Inside $\mathcal{Q}_+$ one can distinguish a five-dimensional subspace H_+ which is the linear envelope of the vector $\mathbf{e}_5$ and of the four multivectors $\mathbf{e}_\alpha \wedge \mathbf{e}_\beta \wedge \mathbf{e}_\lambda \wedge \mathbf{e}_5$ $(0 \leq \alpha < \beta < \lambda \leq 3)$. It is straightforward to check that this subspace is invariant under SO(3,2) transformations produced by the generators M_{AB} defined in (22).

One could think that the direct sum $\mathcal{Q} = \mathcal{Q}_0 \oplus \mathcal{Q}_+$ is the desired Clifford algebra that can be identify with an appropriate subalgebra inside the space of rank (1,1) spinor tensors. However, since the latter space is a tensor product of the type $S_n \otimes \widetilde{S}_n$, where n is the dimension of the spinors, for this to be possible, there should exist four elements in $\mathcal{Q}$ that would produce the same effect as translation generators when commuted with $\mathbf{e}_A$. Denoting these four unknown elements as Θ_μ $(\mu = 0, 1, 2, 3)$, one should require that

$$\Theta_\mu \circ \mathbf{e}_\alpha - \mathbf{e}_\alpha \circ \Theta_\mu = -\kappa \eta_{\mu\alpha} \mathbf{e}_5 \quad \text{and} \quad \Theta_\mu \circ \mathbf{e}_5 - \mathbf{e}_5 \circ \Theta_\mu = 0. \tag{23}$$

One can verify that the first of these equations is satisfied by the operator

$$\Theta_\mu = -\tfrac{1}{2}\,\kappa\, \mathbf{e}_\mu \wedge \mathbf{e}_5\,. \tag{24}$$

However, this operator does not satisfy the second equation in (23). A way out is to slightly modify the definition of multiplication $\circ$ by replacing the inner product h, which is non-degenerate but is not preserved by parallel transport, with the inner product g, which is degenerate on V_5 but is covariantly constant. Let us recall that the inner product g for five-vectors is defined as follows: for any $\mathbf{u} \in V_5$ and $\mathbf{v} \in V_5$ one has $g(\mathbf{u}, \mathbf{v}) = g(\mathbf{U}, \mathbf{V})$, where $\mathbf{U}$ and $\mathbf{V}$ are the four-vectors that correspond to $\mathbf{u}$ and $\mathbf{v}$. It is evident that one has $g(\mathbf{u}, \mathbf{v}) = h(\mathbf{u}^{\mathcal{Z}}, \mathbf{v}^{\mathcal{Z}})$, which means that nothing is changed for the multivectors from $\mathcal{Q}_0$, whereas the product $\circ$ of any two multivectors from $\mathcal{Q}_+$ will now be identically zero. With thus defined multiplication $\circ$, operator (24) obeys the second equation in (23) as well.

It also makes sense to alter accordingly the definition of the inner product d for elements of $\mathcal{Q}$ by replacing everywhere h with g. One can also change the overall normalization of d to make the latter correspond to the normalization of the inner product d in $S_8 \otimes \widetilde{S}_8$. As a result, nothing will change for the elements of $\mathcal{Q}_0$ except for overall normalization, whereas the inner product d of *any* element of $\mathcal{Q}_+$ with any multivector from $\mathcal{Q}$ will now be identically zero.

Let us now observe that to each multivector $\mathbf{a} \wedge \mathbf{b} \wedge \mathbf{c} \wedge \mathbf{d}$ made out of five-vectors one can put into correspondence a five-vector 1-form $\widetilde{\mathbf{w}}$ by requiring that

$$< \widetilde{\mathbf{w}}, \mathbf{u} > = \ \epsilon(\mathbf{a}, \mathbf{b}, \mathbf{c}, \mathbf{d}, \mathbf{u}) \ \text{ for any } \mathbf{u} \in V_5,$$

where ϵ is the Levi-Civita five-tensor. It is easy to see that if the latter tensor is normalized properly, the 1-form corresponding to the multivector $\mathbf{e}_0 \wedge \mathbf{e}_1 \wedge \mathbf{e}_2 \wedge \mathbf{e}_3$ is exactly the fifth element $\widetilde{\mathbf{o}}^5$ of the basis of five-vector 1-forms dual to the basis $\mathbf{e}_A$. Since the latter is a regular basis, the one-dimensional subspace of all the elements proportional to $\widetilde{\mathbf{o}}^5$ is exactly the subspace $\widetilde{\mathcal{E}}$ of $\widetilde{V}_5$. This means that the five-dimensional subspace H_0 introduced above is isomorphic to the direct sum of $\mathcal{Z}$ and $\widetilde{\mathcal{E}}$.

In a similar manner, the five-vector 1-forms corresponding to the four multivectors $\mathbf{e}_\alpha \wedge \mathbf{e}_\beta \wedge \mathbf{e}_\lambda \wedge \mathbf{e}_5$ are exactly the first four elements $\widetilde{\mathbf{o}}^\mu$ of the mentioned basis of five-vector 1-forms dual to the basis $\mathbf{e}_A$. It is easy to see that the four-dimensional linear envelope of the 1-forms $\tilde{\mathbf{o}}^\mu$ is exactly the subspace $\widetilde{\mathcal{Z}}$ of $\widetilde{V}_5$, so the five-dimensional subspace H_+ introduced above is isomorphic to $\widetilde{\mathcal{Z}} \oplus \mathcal{E}$.

We therefore see that our solution to the problem of establishing a relation between five-vectors and spinors that would be invariant under parallel transport is merely a particular realization of the Clifford algebra $\mathcal{Q}$ with 8×8 matrices, or more precisely, as a certain subalgebra in the space of tensors of rank (1,1) over the space of eight-dimensional spinors. The elements of the subalgebra $\mathcal{Q}_0$ and of the ideal $\mathcal{Q}_+$ are represented by the matrices $(\mathbf{1} \otimes U)$ and $(\sigma_+ \otimes U)$, respectively, where U is any matrix from $S_4 \otimes \widetilde{S}_4$. More precisely, one has the following correspondences:

scalar *1*	unity matrix $\mathbf{1} \otimes \mathbf{1} \otimes \mathbf{1}$
vector $\mathbf{e}_\alpha$	matrix $\mathbf{1} \otimes \Gamma_\alpha$
bivector $\mathbf{e}_\alpha \wedge \mathbf{e}_\beta$	matrix $\mathbf{1} \otimes \Gamma_{[\alpha}\Gamma_{\beta]}$
trivector $\mathbf{e}_\alpha \wedge \mathbf{e}_\beta \wedge \mathbf{e}_\mu$	matrix $\mathbf{1} \otimes \Gamma_{[\alpha}\Gamma_\beta\Gamma_{\mu]}$
multivector $\mathbf{e}_0 \wedge \mathbf{e}_1 \wedge \mathbf{e}_2 \wedge \mathbf{e}_3$	matrix $\mathbf{1} \otimes \Gamma_5$
vector $\mathbf{e}_5$	matrix $\sigma_+ \otimes \Gamma_5$
bivector $\mathbf{e}_\alpha \wedge \mathbf{e}_5$	matrix $\sigma_+ \otimes \Gamma_\alpha\Gamma_5$
trivector $\mathbf{e}_\alpha \wedge \mathbf{e}_\beta \wedge \mathbf{e}_5$	matrix $\sigma_+ \otimes \Gamma_{[\alpha}\Gamma_{\beta]}\Gamma_5$
multivector $\mathbf{e}_\alpha \wedge \mathbf{e}_\beta \wedge \mathbf{e}_\mu \wedge \mathbf{e}_5$	matrix $\sigma_+ \otimes \Gamma_{[\alpha}\Gamma_\beta\Gamma_{\mu]}\Gamma_5$
multivector $\mathbf{e}_0 \wedge \mathbf{e}_1 \wedge \mathbf{e}_2 \wedge \mathbf{e}_3 \wedge \mathbf{e}_5$	matrix $-\sigma_+ \otimes \mathbf{1} \otimes \mathbf{1}$

(25)

As one can see, the subspaces H_0 and H_+ are represented by the subspaces G_0 and G_+, respectively.

The considered realization has two seemingly unattractive features. The first one is that only half the elements of the Clifford algebra $S_8 \otimes \widetilde{S}_8$ are used to represent the elements of $\mathcal{Q}$, while the other half seems to be irrelevant and unnecessary. This is actually not so. One way to see this is to observe that in applications one calculates expressions like

$$\overline{\Psi} U \Psi \quad \text{or} \quad \overline{\Psi} U (\partial \Psi) \quad \text{or} \quad (\partial \overline{\Psi}) U \Psi, \tag{26}$$

where $U \in \mathcal{Q}$ and both U and ∂ may carry vector indices. These expressions are nothing but the inner products d of the elements $\Psi \otimes \overline{\Psi}$ or $\partial\Psi \otimes \overline{\Psi}$ or $\Psi \otimes \partial\overline{\Psi}$ with the element U. And if there were no other elements except for those of $\mathcal{Q}$, all expressions of the form (26) with $U \in \mathcal{Q}_+$ would be identically zero. So the elements in $S_8 \otimes \widetilde{S}_8$ not belonging to $\mathcal{Q}$ *are* needed.

To understanding the structure of $S_8 \otimes \widetilde{S}_8$ more clearly, one may recall that the latter is a Clifford algebra associated with the spinors corresponding to a six-dimensional pseudo-Euclidean space. This means that there should exist six constituents of $S_8 \otimes \widetilde{S}_8$, similar to the four constituents Γ_α in $S_4 \otimes \widetilde{S}_4$, that are invariant under Dirac and change conjugations defined by matrices (21) and that satisfy certain anticommutation relations similar to (9). For our purposes, a convenient set of constituents is

$$\Upsilon_\alpha = \mathbf{1} \otimes \Gamma_\alpha \; (\alpha = 0, 1, 2, 3), \quad \Upsilon_5 = \sigma_1 \otimes \Gamma_5, \quad \text{and } \Upsilon_6 = i\sigma_2 \otimes \Gamma_5. \tag{27}$$

It is straightforward to check that these six elements satisfy the following anticommutation relations:

$$\{\Upsilon_A, \Upsilon_B\} = -2\,\eta_{AB}\cdot \mathbb{1}, \;\; \{\Upsilon_A, \Upsilon_6\} = 0, \quad \text{and} \quad \{\Upsilon_6, \Upsilon_6\} = 2\cdot \mathbb{1}$$

where $\mathbb{1}$ is now the unity 8×8 matrix. Let us introduce the notations:

$$\ddot{\Gamma}_\alpha \equiv \Upsilon_\alpha\,, \ddot{\Gamma}_5 \equiv \tfrac{1}{2}\,(\Upsilon_5 + \Upsilon_6), \text{and}\, \ddot{\Gamma}_6 \equiv \tfrac{1}{2}\,(\Upsilon_5 - \Upsilon_6). \tag{28}$$

As Υ's, these constituents are invariant under Dirac and change conjugations. And they satisfy the relations

$$\{\ddot{\Gamma}_\alpha, \ddot{\Gamma}_\beta\} = -2\,\eta_{\alpha\beta}\cdot \mathbb{1}, \;\; \{\ddot{\Gamma}_5, \ddot{\Gamma}_5\} = \{\ddot{\Gamma}_6, \ddot{\Gamma}_6\} = 0, \quad \text{and} \quad \{\ddot{\Gamma}_5, \ddot{\Gamma}_6\} = -\mathbb{1}.$$

Accordingly, one has

$$\mathrm{trace}\{\ddot{\Gamma}_\alpha\ddot{\Gamma}_\beta\} = -8\,\eta_{\alpha\beta}, \; \mathrm{trace}\{\ddot{\Gamma}_5\ddot{\Gamma}_5\} = \mathrm{trace}\{\ddot{\Gamma}_6\ddot{\Gamma}_6\} = 0, \text{ and } \mathrm{trace}\{\ddot{\Gamma}_5\ddot{\Gamma}_6\} = -4.$$

The latter relations mean that the inner product d is nondegenerate on $S_8 \otimes \widetilde{S}_8$, but *is* degenerate on $\mathcal{Q}$ for the reason that $\ddot{\Gamma}_5$ is a *null* vector orthogonal to $\ddot{\Gamma}_\alpha$. This means that—at least, within the realization considered—the only way one can fix the size of $\ddot{\Gamma}_5$ in an invariant way is to specify how it is expressed in terms of the time-like vector Υ_5 and space-like vector Υ_6. Actually, the only place where the size of $\ddot{\Gamma}_5$ becomes important is in the definition of the constant κ in equations like

$$\nabla_\mu\ddot{\Gamma}_\alpha = \Theta_\mu\ddot{\Gamma}_\alpha - \ddot{\Gamma}_\alpha\Theta_\mu = -\kappa\eta_{\mu\alpha}\ddot{\Gamma}_5.$$

In the following, it will always be assumed that $\ddot{\Gamma}_5$ is expressed in terms of a normalized time-like vector and a normalized space-like vector according to the second formula in (28).

In agreement with table (25), the subalgebra $\mathcal{Q}_0$ is made up by the elements 1, $\ddot{\Gamma}_\alpha$, $\ddot{\Gamma}_{[\alpha}\,\ddot{\Gamma}_{\beta]}$, $\ddot{\Gamma}_{[\alpha}\,\ddot{\Gamma}_\beta\,\ddot{\Gamma}_{\lambda]}$, and $\ddot{\Gamma}_0\,\ddot{\Gamma}_1\,\ddot{\Gamma}_2\,\ddot{\Gamma}_3$. The ideal $\mathcal{Q}_+$ is made up by the elements $\ddot{\Gamma}_5$, $\ddot{\Gamma}_\alpha\,\ddot{\Gamma}_5$, $\ddot{\Gamma}_{[\alpha}\,\ddot{\Gamma}_{\beta]}\,\ddot{\Gamma}_5$, $\ddot{\Gamma}_{[\alpha}\,\ddot{\Gamma}_\beta\,\ddot{\Gamma}_{\lambda]}\,\ddot{\Gamma}_5$, and $\ddot{\Gamma}_0\,\ddot{\Gamma}_1\,\ddot{\Gamma}_2\,\ddot{\Gamma}_3\,\ddot{\Gamma}_5$. One can easily see that the ideal whose elements are represented by the matrices $\sigma_-\otimes U$, where $U \in S_4\otimes\widetilde{S}_4$, and which is a "mirror image" of the ideal $\mathcal{Q}_+$, is made up by the elements $\ddot{\Gamma}_6$, $\ddot{\Gamma}_\alpha\,\ddot{\Gamma}_6$, $\ddot{\Gamma}_{[\alpha}\,\ddot{\Gamma}_{\beta]}\,\ddot{\Gamma}_6$, $\ddot{\Gamma}_{[\alpha}\,\ddot{\Gamma}_\beta\,\ddot{\Gamma}_{\lambda]}\,\ddot{\Gamma}_6$, and $\ddot{\Gamma}_0\,\ddot{\Gamma}_1\,\ddot{\Gamma}_2\,\ddot{\Gamma}_3\,\ddot{\Gamma}_6$. Finally, the subspace whose elements are represented by the matrices $\sigma_3 \otimes U$ is made up by the elements $\ddot{\Gamma}_{[5}\,\ddot{\Gamma}_{6]}$, $\ddot{\Gamma}_\alpha\,\ddot{\Gamma}_{[5}\,\ddot{\Gamma}_{6]}$, $\ddot{\Gamma}_{[\alpha}\,\ddot{\Gamma}_{\beta]}\,\ddot{\Gamma}_{[5}\,\ddot{\Gamma}_{6]}$, $\ddot{\Gamma}_{[\alpha}\,\ddot{\Gamma}_\beta\,\ddot{\Gamma}_{\lambda]}\,\ddot{\Gamma}_{[5}\,\ddot{\Gamma}_{6]}$, and $\ddot{\Gamma}_0\,\ddot{\Gamma}_1\,\ddot{\Gamma}_2\,\ddot{\Gamma}_3\,\ddot{\Gamma}_{[5}\,\ddot{\Gamma}_{6]}$.

The other seemingly unattractive feature of the realization considered is that 4-spinors are replaced with 8-spinors. This naturally brings up the question about the meaning of the additional four components of the spinor. One way to avoid this problem is to suppose that the considered 8-spinors are made out of two identical 4-spinors, say,

$$\Psi = \begin{pmatrix} \psi \\ \psi \end{pmatrix} \tag{29}$$

or

$$\Psi = \begin{pmatrix} \psi \\ i\psi \end{pmatrix}. \tag{30}$$

In the following, this type of 8-spinors will be referred to as *double* 4-spinors.

4. Lagrangian Density

We are now sufficiently equipped to examine how the Lagrangian density for a spinor field can be expressed in the language of five-dimensional tangent vectors. Let us proceed from the fact that the standard term with the derivative in the Lagrangian density for a 4-spinor field in flat space-time has the form

$$\tfrac{i}{2}\,\overline{\psi}\gamma^\mu(\nabla_\mu\psi) - \tfrac{i}{2}\,(\nabla_\mu\overline{\psi})\gamma^\mu\psi = \tfrac{i}{2}\,\overline{\psi}\gamma^\mu(\nabla_\mu\psi) + \textit{Dirac conjugate}, \tag{31}$$

where $\gamma^\mu = g^{\mu\nu}\gamma_\nu$ are ordinary 4×4 Dirac's gamma matrices. In the language of four-vectors, the above density should be the $\mathcal{L}_{0123}$ component of a four-vector 4-form. Our idea is to find an expression for the latter that would make explicit its geometric structure and then to present the Lagrangian as a five-vector form by replacing the gammas and derivatives that carry four-vector indices with their five-vector counterparts. One way to find the desired expression for $\mathcal{L}_{0123}$ is to use the identity:

$$i\gamma^\mu\nabla_\mu = 4\,\gamma_5\gamma_{[0}\gamma_1\gamma_2\nabla_{3]},$$

where $\gamma_5 = -i\gamma_0\gamma_1\gamma_2\gamma_3$. By substituting the expression in the right-hand side into the first term in formula (31) and using Dirac conjugation to find a similar expression for the second term, one obtains

$$\mathcal{L}_{0123} = 2\,\overline{\psi}\,\gamma_5\gamma_{[0}\gamma_1\gamma_2\,\nabla_{3]}\psi + \textit{Dirac conjugate}\,. \tag{32}$$

Let us now observe that the relation between four-dimensional tangent vectors and spinors can be establish in a manner similar to how it has been done above for five-vectors: by constructing out of four-vectors a formal Clifford algebra similar to $\mathcal{Q}_0$ and identifying its complexification with $S_4 \otimes \widetilde{S}_4$. Under such identification, V_4 is taken to be isomorphic to some appropriate four-dimensional subspace in the space $G^{(5)}$ introduced earlier, and one can always choose a basis in S_4 in such a way that this four-dimensional subspace would be the linear envelope of the four elements in $S_4 \otimes \widetilde{S}_4$ with the components $(\Gamma_\alpha)^i_j$ given by formulae (5). Considering now that $\gamma_\mu = i\,\Gamma_\mu\Gamma_5 = \gamma_5\,\Gamma_\mu$ and $\gamma_5\gamma_5 = 1$, one can present the right-hand side of equation (32) in the form where its geometric structure becomes transparent[3]:

$$-2\,\overline{\psi}\,\Gamma_{[0}\Gamma_1\Gamma_2\nabla_{3]}\psi + \textit{Dirac conjugate}\,. \tag{33}$$

Let us now try to present this expression as the $\mathcal{L}_{01235}$ component of a five-vector 5-form. First of all, since V_4 is isomorphic to $\mathcal{Z}$, one can interpret the Γ's in expression

[3]The identity $i\gamma^\mu\nabla_\mu = -4\,\Gamma_{[0}\Gamma_1\Gamma_2\nabla_{3]}$ can be understood better if one recalls that the components of any four-vector 4-form $A_{\alpha\beta\sigma\tau}$ can be presented as $A\cdot\epsilon_{\alpha\beta\sigma\tau}$, where $A = -(1/4!)\,\epsilon^{\alpha\beta\sigma\tau}A_{\alpha\beta\sigma\tau}$. Applying this to the expression between $\overline{\psi}$ and ψ in the first term in (33), one obtains:

$$-\tfrac{1}{4!}\,\epsilon^{\alpha\beta\sigma\tau}\cdot(-2)\,\Gamma_{[\alpha}\Gamma_\beta\Gamma_\sigma\nabla_{\tau]} = \tfrac{i}{2}\{\tfrac{1}{3!}\,\Gamma_\alpha\Gamma_\beta\Gamma_\sigma\,\epsilon^{\alpha\beta\sigma\mu}\}\cdot\nabla_\mu = \tfrac{i}{2}\,\gamma^\mu\cdot\nabla_\mu.$$

So γ^μ are images of the basis four-vector 1-forms $\widetilde{\mathbf{O}}^\mu$ relative to the duality isomorphism between $\widetilde{V}_4$ and the space of trivectors $V_4\wedge V_4\wedge V_4$ defined by the Levi-Civita four-tensor—up to the factor i needed to make the expression in the curly brackets real.

(33) as images of the first four elements of a regular five-vector basis. Using the notation $\ddot{\Gamma}_\mu \equiv 1 \otimes \Gamma_\mu$ introduced earlier, one can present the above density as

$$-\overline{\Psi}\,\ddot{\Gamma}_{[0}\ddot{\Gamma}_1\ddot{\Gamma}_2\nabla_{3]}\Psi + \textit{Dirac conjugate}\,, \tag{34}$$

where Ψ is given by formula (29) and $\overline{\Psi}$ is calculated by using the first of the matrices (21). One should then do something so that the above expression would acquire one more index. The simplest way to achieve this is to replaced ∇_α with the operator D_{AB}. Let us first suppose that there are no gauge fields. In this case the components of the bivector derivative in a normalized five-vector basis equal[4,5]

$$\mathsf{D}_{\lambda 5}\Psi = \kappa^{-1}\,\partial_\lambda\Psi - \tfrac{1}{2}\,\ddot{\Gamma}_\lambda\ddot{\Gamma}_5\Psi \quad \text{and} \quad \mathsf{D}_{\mu\nu}\Psi = -\tfrac{1}{2}\,\ddot{\Gamma}_\mu\ddot{\Gamma}_\nu\Psi, \tag{35}$$

where we have used the notation $\ddot{\Gamma}_5 \equiv \sigma_+ \otimes \Gamma_5$. One can then make the following calculations:

$$\begin{aligned}
&5\kappa\overline{\Psi}\ddot{\Gamma}_{[0}\ddot{\Gamma}_1\ddot{\Gamma}_2\mathsf{D}_{35]}\Psi \\
&= \kappa\overline{\Psi}(\ddot{\Gamma}_{[0}\ddot{\Gamma}_1\ddot{\Gamma}_2\mathsf{D}_{3]5} - \ddot{\Gamma}_{[0}\ddot{\Gamma}_1\ddot{\Gamma}_2\mathsf{D}_{|5|3]} + \ddot{\Gamma}_{[0}\ddot{\Gamma}_1\ddot{\Gamma}_{|5|}\mathsf{D}_{23]} - \ddot{\Gamma}_{[0}\ddot{\Gamma}_{|5|}\ddot{\Gamma}_1\mathsf{D}_{23]} + \ddot{\Gamma}_5\ddot{\Gamma}_{[0}\ddot{\Gamma}_1\mathsf{D}_{23]})\Psi \\
&= \kappa\overline{\Psi}(2\,\ddot{\Gamma}_{[0}\ddot{\Gamma}_1\ddot{\Gamma}_2\mathsf{D}_{3]5} + 3\,\ddot{\Gamma}_5\ddot{\Gamma}_{[0}\ddot{\Gamma}_1\mathsf{D}_{23]})\Psi \\
&= \overline{\Psi}(2\,\ddot{\Gamma}_{[0}\ddot{\Gamma}_1\ddot{\Gamma}_2\,\partial_{3]} - \kappa\,\ddot{\Gamma}_{[0}\ddot{\Gamma}_1\ddot{\Gamma}_2\ddot{\Gamma}_{3]}\ddot{\Gamma}_5 - \tfrac{3}{2}\kappa\,\ddot{\Gamma}_5\ddot{\Gamma}_{[0}\ddot{\Gamma}_1\ddot{\Gamma}_2\ddot{\Gamma}_{3]})\Psi \\
&= 2\,\overline{\Psi}\ddot{\Gamma}_{[0}\ddot{\Gamma}_1\ddot{\Gamma}_2\,\partial_{3]}\Psi - \tfrac{5}{2}\kappa\,\overline{\Psi}\ddot{\Gamma}_{[0}\ddot{\Gamma}_1\ddot{\Gamma}_2\ddot{\Gamma}_{3]}\ddot{\Gamma}_5\Psi = 2\,\overline{\Psi}\ddot{\Gamma}_{[0}\ddot{\Gamma}_1\ddot{\Gamma}_2\,\partial_{3]}\Psi + \tfrac{5}{2}\kappa\,\overline{\Psi}(\sigma_+ \otimes 1)\Psi \\
&= 4\,\overline{\psi}\,\Gamma_{[0}\Gamma_1\Gamma_2\,\partial_{3]}\psi + \tfrac{5}{2}\kappa\,\overline{\psi}\psi = -\overline{\psi}(i\gamma^\mu\partial_\mu - m)\psi,
\end{aligned}$$

where $m = \frac{5}{2}\kappa$. We therefore see that the standard term with the derivative in the Lagrangian density can be obtained from the following five-vector 5-form:

$$\mathcal{L}_{01235} = -\tfrac{5}{2}\kappa\,\overline{\Psi}\ddot{\Gamma}_{[0}\ddot{\Gamma}_1\ddot{\Gamma}_2\mathsf{D}_{35]}\Psi + \textit{Dirac conjugate}\,. \tag{36}$$

In addition, one obtains a mass term that technically results from a nonzero connection coefficient in $\mathsf{D}_{\lambda 5}\Psi$ and from $\mathsf{D}_{\mu\nu}\Psi$. Unlike the mass that the field ψ may acquire owing to its ability to interact with other physical fields, mass m is not associated with any explicit interaction and can be termed as the *original* mass.

In the presence of bivector gauge fields, instead of (35) one has

$$\mathsf{D}_{\lambda 5}\Psi = \kappa^{-1}\,\partial_\lambda\Psi - \tfrac{1}{2}\,\ddot{\Gamma}_\lambda\ddot{\Gamma}_5\Psi - \kappa^{-1}\,ieA_\lambda\Psi \quad \text{and} \quad \mathsf{D}_{\mu\nu}\Psi = \tfrac{1}{2}\,\ddot{\Gamma}_\mu\ddot{\Gamma}_\nu\Psi - \kappa^{-1}\,ieE_{\mu\nu}\Psi,$$

where $A_\mu = A^a_\mu \cdot t^a$ can be regarded as ordinary gauge fields (t^a are generators of the corresponding gauge group) and $E_{\mu\nu} = E^a_{\mu\nu} \cdot t^a$ are antisymmetric tensor fields that also mediate interaction between matter particles (for more details see section H of ref.[6]; see also

---

[4]The bivector derivative for double 4-spinor fields should be defined in a manner similar to how it is defined for five-vector fields: It should be such that at any five-vector affine connection $\square$ with respect to which the tensors of the inner products d and c are covariantly constant, the five-vector covariant derivative of such a field could be expressed linearly in terms of its bivector derivatives.

[5]Strictly speaking, the components $\mathcal{L}_{01235}$ should be evaluated in a *passive* regular five-vector basis, where the fifth basis five-vector corresponds to the identity operator. However, the difference in normalization would result only in an additional dimensional factor that would appear in every term in the Lagrangian and therefore can be absorbed into the overall normalization of $\mathcal{L}$. To simplify the presentation, I omit this detail.

ref.[8]). Owing to the presence of these gauge fields, the expression for $5\kappa\overline{\Psi}\ddot{\Gamma}_{[0}\ddot{\Gamma}_1\ddot{\Gamma}_2\mathsf{D}_{35]}\Psi$ acquires the following additional terms:

$$\begin{aligned}-2ie\,\overline{\Psi}\ddot{\Gamma}_{[0}\ddot{\Gamma}_1\ddot{\Gamma}_2A_{3]}\Psi - 3ie\,\overline{\Psi}\ddot{\Gamma}_5\ddot{\Gamma}_{[0}\ddot{\Gamma}_1E_{23]}\Psi&\\ = -4ie\,\overline{\psi}\Gamma_{[0}\Gamma_1\Gamma_2\,A_{3]}\psi - 3ie\,\overline{\psi}\Gamma_5\Gamma_{[0}\Gamma_1E_{23]}\psi&\\ = -e\,\overline{\psi}\gamma^\mu t^a\psi\cdot A^a_\mu - \tfrac{1}{2}e\,\overline{\psi}\,\sigma^{\mu\nu}t^a\psi\cdot E^a_{\mu\nu}\,,&\end{aligned}$$

where $\sigma^{\mu\nu} = \frac{1}{2i}(\gamma^\mu\gamma^\nu - \gamma^\nu\gamma^\mu)$. Therefore, the five-vector 5-form (36) results in the following Lagrangian density:

$$\mathcal{L} = \tfrac{i}{2}\,\overline{\psi}\gamma^\mu(\partial_\mu\psi) - \tfrac{i}{2}\,(\partial_\mu\overline{\psi})\gamma^\mu\psi - m\overline{\psi}\psi + e\,\overline{\psi}\,\gamma^\mu t^a\psi\cdot A^a_\mu + \tfrac{1}{2}e\,\overline{\psi}\,\sigma^{\mu\nu}t^a\psi\cdot E^a_{\mu\nu}. \quad (37)$$

One should now get a clear understanding of the physical assumptions that the Lagrangian density obtained corresponds to. As it has been said in the beginning of this chapter, one of the options for generalizing the concept of ordinary covariant derivative is to replace the latter with the so-called five-vector covariant derivative $\square_A$. In the case of four-vector fields, the latter can be presented as follows:

$$\square_\mu = \nabla_\mu + \mathbf{M}_{\alpha\beta}\,S^{\alpha\beta}{}_\mu \quad\text{and}\quad \square_5 = \mathbf{M}_{\alpha\beta}\,S^{\alpha\beta}{}_5, \quad (38)$$

where ∇ is the torsion-free ordinary covariant derivative with respect to which the metric tensor is covariantly constant and which is uniquely determined by space-time metric; $\mathbf{M}$ is a local operator whose argument is a four-vector bivector and whose components in an arbitrary four-vector basis are $(M_{\alpha\beta})^\mu_\nu = \delta^\mu_\beta g_{\alpha\nu} - \delta^\mu_\alpha g_{\beta\nu}$; and $S^{\alpha\beta}{}_A$ are components of a five-vector 1-form whose values are four-vector bivectors and which is the five-vector generalization of the so-called contorsion tensor (see section A of ref.[6]). It is important to understand that derivative $\square_A$ can be introduced in any case, and the only question is whether or not the *real* space-time can have nonzero five-vector torsion. If it does not, then $\square_5 = 0$ and $\square_\mu$ is completely equivalent to ∇_μ.

Another way of presenting the five-vector covariant derivative is via the bivector derivative D_{AB}:

$$\square_\mu = \mathsf{D}_{\mu 5} + \mathsf{D}_{\alpha\beta}\,S^{\alpha\beta}{}_\mu \quad\text{and}\quad \square_5 = \mathsf{D}_{\alpha\beta}\,S^{\alpha\beta}{}_5, \quad (39)$$

which in the case of four-vector fields is absolutely equivalent to representation (38) and is a consequence of the definition of $\square_A$ and D_{AB}. In the case of five-vector fields, relation (39) is *not* a consequence but is an *assumption* that one makes to define the five-vector covariant derivative at non-zero torsion. For this to be possible, the bivector derivative should be defined in such a way that $\mathsf{D}_{\mu 5}$ would equal ∇_μ. The same requirement was imposed above on the bivector derivative of spinor fields.

The situation is more interesting for the fields whose values are abstract vectors or tensors associated with internal gauge symmetries. In this case in flat space-time

$$\square_\mu = \partial_\mu - ie\,B^a_\mu\cdot t^a \quad\text{and}\quad \square_5 = -\,ie\,B^a_5\cdot t^a, \quad (40)$$

where B^a_μ can be regarded as ordinary gauge fields, t^a are generators of the corresponding gauge group, and B^a_5 are new (scalar) gauge fields.[6] As in the case of four-vector fields,

[6] For simplicity, in equations (40) and (41) I omit the connection coefficients that could be associated with the possible vector or spinor indices that the field may carry.

the derivative $\square_A$ can be introduced in any case, and the question is whether or not the additional gauge fields B^a_5 exist in the real world. However, this is not all there is to it. Another question that one may ask in this case is whether the fields B^a_A are elementary or composite. What the latter means is a situation where for the considered type of fields the bivector derivative, defined according to the formulae

$$\mathsf{D}_{\mu 5} = \partial_\mu - ie\, A^a_\mu \cdot t^a \quad \text{and} \quad \mathsf{D}_{\mu\nu} = -\,ie\, E^a_{\mu\nu} \cdot t^a, \tag{41}$$

is a *more fundamental* object, while the five-vector covariant derivative is a subordinate object, which is completely determined by D_{AB} and by five-vector torsion according to relations (39). It is evident that Lagrangian density (37) corresponds to this latter situation.

There is also a third question that one may ask in the case where the fields B^a_A are considered composite: Does torsion have any direct effect on the dynamics of the matter fields that interact with the bivector gauge fields A^a_μ and $E^a_{\mu\nu}$? It is evident that Lagrangian density (37) corresponds to the case where the answer to this latter question is 'no'.

Let us now consider a situation where the derivative ∇ in density (33) is replaced with the five-vector covariant derivative. In accordance with what has been said above, in doing so one should assume that the five-vector gauge fields B^a_A are elementary. A straightforward substitution of $\square$ for ∇ in expression (34) yields

$$-\overline{\Psi}\,\ddot{\Gamma}_{[A}\ddot{\Gamma}_B\ddot{\Gamma}_C\,\square_{D]}\Psi + \textit{Dirac conjugate.} \tag{42}$$

This is a five-vector 4-form. And since the Lagrangian density is integrated over a four-dimensional volume, only the $\breve{\mathcal{Z}}$-component of the above expression will give contribution to the action (see ref.[5]), which means that none of the indices can have the value 5. Since in any regular five-vector basis $\square_\mu = \nabla_\mu$ and

$$\nabla_\mu\Psi = \partial_\mu\Psi - \tfrac{1}{2}\kappa\,\ddot{\Gamma}_\mu\ddot{\Gamma}_5\Psi - \tfrac{1}{2}\,\ddot{\Gamma}_\alpha\ddot{\Gamma}_\beta\, S^{\alpha\beta}{}_\mu\Psi - ieB_\mu\Psi,$$

expression (42) equals

$$-\,\overline{\Psi}\ddot{\Gamma}_{[0}\ddot{\Gamma}_1\ddot{\Gamma}_2\,\partial_{3]}\Psi + \tfrac{1}{2}\kappa\,\overline{\Psi}\ddot{\Gamma}_{[0}\ddot{\Gamma}_1\ddot{\Gamma}_2\ddot{\Gamma}_{3]}\ddot{\Gamma}_5\Psi$$
$$+\,\tfrac{1}{2}\overline{\Psi}\ddot{\Gamma}_{[0}\ddot{\Gamma}_1\ddot{\Gamma}_2 S^{\alpha\beta}{}_{3]}\ddot{\Gamma}_\alpha\ddot{\Gamma}_\beta\Psi + ie\,\overline{\Psi}\ddot{\Gamma}_{[0}\ddot{\Gamma}_1\ddot{\Gamma}_2 B_{3]}\Psi + \textit{Dirac conjugate.}$$

Performing calculations similar to those made above for the bivector derivative, one obtains the Lagrangian

$$\tfrac{i}{2}\,\overline{\psi}\gamma^\mu(\partial_\mu\psi) - \tfrac{i}{2}\,(\partial_\mu\overline{\psi})\gamma^\mu\psi - \kappa\,\overline{\psi}\psi - \tfrac{1}{2}\,\overline{\psi}\,\gamma^\mu\gamma^5\psi\cdot\epsilon_{\mu\alpha\beta\rho}\,S^{\alpha\beta\rho} + e\,\overline{\psi}\,\gamma^\mu t^a\psi\cdot B^a_\mu. \tag{43}$$

As was to be expected, the latter coincides with the density obtained from the four-vector 4-form (33), except for the term with the original mass, which now equals κ. Apparently, the only way one can make the scalar gauge fields B^a_5 give contribution to the action is to alter density (42) in such a way that it would become a five-vector 5-form. An evident candidate is

$$i\overline{\Psi}\,\ddot{\Gamma}_{[A}\ddot{\Gamma}_B\ddot{\Gamma}_C\ddot{\Gamma}_D\,\square_{E]}\Psi + \textit{Dirac conjugate.}$$

Unfortunately, this Lagrangian produces the wrong kind of kinematic term: $i\overline{\psi}\Gamma^\mu\partial_\mu\psi$ instead of $i\overline{\psi}\gamma^\mu\partial_\mu\psi$, and a term with the scalar gauge field of the form $e\,\overline{\psi}\gamma_5 t^a\psi\cdot B^a_5$, which leaves one no hope that the field B^a_5 could play the role of the Higgs boson.

It is not difficult to see that the right kind of kinematic term and a suitable term with the scalar gauge field follow from a density proportional to

$$\overline{\psi}\,\Gamma_5\Gamma_{[A}\Gamma_B\Gamma_C\Gamma_D\,\square_{E]}\,\psi.$$

One should now find such elements in $S_8 \otimes \widetilde{S}_8$ that would produce the above expression when properly combined with $\square$ and inserted between $\overline{\Psi}$ and Ψ. By trial and error one can arrive at the combination

$$\ddot{\Gamma}_{[6}\ddot{\Gamma}_A\ddot{\Gamma}_B\ddot{\Gamma}_C\ddot{\Gamma}_D\,\square_{E]},$$

where in accordance with defenitions (28), we used the notation $\ddot{\Gamma}_6 \equiv \sigma_- \otimes \Gamma_5$, and it is taken that $\square_6$ is identically zero. Inserting this between $\overline{\Psi}$ and Ψ, one obtains the density

$$\overline{\Psi}\,\ddot{\Gamma}_{[6}\ddot{\Gamma}_0\ddot{\Gamma}_1\ddot{\Gamma}_2\ddot{\Gamma}_3\,\square_{5]}\Psi = \tfrac{1}{3}\overline{\Psi}(\sigma_3 \otimes \Gamma_{[0}\Gamma_1\Gamma_2)\,\square_{3]}\Psi - \tfrac{1}{6}\overline{\Psi}(\sigma_- \otimes \mathbf{1} \otimes \mathbf{1})\,\square_5\Psi, \qquad (44)$$

which for the Ψ given by formula (29) results in a *zero* kinematic term. A way out is to suppose that in this case Ψ is given by formula (30). Multiplying the expression in the left-hand side of (44) by $-3i$ to get the right normalization of the kinematic term, one arrives at the following Lagrangian density:

$$-3i\overline{\Psi}\,\ddot{\Gamma}_{[6}\ddot{\Gamma}_A\ddot{\Gamma}_B\ddot{\Gamma}_C\ddot{\Gamma}_D\,\square_{E]}\Psi + \textit{Dirac conjugate}. \qquad (45)$$

Making use of equation (44) and considering that

$$\square_\mu\Psi = \partial_\mu\Psi - \tfrac{1}{2}\kappa\,\ddot{\Gamma}_\mu\ddot{\Gamma}_5\Psi - \tfrac{1}{2}\,\ddot{\Gamma}_\alpha\ddot{\Gamma}_\beta\,S^{\alpha\beta}_{\ \ \mu}\Psi - ieB_\mu\Psi$$

and

$$\square_5\Psi = -\tfrac{1}{2}\,\ddot{\Gamma}_\alpha\ddot{\Gamma}_\beta\,S^{\alpha\beta}_{\ \ 5}\Psi - ieB_5\Psi,$$

one obtains the following expression for $\mathcal{L}_{01235}$:

$$-2\,\overline{\psi}\,\Gamma_{[0}\Gamma_1\Gamma_2\,\partial_{3]}\psi + \tfrac{i}{2}\kappa\,\overline{\psi}\,\Gamma_{[0}\Gamma_1\Gamma_2\Gamma_{3]}\Gamma_5\psi + \overline{\psi}\,\Gamma_{[0}\Gamma_1\Gamma_2 S^{\alpha\beta}_{\ \ 3]}\Gamma_\alpha\Gamma_\beta\psi$$
$$+\,2ie\,\overline{\psi}\,\Gamma_{[0}\Gamma_1\Gamma_2 B_{3]}\psi - \tfrac{i}{4}\,\overline{\psi}\,\Gamma_\alpha\Gamma_\beta\,\psi\cdot S^{\alpha\beta}_{\ \ 5} + \tfrac{1}{2}e\,\overline{\psi}\,B_5\psi + \textit{Dirac conjugate}.$$

Performing calculations similar to those made earlier, one obtains the Lagrangian density:

$$\mathcal{L} = \tfrac{i}{2}\,\overline{\psi}\gamma^\mu(\partial_\mu\psi) - \tfrac{i}{2}\,(\partial_\mu\overline{\psi})\gamma^\mu\psi - \tfrac{1}{2}\,\overline{\psi}\,\gamma^\mu\gamma^5\psi\cdot\epsilon_{\mu\alpha\beta\rho}\,S^{\alpha\beta\rho}$$
$$-\,\tfrac{1}{2}\,\overline{\psi}\sigma_{\alpha\beta}\psi\cdot S^{\alpha\beta}_{\ \ 5} + e\,\overline{\psi}\,\gamma^\mu t^a\psi\cdot B^a_\mu + e\,\overline{\psi}\,t^a\psi\cdot B^a_5.$$

As one can see, the term with the original mass has disappeared. But now one has a term with the fifth component of five-vector torsion and a term with the scalar gauge field B^a_5.

5. Conclusion

Driven by the desire to find some clues that would indicate how the notion of the covariant derivative should be generalized within the formalism of five-dimensional tangent vectors, I have examined the possibility of establishing a relation between five-vectors and spinors that would be preserved by parallel transport and that would reproduce the conventional relation between spinors and four-dimensional tangent vectors. It turned out that such a relation cannot exist between five-vectors and four-dimensional spinors; the lowest dimension of spinors for which such a relation can be established is eight. After discussing the necessary technical details, I have used my findings to construct a generalization of the Lagrangian density for a spinor field that reproduces the conventional kinematic term and, in addition, yields terms that describe the interaction of the spinor field with bivector or five-vector gauge fields and with five-vector torsion. Though the results obtained do show one how to generalize the Lagrangian density, basing on them it does not seem possible to give definite preference to one particular way of generalizing the covariant derivative.

References

[1] A.B. Krasulin, Five-dimensional tangent vectors in space-time. I. *Introduction and formal theory*, Moscow, see e.g., math-ph/9805004 at arxiv.org.

[2] A.B. Krasulin, Five-dimensional tangent vectors in space-time. II. *Differential-geometric approach*, Moscow, see e.g., math-ph/9805025 at arxiv.org.

[3] A.B. Krasulin, Five-dimensional tangent vectors in space-time. III. *Some applications*, Moscow, see e.g., math-ph/9807004 at arxiv.org

[4] A.B. Krasulin, Five-dimensional tangent vectors in space-time. IV. *Generalization of exterior calculus*, Moscow, see e.g., math-ph/9808006 at arxiv.org.

[5] A.B. Krasulin, Five-dimensional tangent vectors in space-time. V. *Generalization of covariant derivative*, Moscow, see e.g., math-ph/9808014 at arxiv.org.

[6] A.B. Krasulin, Five-dimensional tangent vectors in space-time. VI. *Bivector derivative and its application*, Moscow, see e.g., 1006.3010 at arxiv.org.

[7] L. Schwartz, *Analyse Mathématique*, Hermann, 1967.

[8] A. Krasulin, Bivector gauge fields from classical electrodynamics, In: Dvoeglazov, V.V. (ed.): *Photon and Poincare group*, 1999, p.218-230.

In: Einstein and Others: Unification
Editor: Valeriy V. Dvoeglazov
ISBN: 978-1-63463-276-8

Chapter 8

HOW TO CONSTRUCT NEUTRAL PARTICLE STATES FOR HIGH SPINS?

Valeriy V. Dvoeglazov*
UAF, Universidad de Zacatecas, Zacatecas, México

Abstract

We construct self/anti-self charge conjugate (Majorana-like) states for the $(1/2, 0) \oplus (0, 1/2)$ representation of the Lorentz group, and their analogs for higher spins within the quantum field theory. The problem of the basis rotations and that of the selection of phases in the Dirac-like and Majorana-like field operators are considered. The discrete symmetries properties (P, C, T) are studied. The corresponding dynamical equations are presented. In the $(1/2, 0) \oplus (0, 1/2)$ representation they obey the Dirac-like equation with eight components, which has been first introduced by Markov. Thus, the Fock space for corresponding quantum fields is doubled (as shown by Ziino). The particular attention has been paid to the questions of chirality and helicity (two concepts which are frequently confused in the literature) for Dirac and Majorana states. We further review several experimental consequences which follow from the previous works of M. Kirchbach et al. on neutrinoless double beta decay, and G. J. Ni et al. on meson lifetimes.

Many authors introduced *self/anti-self* charge-conjugate 4-spinors (including in the momentum representation), see, e. g., [1–4]. These spinors have been studied in Refs. [5–9]. Authors found corresponding dynamical equations, gauge transformations and other specific features of them. On using $C = -e^{i\theta}\gamma^2\mathcal{K}$, the anti-linear operator of charge conjugation ($\mathcal{K}$ is the complex conjugation operator), we define the *self/anti-self* charge-conjugate 4-spinors in the momentum space $C\lambda^{S,A}(\mathbf{p}) = \pm\lambda^{S,A}(\mathbf{p})$, $C\rho^{S,A}(\mathbf{p}) = \pm\rho^{S,A}(\mathbf{p})$. The Wigner matrix is $\Theta_{[1/2]} = -i\sigma_2$, and the 2-spinors ϕ_L, ϕ_R can be boosted with $\Lambda_{L,R}$ matrices.

Such definitions of 4-spinors differ, of course, from the original Majorana definition in the x-representation $\nu(x) = \frac{1}{\sqrt{2}}(\Psi_D(x) + \Psi_D^c(x))$, $C\nu(x) = \nu(x)$ that represents the field with the positive real $C-$ parity only.

*E-mail address: vdvoeglazov@yahoo.com.mx

The rest λ and ρ spinors can be defined in analogious way with the Dirac spinors. Moreover, $\rho^S_{\uparrow\downarrow}(\mathbf{0}) = \mp i\lambda^A_{\downarrow\uparrow}(\mathbf{0})$, $\rho^A_{\uparrow\downarrow}(\mathbf{0}) = \pm i\lambda^S_{\downarrow\uparrow}(\mathbf{0})$. Thus, in this basis with the appropriate normalization ("mass dimension") the explicite forms of the 4-spinors of the second kind $\lambda^{S,A}_{\uparrow\downarrow}(\mathbf{p})$ (and the corresponding expressions for $\rho^{S,A}_{\uparrow\downarrow}(\mathbf{p})$) are:

$$\lambda^S_\uparrow(\mathbf{p}) = \frac{1}{2\sqrt{E+m}}\begin{pmatrix} ip_l \\ i(p^-+m) \\ p^-+m \\ -p_r \end{pmatrix}, \lambda^S_\downarrow(\mathbf{p}) = \frac{1}{2\sqrt{E+m}}\begin{pmatrix} -i(p^++m) \\ -ip_r \\ -p_l \\ (p^++m) \end{pmatrix},$$

$$\lambda^A_\uparrow(\mathbf{p}) = \frac{1}{2\sqrt{E+m}}\begin{pmatrix} -ip_l \\ -i(p^-+m) \\ (p^-+m) \\ -p_r \end{pmatrix}, \lambda^A_\downarrow(\mathbf{p}) = \frac{1}{2\sqrt{E+m}}\begin{pmatrix} i(p^++m) \\ ip_r \\ -p_l \\ (p^++m) \end{pmatrix}.$$

As claimed in [4], λ and ρ 4-spinors are *not* the eigenspinors of the helicity. Moreover, λ and ρ are *not* the eigenspinors of the parity, as opposed to the Dirac case ($P = \gamma^0 R$, where $R = (\mathbf{x} \to -\mathbf{x})$). The indices $\uparrow\downarrow$ should be referred to the chiral helicity quantum number introduced in the 60s, $\eta = -\gamma^5 h$, Ref. [10]. The normalizations of the spinors $\lambda^{S,A}_{\uparrow\downarrow}(\mathbf{p})$ and $\rho^{S,A}_{\uparrow\downarrow}(\mathbf{p})$ have been given in the previous works [4].

The dynamical coordinate-space equations are:

$$i\gamma^\mu\partial_\mu\lambda^S(x) - m\rho^A(x) = 0\,,\; i\gamma^\mu\partial_\mu\rho^A(x) - m\lambda^S(x) = 0\,,$$
$$i\gamma^\mu\partial_\mu\lambda^A(x) + m\rho^S(x) = 0\,,\; i\gamma^\mu\partial_\mu\rho^S(x) + m\lambda^A(x) = 0\,.$$

These are *not* the Dirac equation. However, they can be written in the 8-component form. One can also re-write the equations into the two-component form. Thus, one obtains the equations of Ref. [11]. Similar formulations have been presented by M. Markov [12], and by A. Barut and G. Ziino [3]. The group-theoretical basis for such doubling has been given in the papers by Gelfand, Tsetlin and Sokolik [13], who first presented the theory in the 2-dimensional representation of the inversion group in 1956 (later called as "the Bargmann-Wightman-Wigner-type quantum field theory" in 1993). The Lagrangian is

$$\mathcal{L} = \frac{i}{2}\left[\bar\lambda^S\gamma^\mu\partial_\mu\lambda^S - (\partial_\mu\bar\lambda^S)\gamma^\mu\lambda^S + \bar\rho^A\gamma^\mu\partial_\mu\rho^A - (\partial_\mu\bar\rho^A)\gamma^\mu\rho^A \right.$$
$$\left. + \bar\lambda^A\gamma^\mu\partial_\mu\lambda^A - (\partial_\mu\bar\lambda^A)\gamma^\mu\lambda^A + \bar\rho^S\gamma^\mu\partial_\mu\rho^S - (\partial_\mu\bar\rho^S)\gamma^\mu\rho^S\right]$$
$$- m(\bar\lambda^S\rho^A + \bar\rho^A\lambda^S - \bar\lambda^A\rho^S - \bar\rho^S\lambda^A).$$

The connection with the Dirac spinors has been found [6, 8]. We can see that the two sets are connnected by the unitary transformations, and this represents itself the rotation of the spin-parity basis.

It was shown [6] that the covariant derivative (and, hence, the interaction) can be introduced in this construct in the following way $\partial_\mu \to \nabla_\mu = \partial_\mu - ig\text{Ł}^5 A_\mu$, where $\text{Ł}^5 = \text{diag}(\gamma^5 \quad -\gamma^5)$, the 8×8 matrix. In other words, with respect to the transformations

$$\lambda'(x) \to (\cos\alpha - i\gamma^5\sin\alpha)\lambda(x)\,,\; \bar\lambda'(x) \to \bar\lambda(x)(\cos\alpha - i\gamma^5\sin\alpha)\,,$$
$$\rho'(x) \to (\cos\alpha + i\gamma^5\sin\alpha)\rho(x)\,,\; \bar\rho'(x) \to \bar\rho(x)(\cos\alpha + i\gamma^5\sin\alpha)$$

the spinors retain their properties to be self/anti-self charge conjugate spinors and the proposed Lagrangian [6] remains to be invariant. This tells us that while the self/anti-self charge conjugate states have zero eigenvalues of the ordinary (scalar) charge operator but they can possess the axial charge [3]. Next, due to the fact that the transformations

$$\lambda'_S(\mathbf{p}) = \begin{pmatrix} \Xi & 0 \\ 0 & \Xi \end{pmatrix} \lambda_S(\mathbf{p}) \equiv \lambda^*_A(\mathbf{p}), \lambda''_S(\mathbf{p}) = \begin{pmatrix} i\Xi & 0 \\ 0 & -i\Xi \end{pmatrix} \lambda_S(\mathbf{p}) \equiv -i\lambda^*_S,$$

$$\lambda'''_S(\mathbf{p}) = i \begin{pmatrix} 0 & \Xi \\ \Xi & 0 \end{pmatrix} \lambda_S(\mathbf{p}) \equiv i\gamma^0\lambda^*_A(\mathbf{p}), \lambda^{IV}_S(\mathbf{p}) = \begin{pmatrix} 0 & \Xi \\ -\Xi & 0 \end{pmatrix} \lambda_S(\mathbf{p}) \equiv \gamma^0\lambda^*_S,$$

with the 2×2 matrix Ξ defined in Ref. [4], $\Xi\Lambda_{R,L}(\mathbf{p} \leftarrow \mathbf{0})\Xi^{-1} = \Lambda^*_{R,L}(\mathbf{p} \leftarrow \mathbf{0})$, and corresponding transformations for λ^A, do *not* change the properties of bispinors to be in the self/anti-self charge-conjugate spaces, the Majorana-like field operator ($b^\dagger \equiv a^\dagger$) admits additional phase (and, in general, normalization) transformations $\nu^{ML\,\prime}(x^\mu) = [c_0 + i(\tau \cdot \mathbf{c})]\,\nu^{ML\,\dagger}(x^\mu)$, where c_α are arbitrary parameters. The τ matrices are defined over the field of 2×2 matrices. One can parametrize $c_0 = \cos\phi$ and $\mathbf{c} = \mathbf{n}\sin\phi$ and, thus, define the $SU(2)$ group of phase transformations.

The Dirac-like and Majorana-like field operators can be built from both $\lambda^{S,A}(\mathbf{p})$ and $\rho^{S,A}(\mathbf{p})$, or their combinations. It is interesting to note that $\left[\nu^{ML}(x^\mu) \pm C\nu^{ML\,\dagger}(x^\mu)\right]/2$ lead naturally to the Ziino-Barut scheme of massive chiral fields, Ref. [3], if they are composed from $\lambda^{S,A}$ spinors. Recently, the interest to these models raised again [9, 14].

Ahluwalia [4] claimed "Incompatibility of Self-Charge Conjugation with Helicity Eignestates and Gauge Interactions". I showed that the gauge interactions of λ and ρ 4-spinors are different. Z. Q. Shi and G. J. Ni promote a very extreme standpoint. Namely, ""the spin states, the helicity states and the chirality states of fermions in Relativistic Quantum Mechanics ... are entirely different: a spin state is helicity degenerate; a helicity state can be expanded as linear combination of the chirality states; the polarization of fermions in flight must be described by the helicity states" (see also the Conclusion Section [15]). In fact, they showed experimental consequences of their statement: "the lifetime of RH polarized fermions is always greater than of LH ones with the same speed in flight". However, we show below that the helicity, chiral helicity and chirality operators are connected by the unitary transformations. M. Markov wrote long ago [12] *two* Dirac equations with opposite signs at the mass term $[i\gamma^\mu\partial_\mu - m]\,\Psi_1(x) - 0$, $[i\gamma^\mu\partial_\mu + m]\,\Psi_2(x) = 0$. In fact, he studied all properties of this relativistic quantum model . Next, he added and subtracted these equations. $i\gamma^\mu\partial_\mu\chi(x) - m\eta(x) = 0$, $i\gamma^\mu\partial_\mu\eta(x) - m\chi(x) = 0$, Thus, χ and η solutions can be presented as some superpositions of the Dirac 4-spinors $u-$ and $v-$. These equations, of course, can be identified with the equations for λ and ρ we presented above. Barut and Ziino [3] proposed yet another model. They considered γ^5 operator as the operator of charge-conjugation. Thus, the charge-conjugated Dirac equation has the different sign comparing with the ordinary formulation $[i\gamma^\mu\partial_\mu + m]\Psi^c_{BZ} = 0$, and the so-defined charge conjugation applies to the whole system, fermions+electromagnetic field in the covariant derivative. The concept of the doubling of the Fock space has been developed in Ziino works (cf. [13, 16]).

Let us analize the above statements. The helicity operator is $\hat{h} = (\sigma \cdot \mathbf{p}) \otimes I$ However, we can do the equivalence transformation of the helicity $\hat{h}$-operator with the unitary matrix.

It is known [17] that there exists $\mathcal{U}_1$ such that $\mathcal{U}_1(\sigma\cdot\mathbf{a})\mathcal{U}_1^{-1} = \sigma_3|\mathbf{a}|$. In the case of the momentum vector $(\mathbf{n}\equiv\hat{\mathbf{p}}=\mathbf{p}/|\mathbf{p}|)$, one has

$$\mathcal{U}_1 = \begin{pmatrix} 1 & p_l/(p+p_3) \\ -p_r/(p+p_3) & 1 \end{pmatrix}, \; U_1 = \begin{pmatrix} \mathcal{U}_1 & 0 \\ 0 & \mathcal{U}_1 \end{pmatrix},$$

$$U_1 \hat{h} U_1^{-1} = \frac{|\mathbf{n}|}{2}\begin{pmatrix} \sigma_3 & 0 \\ 0 & \sigma_3 \end{pmatrix}.$$

Then, applying other unitary matrix U_3:

$$\begin{pmatrix} 1&0&0&0\\0&0&1&0\\0&1&0&0\\0&0&0&1 \end{pmatrix}\begin{pmatrix} \sigma_3 & 0\\ 0 & \sigma_3\end{pmatrix}\begin{pmatrix} 1&0&0&0\\0&0&1&0\\0&1&0&0\\0&0&0&1 \end{pmatrix} = \begin{pmatrix} 1&0&0&0\\0&1&0&0\\0&0&-1&0\\0&0&0&-1 \end{pmatrix}.$$

we transform to the basis, where the helicity operator is equal (within the factor $\frac{1}{2}$) to γ^5, the chirality operator. The author of [10] and others introduced the *chiral* helicity $\eta = -\gamma_5 h$, which is equal (within the sign and the factor $\frac{1}{2}$) to the well-known matrix α multiplied by $\mathbf{n}$. Again, $U_1(\alpha\cdot\mathbf{n})U_1^{-1} = \alpha_3|\mathbf{n}|$, with the same matrix U_1. And applying the second unitary transformation:

$$U_2\alpha_3 U_2^\dagger = \begin{pmatrix} 1&0&0&0\\0&0&0&1\\0&0&1&0\\0&1&0&0 \end{pmatrix}\alpha_3\begin{pmatrix} 1&0&0&0\\0&0&0&1\\0&0&1&0\\0&1&0&0 \end{pmatrix} = \begin{pmatrix} 1&0&0&0\\0&1&0&0\\0&0&-1&0\\0&0&0&-1 \end{pmatrix}$$

we again come to the γ_5 matrix. The determinants are: $DetU_1 = 1 \neq 0$, $DetU_{2,3} = -1 \neq 0$. Thus, helicity, chirality and chiral helicity are connected by the unitary transformations. It is *not* surprising to have such a situation because the different helicity 2-spinors can be connected *not only* by the anti-linear transformation [4, 18], $\xi_h = (-1)^{1/2+h}e^{i\alpha_h}\Theta_{[1/2]}\mathcal{K}\xi_{-h}$, but the unitary transformation too. For instance, when we parametrize the 2-spinors as in [19] we obtain

$$\xi_\downarrow = U\xi_\uparrow = e^{i(\beta-\alpha)}\begin{pmatrix} 0 & e^{-i\phi}\\ -e^{i\phi} & 0\end{pmatrix}\xi_\uparrow,\; \xi_\uparrow = U^\dagger\xi_\downarrow = e^{i(\alpha-\beta)}\begin{pmatrix} 0 & -e^{-i\phi}\\ e^{i\phi} & 0\end{pmatrix}\xi_\downarrow .$$

In fact, Ahluwalia [4] and da Rocha [14] claimed that a 4-spinor can be an eigenspinor of the *chiral helicity*, and, at the same time, it is *not!* an eigenspinor of the helicity operator, and that the physical results would depend on this. Similar claims have been made by Shi and Ni [15]. These statements may signify that rotations have physical significance on the fundamental level.

Next, R. da Rocha et al. write [14]: "Now let one denotes the eigenspinors of the Dirac operator for particles and antiparticles respectively by $u_\pm(\mathbf{p})$ and $v_\pm(\mathbf{p})$. The subindex $\pm$ regards the eigenvalues of the helicity operator $(\sigma\cdot\widehat{\mathbf{p}})\otimes I$. $P^2 = 1$ in this case. The action of C on these spinors is given by $C(u_\pm(\mathbf{p})) = \mp v_\mp(\mathbf{p})$, $C(v_\pm(\mathbf{p})) = \pm u_\mp(\mathbf{p})$ [in textbooks [24]], which implies that $\{C,P\} = 0$, [anticommutator]. On another hand the parity operator P acts on ELKO by $P\lambda^S_{\mp,\pm}(\mathbf{p}) = \pm i\,\lambda^A_{\pm,\mp}(\mathbf{p})$, $P\lambda^A_{\mp,\pm}(\mathbf{p}) = \mp i\,\lambda^S_{\pm,\mp}(\mathbf{p})$, and it follows that $[C, P]_- = 0$ [when acting on the Majorana-like states]." In the earlier papers of the 50's-60's, Ref. [25] it is this case which has been attributed to the $Q = 0$ eigenvalues (the truly neutral particles). This may be compared with the results of Refs. [4, 7, 26], where the same statements have been done on the quantum-field level even at the earlier time comparing with Ref. [14]. The notation for the 4-spinors used in the cited papers is slightly different. The acronym "ELKO" is (**almost**) the synonym for the self/anti-self charge conjugated states (the Majorana-like spinors) used before. So, one can see the difference in the da Rocha formulas comparing with previous results on the classical level? In Refs. [6, 7, 26] the explicite forms of the $\lambda-$ and $\rho-$ 2-spinors have been given in the basis $\hat{S}_3\xi(\mathbf{0}) = \pm\frac{1}{2}\xi(\mathbf{0})$. The corresponding properties with respect to the parity (on the classical level) are different: $\gamma^0\lambda^S_{\uparrow\downarrow}(p^{\mu'}) = \pm i\lambda^S_{\downarrow\uparrow}(p^\mu)$, $\gamma^0\lambda^A_{\uparrow\downarrow}(p^{\mu'}) = \mp i\lambda^A_{\downarrow\uparrow}(p^\mu)$ therein. It is easy to find the correspondence between "the new notation", Refs. [14, 27] and the previous one. Namely, $\lambda^{S,A}_\uparrow \to \lambda^{S,A}_{-,+}$, $\lambda^{S,A}_\downarrow \to \lambda^{S,A}_{+,-}$. However, the difference is also in the choice of the basis for the 2-spinors. As in Ref. [28], Ahluwalia, Grumiller and da Rocha have chosen the well-known helicity basis (cf. [19, 29]). It was shown in my paper [28] that the helicity-basis 4-spinors satisfy the same Dirac equation, the parity matrix can be defined in the similar fashion as in the spinorial basis (according to the Itzykson-Zuber textbook [24]), but the helicity-basis 4-spinors are *not* the eigenspinors of the parity, This is in full agreement with the claims made in the 4th volume of the Landau course of theoretical physics and with the fact that $[\hat{h}, \hat{P}]_+ = 0$, Ref. [30]. In this basis, the parity transformation lead to the properties:

$$R\phi^-_L(\mathbf{0}) = -ie^{i(\theta_2-\theta_1)}\phi^+_L(\mathbf{0})\,,\; R\phi^+_L(\mathbf{0}) = -ie^{i(\theta_1-\theta_2)}\phi^-_L(\mathbf{0}),$$
$$R\Theta(\phi^-_L(\mathbf{0}))^* = -ie^{-2i\theta_2}\phi^-_L(\mathbf{0})\,,\; R\Theta(\phi^+_L(\mathbf{0}))^* = +ie^{-2i\theta_1}\phi^+_L(\mathbf{0}).$$

This opposes to the spinorial basis, where, for instance: $R\phi^\pm_L(\mathbf{0}) = \phi^\pm_L(\mathbf{0})$. Further calculations are straightforward, and, indeed, under certain choice of the phases in 2-spinors they can lead to the commutation of C and P operators $[C, P]_- = 0$ when acting on the "ELKO" states. This case corresponds to the neutral particles [25].

Several formalisms have been used for higher spin fields, e. g., [20,21]. The $2(2S+1)$ formalism gives the equations which are in some sense on an equal footing with the Dirac equation. For instance, for the spin-1 field the equation is $[\gamma_{\mu\nu}p_\mu p_\nu - m^2]\Psi(x) = 0$, with the $\gamma_{\mu\nu}$ being the 6x6 covariantly-defined matrices. However, it was argued later that the signs before the mass terms should be opposite for charged particles of positive- and negative- frequencies [22, 23]: $[\gamma_{\mu\nu}p_\mu p_\nu - (\frac{i\partial/\partial t}{E})m^2]\Psi(x) = 0$. Hence, Ahluwalia et al. write: "The charge conjugation operation C must be carried through with a little greater care for bosons than for fermions within [this] framework because of $\wp_{u,v} = \pm 1$ factor in the mass term. For the $(1,0)\oplus(0,1)$ case, at the classical level we want

$$C:\left(\gamma_{\mu\nu}\,D^\mu_+\,D^\nu_+ + m^2\right)u(x) = 0 \to \left(\gamma_{\mu\nu}\,D^\mu_-\,D^\nu_- - m^2\right)v(x) = 0,$$

where the local $U(1)$ gauge covariant derivatives are defined as: $D^\mu_+ = \partial^\mu + i\,q\,A^\mu(x)$, $D^\mu_- = \partial^\mu - i\,q\,A^\mu(x)$", Ref. [23]. "For spin-1 ... the requirement of self/anti-self charge conjugacy *cannot* be satisfied. That is, there does not exist a ζ [the phase factors between right- and left- 3-"spinors"] that can satisfy the spin-1 ... requirement" $S^c_{[1]}\,\lambda(p^\mu) = \pm\lambda(p^\mu)$, $S^c_{[1]}\,\rho(p^\mu) = \pm\rho(p^\mu)$". This is due to the fact that

$C^2 = -1$ within this definition of the charge conjugation operator. "We find, however, that the requirement of self/anti-self conjugacy under charge conjugation can be replaced by the requirement of self/anti-self conjugacy under the operation of $\Gamma^5\, S^c_{[1]}$ [precisely, which was used by Weinberg in Ref. [21] due to the different choice of the equation for the negative-frequency 6-"bispinors"], where Γ^5 is the *chirality* operator for the $(1,\,0) \oplus (0,\,1)$ representation space... The requirements $\left[\Gamma^5\, S^c_{[1]}\right] \lambda(p^\mu) \;=\; \pm\,\lambda(p^\mu)\,, \quad \left[\Gamma^5\, S^c_{[1]}\right] \rho(p^\mu) \;=\; \pm\,\rho(p^\mu)$ determine $\zeta^S_\lambda \;=\; +1 \;=\; \zeta^S_\rho$ for the self $\left[\Gamma^5\, S^c_{[1]}\right]$-conjugate spinors $\lambda^S(p^\mu)$ and $\rho^S(p^\mu)$; and $\zeta^A_\lambda \;=\; -1 \;=\; \zeta^A_\rho$ for the anti-self $\left[\Gamma^5\, S^c_{[1]}\right]$-conjugate spinors $\lambda^A(p^\mu)$ and $\rho^A(p^\mu)$."

The covariant equations for the $\lambda-$ and $\rho-$ objects and their explicite forms have been obtained in Ref. [6] in the $(1,0)\oplus(0,1)$ representation, under the certain choice of the phase factors in the definition of left- and right- 3-objects. While we introduced $\Gamma_5 S^c_{[1]}$ self/anti-self objects in the $(1,0)\oplus(0,1)$ representation the situation is rather similar to the $(1/2,0)\oplus(0,1/2)$ representation, because, again, we just have the rotation of the spin-parity basis. If we would like to extend the Nigam-Foldy conclusion, Ref. [25] (about the fact that the commutation $[C,P]_- \;=\; 0$ corresponds to the neutral particles even in the higher spin case), then we should use a) the helicity basis on the classical level, b) the appropriate choice of the phases of the 3-"spinors". The corresponding relations are:

$$R\phi^-_L(\mathbf{0}) \;=\; +e^{i(\theta_- - \theta_+)}\phi^+_L(\mathbf{0})\,,\; R\phi^+_L(\mathbf{0}) = +e^{i(\theta_+ - \theta_-)}\phi^-_L(\mathbf{0}), \tag{1}$$

$$R\Theta(\phi^-_L(\mathbf{0}))^* \;=\; -e^{-2i\theta_-}\,\phi^-_L(\mathbf{0})\,,\; R\Theta(\phi^+_L(\mathbf{0}))^* = -e^{-2i\theta_+}\,\phi^+_L(\mathbf{0}), \tag{2}$$

$$R\phi^0_L(\mathbf{0}) \;=\; -\phi^0_L(\mathbf{0})\,,\; R\Theta(\phi^0_L(\mathbf{0}))^* = +e^{-2i\theta_0}\,\phi^0_L(\mathbf{0}). \tag{3}$$

On using these rules and CFT/QFT it is easy to find C and P properties of the 6-component objects $\lambda^{S,A}$ and $\rho^{S,A}$, and the (anti)commutators $[C,P]_\pm$.

However, on the level of the quantum-field theory (the "secondary" quantization) the situation is self-consistent even in the case of the use of $u-$ and $v-$ objects. We have to introduce the unitary operators for the charge conjugation and the parity in the Fock space $U^c_{[S]}\Psi_{[S]}(x^\mu)(U^c_{[S]})^{-1} \;=\; C_{[S]}\Psi^\dagger_{[S]}(x^\mu)$, $U^s_{[S]}\Psi_{[S]}(x^\mu)(U^s_{[S]})^{-1} \;=\; \gamma^0\Psi_{[S]}(x'^\mu)$. For the spin $S = 1/2$ they can be found in the well-known textbooks [24]. The procedure of their construction can be easily generalized for higher spins. As shown in 1997, Ref. [7,26], we can obtain easily **both** cases (commutation and anti-commutation) on using $\lambda^{S,A}$ 4-spinors, which have been used earlier (in the basis $column(1\;0)\; column(0\;1)$). Presumably, this conclusion holds in the case of higher-spin field operators too.

We presented a review of the formalism for the momentum-space Majorana-like particles in the $(S,0)\oplus(0,S)$ representation of the Lorentz Group. The $\lambda-$ and $\rho-$ 4-spinors satisfy the 8- component analogue of the Dirac equation. Apart, they have different gauge transformations comparing with the usual Dirac 4-spinors. Their helicity, chirality and chiral helicity properties have been investigated in detail. At the same time, we showed that the Majorana-like 4-spinors can be obtained by the rotation of the spin-parity basis. Meanwhile, several authors have claimed that the physical results would be different on using calculations with these Majorana-like spinors. However, several claims made by other researchers concerning with chirality, helicity, chiral helicity should not be considered to be true until the time when experiments confirm them. Usually, it is considered that the rotations (unitary transformations) have *no* any physical consequences on the level of the Lorentz-covariant theories. Next, we discussed the $[C, P]_{\pm} = 0$ dilemma for neutral and charged particles on using the analysis of the basis rotations and phases.

References

[1] E. Majorana, *Nuovo Cimento* 14 (1937) 171.

[2] S. M. Bilenky and B. M. Pontekorvo, *Phys. Repts.* 42 (1978) 224.

[3] A. Barut and G. Ziino, *Mod. Phys. Lett. A* 8 (1993) 1099; G. Ziino, *Int. J. Mod. Phys. A* 11 (1996) 2081.

[4] D. V. Ahluwalia, *Int. J. Mod. Phys. A* 11 (1996) 1855; Incompatibility of Self-Charge Conjugation with Helicity Eignestates and Gauge Interactions. Preprint LANL UR-94-1252 (Los Alamos, 1994).

[5] P. Lounesto, *Clifford Algebras and Spinors.* (Cambridge University Press, 2002), Ch. 11 and 12.

[6] V. V. Dvoeglazov, *Int. J. Theor. Phys.* 34 (1995) 2467; *Nuovo Cim.* 108A (1995) 1467, *Hadronic J.* 20 (1997) 435; *Acta Phys. Polon. B* 29 (1998) 619.

[7] V. V. Dvoeglazov, *Mod. Phys. Lett. A* 12 (1997) 2741.

[8] M. Kirchbach, C. Compean and L. Noriega, *Eur. Phys. J. A* 22 (2004) 149.

[9] R. da Rocha and W. Rodrigues, Jr., *Mod. Phys. Lett. A* 21 (2006) 65, math-ph/0506075.

[10] N. D. Sen Gupta, *Nucl. Phys. B* 4 (1967) 147.

[11] R. P. Feynman and M. Gell-Mann, *Phys. Rev.* 109 (1958) 193.

[12] M. Markov, ZhETF 7 (1937) 579, 603; *Nucl. Phys.* 55 (1964) 130.

[13] I. M. Gelfand and M. L. Tsetlin, *ZhETF* 31 (1956) 1107; G. A. Sokolik, *ZhETF* 33 (1957) 1515.

[14] R. da Rocha, A. E. Bernardini and J. M. Hoff da Silva, *JHEP* 1104 (2011) 110.

[15] Z. Q. Shi and G. J. Ni, *Chin. Phys. Lett.* 19 (2002) 1427; Zhi-Qiang Shi, arXiv:1101.0481v1.

[16] V. V. Dvoeglazov, *Int. J. Theor. Phys.* 37 (1998) 1915.

[17] R. A. Berg, *Nuovo Cim.* 42 (1966) 148.

[18] L. H. Ryder, *Quantum Field Theory.* (Cambridge University Press, Cambridge, 1985).

[19] V. V. Dvoeglazov, *Fizika B* 6 (1997) 111.

[20] V. Bargmann and E. P. Wigner, *Proc. Nat. Acad. Sci. (USA)* 34 (1948) 211.

[21] S. Weinberg, *Phys. Rev.* 133 (1964) B1318; ibid. 134 (1964) B882.

[22] D. L. Weaver, C. L. Hammer and R. H. Good, jr., *Phys. Rev. B* 135 (1964) 241; A. Sankaranarayanan and R. H. Good, jr., *Nuovo Cim.* 36 (1965) 1303.

[23] D. V. Ahluwalia, M. B. Johnson and T. Goldman, *Phys. Lett. B* 316 (1993) 102.

[24] C. Itzykson and J.-B. Zuber, *Quantum Field Theory.* (McGraw-Hill Book Co., 1980).

[25] B. Nigam and L. L. Foldy, *Phys. Rev.* 102 (1956) 1410.

[26] V. V. Dvoeglazov, *J. Phys. Conf. Ser.* 343 (2012) 012033.

[27] D. V. Ahluwalia and D. Grumiller, *Phys. Rev. D* 72 (2005) 067701.

[28] V. V. Dvoeglazov, *Int. J. Theor. Phys.* 43 (2004) 1287.

[29] D. A. Varshalovich et al., *Quantum Theory of Angular Momentum* (World Scientific, 1988).

[30] V. B. Berestetskii et al., *Quantum Electrodynamics.* 2nd Edition (Butterworth-Heinemann, 1982).

In: Einstein and Others: Unification
Editor: Valeriy V. Dvoeglazov

ISBN: 978-1-63463-276-8

Chapter 9

THE CONSTRUCTION OF QUANTUM FIELD OPERATORS: SOMETHING OF INTEREST

Valeriy V. Dvoeglazov
Universidad de Zacatecas, Zacatecas, Zac., México

Abstract

We draw attention to some tune problems in constructions of the quantum-field operators for spins 1/2 and 1. They are related to the existence of negative-energy solutions and acausal solutions of relativistic wave equations. Particular attention is paid to chiral theories, and to the method of the Lorentz boosts.

1. The Dirac Equation

The Dirac equation can be deduced either by the Dirac method [1] (the Hamiltonian should be linear in $\partial/\partial x^i$, and be compatible with $E^2 - \mathbf{p}^2 c^2 = m^2 c^4$); or by the Sakurai method [2] (based on the equation $(E - \sigma \cdot \mathbf{p})(E + \sigma \cdot \mathbf{p})\phi = m^2\phi$); or by the Ryder one [3] (the relation between 2-spinors at rest is $\phi_R(\mathbf{0}) = \pm\phi_L(\mathbf{0})$). Usually, everybody uses the following definition of the field operator [4]:

$$\Psi(x) = \frac{1}{(2\pi)^3} \sum_\sigma \int \frac{d^3\mathbf{p}}{2E_p} [u_\sigma(\mathbf{p}) a_\sigma(\mathbf{p}) e^{-ip\cdot x} + v_\sigma(\mathbf{p}) b_\sigma^\dagger(\mathbf{p})] e^{+ip\cdot x}] , \qquad (1)$$

as given *ab initio*. We begin with the method of constructions of the field operators given in [5]. The Wigner rules [6] of the Lorentz transformations for the $(0, S)$ left- $\phi_L(\mathbf{p})$ and the $(S, 0)$ right- $\phi_R(\mathbf{p})$ spinors are used. The solutions of the Dirac equation can be denoted as $u(\mathbf{p})$ and $v(\mathbf{p}) = \gamma^5 u(\mathbf{p})$, ref. [3, p.53] with $E = \sqrt{\mathbf{p}^2 + m^2} > 0$, $p_0 = \pm E$, $p^\pm = E \pm p_z$, $p_{r,l} = p_x \pm i p_y$. They are the parity eigenstates with the eigenvalues of ± 1. They also describe eigenstates of the charge operator. Thus, in this Section we use the basis for charged particles in the $(S,0)\oplus(0,S)$ representation (in general)

$$u_{+\sigma}(\mathbf{0}) = N(\sigma)\begin{pmatrix}1\\0\\ \cdot\\ \cdot\\ \cdot\\0\end{pmatrix}, u_{\sigma-1}(\mathbf{0}) = N(\sigma)\begin{pmatrix}0\\1\\ \cdot\\ \cdot\\ \cdot\\0\end{pmatrix}, \ldots v_{-\sigma}(\mathbf{0}) = N(\sigma)\begin{pmatrix}0\\0\\ \cdot\\ \cdot\\ \cdot\\1\end{pmatrix} \tag{2}$$

Sometimes, the normalization factor is convenient to choose $N(\sigma) = m^\sigma$ in order the rest spinors to vanish in the massless limit. However, other constructs are possible in the $(1/2,0)\oplus(0,1/2)$ representation.

2. Majorana Spinors in the Momentum Representation

During the 20th century various authors introduced *self/anti-self* charge-conjugated 4-spinors (including in the momentum representation), see [7, 8, 9, 10]. Later several authors [11, 12, 13, 14] studied these spinors, they found corresponding dynamical equations, gauge transformations and other specific features of them. The definitions are:

$$C = e^{i\theta}\begin{pmatrix}0 & 0 & 0 & -i\\ 0 & 0 & i & 0\\ 0 & i & 0 & 0\\ -i & 0 & 0 & 0\end{pmatrix}\mathcal{K} = -e^{i\theta}\gamma^2\mathcal{K} \tag{3}$$

is the anti-linear operator of charge conjugation. $\mathcal{K}$ is the complex conjugation operator. We define the *self/anti-self* charge-conjugate 4-spinors in the momentum space

$$C\lambda^{S,A}(\mathbf{p}) = \pm\lambda^{S,A}(\mathbf{p})\,, C\rho^{S,A}(\mathbf{p}) = \pm\rho^{S,A}(\mathbf{p})\,. \tag{4}$$

Such definitions of 4-spinors differ, of course, from the original Majorana definition in x-representation: $\nu(x) = (\Psi_D(x) + \Psi_D^c(x))/\sqrt{2}$, $C\nu(x) = \nu(x)$ that represents the positive real $C-$ parity field operator. The rest λ spinors are:

$$\lambda_\uparrow^S(\mathbf{0}) = \sqrt{\frac{m}{2}}\begin{pmatrix}0\\i\\1\\0\end{pmatrix}, \lambda_\downarrow^S(\mathbf{0}) = \sqrt{\frac{m}{2}}\begin{pmatrix}-i\\0\\0\\1\end{pmatrix}, \lambda_\uparrow^A(\mathbf{0}) = \sqrt{\frac{m}{2}}\begin{pmatrix}0\\-i\\1\\0\end{pmatrix}, \lambda_\downarrow^A(\mathbf{0}) = \sqrt{\frac{m}{2}}\begin{pmatrix}i\\0\\0\\1\end{pmatrix}. \tag{5}$$

It is easy to derive the explicite forms of the 4-spinors of the second kind $\lambda^{S,A}_{\uparrow\downarrow}(\mathbf{p})$ and the $\rho^{S,A}_{\uparrow\downarrow}(\mathbf{p})$ in this basis. As we showed, the 4-spinors λ and ρ are neither eigenspinors of the helicity nor eigenspinors of the parity (as opposed to the Dirac case). The indices $\uparrow\downarrow$ should be referred to the chiral helicity quantum number introduced in the 60s, $\eta = -\gamma^5 h$ (or $-\gamma_5\mathbf{S}_3$). While $Pu_\sigma(\mathbf{p}) = +u_\sigma(\mathbf{p})\,, Pv_\sigma(\mathbf{p}) = -v_\sigma(\mathbf{p})$, we have $P\lambda^{S,A}(\mathbf{p}) = \rho^{A,S}(\mathbf{p})\,, P\rho^{S,A}(\mathbf{p}) = \lambda^{A,S}(\mathbf{p})$ for the Majorana-like momentum-space 4-spinors on the first quantization level.

The dynamical coordinate-space equations are:

$$i\gamma^\mu\partial_\mu\lambda^S(x) - m\rho^A(x) = 0\,, i\gamma^\mu\partial_\mu\rho^A(x) - m\lambda^S(x) = 0\,, \tag{6}$$

$$i\gamma^\mu\partial_\mu\lambda^A(x) + m\rho^S(x) = 0\,, i\gamma^\mu\partial_\mu\rho^S(x) + m\lambda^A(x) = 0\,. \tag{7}$$

These are *not* the Dirac equation. However, they can be written in the 8-component form or in the two-component form. Similar formulations have been presented by M. Markov [15], and A. Barut and G. Ziino [9]. The group-theoretical basis for such doubling has been given in the papers by Gelfand et al. [16].

The Lagrangian is

$$\mathcal{L} = \frac{i}{2}\Big[\bar{\lambda}^S\gamma^\mu\partial_\mu\lambda^S - (\partial_\mu\bar{\lambda}^S)\gamma^\mu\lambda^S + \bar{\rho}^A\gamma^\mu\partial_\mu\rho^A - (\partial_\mu\bar{\rho}^A)\gamma^\mu\rho^A +$$
$$+\bar{\lambda}^A\gamma^\mu\partial_\mu\lambda^A - (\partial_\mu\bar{\lambda}^A)\gamma^\mu\lambda^A + \bar{\rho}^S\gamma^\mu\partial_\mu\rho^S - (\partial_\mu\bar{\rho}^S)\gamma^\mu\rho^S -$$
$$-m(\bar{\lambda}^S\rho^A - \bar{\lambda}^A\rho^S + \bar{\rho}^A\lambda^S - \bar{\rho}^S\lambda^A)\Big] . \tag{8}$$

The connection with the Dirac spinors has been found. For instance,

$$\begin{pmatrix}\lambda^S_\uparrow(\mathbf{p})\\ \lambda^S_\downarrow(\mathbf{p})\\ \lambda^A_\uparrow(\mathbf{p})\\ \lambda^A_\downarrow(\mathbf{p})\end{pmatrix} = \frac{1}{2}\begin{pmatrix}1 & i & -1 & i\\ -i & 1 & -i & -1\\ 1 & -i & -1 & -i\\ i & 1 & i & -1\end{pmatrix}\begin{pmatrix}u_{+1/2}(\mathbf{p})\\ u_{-1/2}(\mathbf{p})\\ v_{+1/2}(\mathbf{p})\\ v_{-1/2}(\mathbf{p})\end{pmatrix} . \tag{9}$$

Possibilities have been noted for generalizations of the concept of the Fock space, which lead to the "doubling" Fock space [16, 9].

It was shown [13] that the covariant derivative (and, hence, the interaction) can be introduced in this construct in the following way: $\partial_\mu \rightarrow \nabla_\mu = \partial_\mu - ig\text{Ł}^5 A_\mu$, where $\text{Ł}^5 = \mathrm{diag}(\gamma^5 \quad -\gamma^5)$, the 8×8 matrix. With respect to the transformations $\lambda'(x) \rightarrow (\cos\alpha - i\gamma^5\sin\alpha)\lambda(x)$, $\bar{\lambda}'(x) \rightarrow \bar{\lambda}(x)(\cos\alpha - i\gamma^5\sin\alpha)$ the spinors retain their properties to be self/anti-self charge conjugate spinors and the proposed Lagrangian [13, p.1472] remains to be invariant. This tells us that while self/anti-self charge conjugate states have zero eigenvalues of the ordinary (scalar) charge operator but they can possess the axial charge. Next, because the transformations

$$\lambda'_S(\mathbf{p}) = \begin{pmatrix}\Xi & 0\\ 0 & \Xi\end{pmatrix}\lambda_S(\mathbf{p}) \equiv \lambda^*_A(\mathbf{p}), \lambda''_S(\mathbf{p}) = \begin{pmatrix}i\Xi & 0\\ 0 & -i\Xi\end{pmatrix}\lambda_S(\mathbf{p}) \equiv -i\lambda^*_S(\mathbf{p}), \tag{10}$$

$$\lambda'''_S(\mathbf{p}) = \begin{pmatrix}0 & i\Xi\\ i\Xi & 0\end{pmatrix}\lambda_S(\mathbf{p}) \equiv i\gamma^0\lambda^*_A(\mathbf{p}), \lambda^{IV}_S(\mathbf{p}) = \begin{pmatrix}0 & \Xi\\ -\Xi & 0\end{pmatrix}\lambda_S(\mathbf{p}) \equiv \gamma^0\lambda^*_S(\mathbf{p}) \tag{11}$$

with the 2×2 matrix Ξ defined as

$$\Xi = \begin{pmatrix}e^{i\phi} & 0\\ 0 & e^{-i\phi}\end{pmatrix} \quad , \quad \Xi\Lambda_{R,L}(\mathbf{p} \leftarrow \mathbf{0})\Xi^{-1} = \Lambda^*_{R,L}(\mathbf{p} \leftarrow \mathbf{0}) \ , \tag{12}$$

and corresponding transformations for λ^A, do *not* change the properties of bispinors to be in the self/anti-self charge conjugate spaces, the Majorana-like field operator ($b^\dagger \equiv a^\dagger$) admits additional phase (and, in general, normalization) transformations:

$$\nu^{ML\,\prime}(x^\mu) = [c_0 + i(\tau\cdot\mathbf{c})]\,\nu^{ML\,\dagger}(x^\mu) , \tag{13}$$

where c_α are arbitrary parameters. The τ matrices are defined over the field of 2×2 matrices. One can parametrize $c_0 = \cos\phi$ and $\mathbf{c} = \mathbf{n}\sin\phi$ and, thus, define the $SU(2)$ group of phase transformations. One can select the Lagrangian which is composed from the both field operators (with λ spinors and ρ spinors) and which remains to be invariant with respect to this kind of transformations. The conclusion is: it is permitted a non-Abelian construct which is based on the spinors of the Lorentz group only. This is not surprising because both the $SU(2)$ group and $U(1)$ group are the sub-groups of the extended Poincaré group (cf. [3]).

The Dirac-like and Majorana-like field operators can be built from both $\lambda^{S,A}(\mathbf{p})$ and $\rho^{S,A}(\mathbf{p})$, or their combinations. The anticommutation relations are the following ones (due to the *bi-orthonormality*):

$$[a_{\eta'}(\mathbf{p}'), a^\dagger_\eta(\mathbf{p})]_\pm = (2\pi)^3 2E_p \delta(\mathbf{p}-\mathbf{p}')\delta_{\eta,-\eta'}, \tag{14}$$
$$[b_{\eta'}(\mathbf{p}'), b^\dagger_\eta(\mathbf{p})]_\pm = (2\pi)^3 2E_p \delta(\mathbf{p}-\mathbf{p}')\delta_{\eta,-\eta'} \tag{15}$$

Other (anti)commutators are equal to zero: ($[a_{\eta'}(\mathbf{p}'), b^\dagger_\eta(\mathbf{p})] = 0$).

In the Fock space the operations of the charge conjugation and space inversions can be defined through unitary operators such that:

$$U^c_{[1/2]}\Psi(x^\mu)(U^c_{[1/2]})^{-1} = \mathcal{C}_{[1/2]}\Psi^\dagger_{[1/2]}(x^\mu)\,,\; U^s_{[1/2]}\Psi(x^\mu)(U^s_{[1/2]})^{-1} = \gamma^0\Psi(x'^\mu), \tag{16}$$

the time reversal operation, through *an antiunitary* operator

$$\left[V^T_{[1/2]}\Psi(x^\mu)(V^T_{[1/2]})^{-1}\right]^\dagger = S(T)\Psi^\dagger(x''^\mu) \quad , \tag{17}$$

with $x'^\mu \equiv (x^0, -\mathbf{x})$ and $x''^\mu = (-x^0, \mathbf{x})$. We further assume the vacuum state to be assigned an even P- and C-eigenvalue and, then, proceed as in ref. [4]. As a result we have the following properties of creation (annihilation) operators in the Fock space:

$$U^s_{[1/2]}a_\uparrow(\mathbf{p})(U^s_{[1/2]})^{-1} = -ia_\downarrow(-\mathbf{p})\,,\; U^s_{[1/2]}a_\downarrow(\mathbf{p})(U^s_{[1/2]})^{-1} = +ia_\uparrow(-\mathbf{p}) \tag{18}$$
$$U^s_{[1/2]}b^\dagger_\uparrow(\mathbf{p})(U^s_{[1/2]})^{-1} = +ib^\dagger_\downarrow(-\mathbf{p})\; U^s_{[1/2]}b^\dagger_\downarrow(\mathbf{p})(U^s_{[1/2]})^{-1} = -ib_\uparrow(-\mathbf{p})\,, \tag{19}$$

what signifies that the states created by the operators $a^\dagger(\mathbf{p})$ and $b^\dagger(\mathbf{p})$ have very different properties with respect to the space inversion operation, comparing with Dirac states (the case also regarded in [9]):

$$U^s_{[1/2]}|\mathbf{p},\uparrow>^+ = +i|-\mathbf{p},\downarrow>^+, U^s_{[1/2]}|\mathbf{p},\uparrow>^- = +i|-\mathbf{p},\downarrow>^-\,, \tag{20}$$
$$U^s_{[1/2]}|\mathbf{p},\downarrow>^+ = -i|-\mathbf{p},\uparrow>^+, U^s_{[1/2]}|\mathbf{p},\downarrow>^- = -i|-\mathbf{p},\uparrow>^-\,. \tag{21}$$

For the charge conjugation operation in the Fock space we have two physically different possibilities. The first one, *e.g.*,

$$U^c_{[1/2]}a_\uparrow(\mathbf{p})(U^c_{[1/2]})^{-1} = +b_\uparrow(\mathbf{p})\,,\; U^c_{[1/2]}a_\downarrow(\mathbf{p})(U^c_{[1/2]})^{-1} = +b_\downarrow(\mathbf{p})\,, \tag{22}$$
$$U^c_{[1/2]}b^\dagger_\uparrow(\mathbf{p})(U^c_{[1/2]})^{-1} = -a^\dagger_\uparrow(\mathbf{p})\,,\; U^c_{[1/2]}b^\dagger_\downarrow(\mathbf{p})(U^c_{[1/2]})^{-1} = -a^\dagger_\downarrow(\mathbf{p})\,. \tag{23}$$

The action of this operator on the physical states are

$$U^c_{[1/2]}|\mathbf{p},\uparrow>^+ = +|\mathbf{p},\uparrow>^-,\; U^c_{[1/2]}|\mathbf{p},\downarrow>^+ = +|\mathbf{p},\downarrow>^-, \tag{24}$$
$$U^c_{[1/2]}|\mathbf{p},\uparrow>^- = -|\mathbf{p},\uparrow>^+,\; U^c_{[1/2]}|\mathbf{p},\downarrow>^- = -|\mathbf{p},\downarrow>^+\,. \tag{25}$$

The second one acts in the following manner:

$$\widetilde{U}^c_{[1/2]} a_\uparrow(\mathbf{p}) (\widetilde{U}^c_{[1/2]})^{-1} = -b_\downarrow(\mathbf{p})\,,\; \widetilde{U}^c_{[1/2]} a_\downarrow(\mathbf{p}) (\widetilde{U}^c_{[1/2]})^{-1} = -b_\uparrow(\mathbf{p}), \tag{26}$$

$$\widetilde{U}^c_{[1/2]} b^\dagger_\uparrow(\mathbf{p}) (\widetilde{U}^c_{[1/2]})^{-1} = +a^\dagger_\downarrow(\mathbf{p})\,,\; \widetilde{U}^c_{[1/2]} b^\dagger_\downarrow(\mathbf{p}) (\widetilde{U}^c_{[1/2]})^{-1} = +a^\dagger_\uparrow(\mathbf{p}), \tag{27}$$

and, therefore,

$$\widetilde{U}^c_{[1/2]} |\mathbf{p},\uparrow>^+ \;=\; -\,|\mathbf{p},\downarrow>^-\,,\; \widetilde{U}^c_{[1/2]} |\mathbf{p},\downarrow>^+ = -\,|\mathbf{p},\uparrow>^-\,, \tag{28}$$

$$\widetilde{U}^c_{[1/2]} |\mathbf{p},\uparrow>^- \;=\; +\,|\mathbf{p},\downarrow>^+\,,\; \widetilde{U}^c_{[1/2]} |\mathbf{p},\downarrow>^- = +\,|\mathbf{p},\uparrow>^+\,. \tag{29}$$

It is possible a situation when the operators of the space inversion and charge conjugation commute each other in the Fock space [18]. Finally, the time reversal *anti-unitary* operator in the Fock space should be defined in such a way that the formalism to be compatible with the CPT theorem. We obtain for the $\Psi(x^\mu)$ field:

$$V^T a^\dagger_\uparrow(\mathbf{p}) (V^T)^{-1} \;=\; a^\dagger_\downarrow(-\mathbf{p}),\; V^T a^\dagger_\downarrow(\mathbf{p}) (V^T)^{-1} = -a^\dagger_\uparrow(-\mathbf{p})\,, \tag{30}$$

$$V^T b_\uparrow(\mathbf{p}) (V^T)^{-1} \;=\; b_\downarrow(-\mathbf{p}),\; V^T b_\downarrow(\mathbf{p}) (V^T)^{-1} = -b_\uparrow(-\mathbf{p}). \tag{31}$$

Thus, this construct has very different properties with respect to C, P and T comparing with the Dirac construct.

The dependence of the physical results on the choice of the basis is a bit strange thing. Somewhat similar things have been presented in [17] when compared the Dirac-like constructs in the parity basis and the helicity basis. It was shown that the helicity eigenstates of the operator $(\sigma\cdot\mathbf{n})\otimes I)$ are NOT the parity eigenstates, and vice versa, in the helicity basis (cf. with [19]), while they obey the same Dirac equation. The bases are connected by the unitary transformation. The both sets of 4-spinors form the complete system in a mathematical sense.

3. The Spin 1

In refs. [20, 21] the Maxwell-like equations have been derived from the D'Alembert equation. Similar to the spin-1/2 case we have in the spin-1 case:

$$(EI^{(3)} - \mathbf{S}\cdot\mathbf{p})(EI^{(3)} + \mathbf{S}\cdot\mathbf{p})\Psi^{(3)} - \mathbf{p}(\mathbf{p}\cdot\Psi^{(3)}) = m^2\Psi^{(3)}\,. \tag{32}$$

Here they are:

$$\nabla\times\mathbf{E} = -\frac{1}{c}\frac{\partial\mathbf{B}}{\partial t} + \nabla Im\chi\,,\; \nabla\cdot\mathbf{E} = -\frac{1}{c}\frac{\partial}{\partial t}Re\chi\,, \tag{33}$$

$$\nabla\times\mathbf{B} = \frac{1}{c}\frac{\partial\mathbf{E}}{\partial t} + \nabla Re\chi\,,\; \nabla\cdot\mathbf{B} = \frac{1}{c}\frac{\partial}{\partial t}Im\chi\,. \tag{34}$$

Of course, similar equations can be obtained in the massive case $m \neq 0$, i.e., within the Proca-like theory. When $\chi = 0$ we recover the common-used Maxwell equations.

Otherwise, we can start with ($c = \hbar = 1$) $\partial\mathbf{E}/\partial t = curl\mathbf{B}$, $\partial\mathbf{B}/\partial t = -curl\mathbf{E}$. Since the spin-1 matrices can be presented in the form: $(\mathbf{S}^i)^{jk} = -i\epsilon^{ijk}$, we have

$$\frac{\partial(\mathbf{E}+i\mathbf{B})^i}{\partial t} + (\mathbf{S}\cdot\nabla)^{ik}(\mathbf{E}+i\mathbf{B})^k = 0, \tag{35}$$

$$\frac{\partial(\mathbf{E}-i\mathbf{B})^i}{\partial t} - (\mathbf{S}\cdot\nabla)^{ik}(\mathbf{E}-i\mathbf{B})^k = 0. \tag{36}$$

Finally, on using that $\hat{\mathbf{p}} = -i\hbar\nabla$ we have

$$i\frac{\partial\phi}{\partial t} = (\mathbf{S}\cdot\hat{\mathbf{p}})\phi, \quad i\frac{\partial\xi}{\partial t} = -(\mathbf{S}\cdot\hat{\mathbf{p}})\xi. \tag{37}$$

We can show that these equations can also be considered as the massless limit of the Weinberg $2(2S+1)$ quantum-field equation for spin-1. Meanwhile, we can calculate the determinants of the above equations, $Det[E \mp (\mathbf{S}\cdot\mathbf{p})] = 0$, and we can find that we have both the causal $E = \pm|\mathbf{p}|$ and the "acausal" $E = 0$ solutions.

The Weinberg formalism is based on the relativity postulates [6]. In such a way he proceeds to form the $2(2S+1)$-component object $\Psi = column(\Phi_\sigma \quad \Xi_\sigma)$ transformed according to the Wigner rules. Weinberg gives the higher-spin equations:

$$[\gamma^{\mu_1\mu_2\ldots\mu_{2s}}\partial_{\mu_1}\partial_{\mu_2}\ldots\partial_{\mu_{2s}} + m^{2s}]\Psi(x) = 0, \tag{38}$$

with the Barut-Muzinich-Williams covariantly-defined matrices [22, 23]. For the spin-1 they are:

$$\gamma_{44} = \begin{pmatrix} 0 & 1 \\ 1 & 0 \end{pmatrix}, \quad \gamma_{i4} = \gamma_{4i} = \begin{pmatrix} 0 & iS_i \\ -iS_i & 0 \end{pmatrix}, \tag{39}$$

$$\gamma^{ij} = \begin{pmatrix} 0 & \delta_{ij} - S_iS_j - S_jS_i \\ \delta_{ij} - S_iS_j - S_jS_i & 0 \end{pmatrix}. \tag{40}$$

Later Sankaranarayanan and Good considered another version of this theory [23] (see also [25]). For the $S = 1$ case they introduced the Weaver-Hammer-Good sign operator, ref. [24], $m^2 \rightarrow m^2\,(i\partial/\partial t)/E$, which led to the different parity properties of an antiparticle with respect to a *boson* particle. Next, Tucker and Hammer *et al* [26] introduced another higher-spin equations. In the spin-1 case it is:

$$[\gamma_{\mu\nu}\partial_\mu\partial_\nu + \partial_\mu\partial_\mu - 2m^2]\Psi^{(s=1)} = 0. \tag{41}$$

One can add the Klein-Gordon equation with arbitrary multiple factor to the Weinberg equation. The new equation is written:

$$[\gamma_{\alpha\beta}p_\alpha p_\beta + Ap_\alpha p_\alpha + Bm^2]\Psi = 0. \tag{42}$$

It has solutions with relativistic dispersion relations $E^2 - \mathbf{p}^2 = m^2$ provided that $B/(A+1) = 1$, or $B/(A-1) = 1$. This can be proven by considering the algebraic equation $Det[\gamma_{\alpha\beta}p_\alpha p_\beta + Ap_\alpha p_\alpha + Bm^2] = 0$. It is of the 12th order in p_μ. Solving it with respect to energy one concludes that **unlike the Maxwell equations there are NO any $E = 0$ solutions.** The solutions in the momentum representation have been explicitly presented in [25] in the standard representation of $\gamma_{\mu\nu}$ matrices. If the 6-component $v(\mathbf{p})$ are defined in such way, we inevitably should get the additional energy-sign operator, refs. [24, 23], $\epsilon = i\partial_t/E = \pm 1$ in the dynamical equation, and the different parities of the corresponding boson and antiboson, $\hat{P}u_{1\sigma}(\mathbf{p}) = +u_{1\sigma}(\mathbf{p})$ and $\hat{P}v_{1\sigma}(\mathbf{p}) = -v_{1\sigma}(\mathbf{p})$.

4. The Construction of Field Operators

The method for constructions of scalar field operators has been given in [5]. In the 4-spinor case we have more components $(\hat{k}+m)\psi(k)|_{k^2=m^2}=0$. However, we can use again

$$\psi(x) = \frac{1}{(2\pi)^{3/2}}\int dk\, e^{ikx}\delta(k^2-m^2)(\theta(k_0)+\theta(-k_0))\psi(k) = $$
$$= \frac{1}{(2\pi)^3}\int\frac{d^3\mathbf{k}}{2E}\left[e^{ikx}\theta(k_0)\psi(k)+e^{-ikx}\theta(k_0)\psi(-k)\right], \quad (43)$$

where $k_0 = E = \sqrt{\mathbf{k}^2+m^2}$ is positive in this case, $(\hat{k}+m)\psi^+(\mathbf{k})=0$, $(-\hat{k}+m)\psi^-(\mathbf{k})=0$. **Please note that the momentum-space Dirac equations have solutions $k_0=\pm\sqrt{\mathbf{k}^2+m^2}$, both for $u-$ and $v-$ spinors. This can be checked by calculating the determinants. Usually, one chooses $k_0=E$ in the $u-$ and in the $v-$. This is because on the classical level (better to say, on the first quantization level) the negative-energy $u-$ can be transformed into the positive-energy $v-$, and vice versa. This is not precisely so, if we go to the secondary quantization level. The introduction of creation/annihilation noncommutating operators gives us more possibilities in constructing generalized theories even on the basis of the Dirac equation.**

Various-type field operators are possible in the $(1/2,1/2)$ representation. During the calculations below we have to present $1=\theta(k_0)+\theta(-k_0)$ (as previously) in order to get positive- and negative-frequency parts.

$$A_\mu(x) = \frac{1}{(2\pi)^3}\int d^4k\,\delta(k^2-m^2)e^{+ik\cdot x}A_\mu(k) = \quad (44)$$
$$= \frac{1}{(2\pi)^3}\int\frac{d^4k}{2E}[\delta(k_0-E_k)+\delta(k_0+E_k)][\theta(k_0)+\theta(-k_0)]e^{+ik\cdot x}A_\mu(k) =$$
$$= \frac{1}{(2\pi)^3}\sum_\lambda\int\frac{d^3\mathbf{k}}{2E_k}[\epsilon_\mu(k,\lambda)a_\lambda(k)e^{+ik\cdot x}+\epsilon_\mu(-k,\lambda)a_\lambda(-k)e^{-ik\cdot x}].$$

In general, due to theorems for integrals and for distributions the presentation $1=\theta(k_0)+\theta(-k_0)$ is possible because we use this in the integrand. However, remember, that we have the $k_0=E=0$ solution of the Maxwell equations. Moreover, it has the experimental confirmation (for instance, the stationary magnetic field $curl\mathbf{B}=0$). Meanwhile the θ function is NOT defined in $k_0=0$. Do we not loose this solution in the above construction of the quantum field operator? Of course, the same arguments can be applied in the construction of the quantum field operator for $F_{\mu\nu}$.

Next, we should transform the second part to $\epsilon^*_\mu(k,\lambda)b^\dagger_\lambda(k)$. In such a way we obtain the charge-conjugate states. In the Dirac case we should assume the following relation in the field operator:

$$\sum_\lambda v_\lambda(k)b^\dagger_\lambda(k)=\sum_\lambda u_\lambda(-k)a_\lambda(-k)\,. \quad (45)$$

Hence, $b^\dagger_\mu(k)=i(\sigma\cdot\mathbf{n})_{\mu\lambda}a_\lambda(-k)$. In the $(1,0)\oplus(0,1)$ representation we have somewhat different situation. Namely, $a_\mu(k)=[1-2(\mathbf{S}\cdot\mathbf{n})^2]_{\mu\lambda}a_\lambda(-k)$. This signifies that in order to construct the field operator (which was used in [25]) we need additional postulates. We can set for the 4-vector field operator: $\sum_\lambda \epsilon_\mu(-k,\lambda)a_\lambda(-k) = \sum_\lambda \epsilon^*_\mu(k,\lambda)b^\dagger_\lambda(k)$. However, in the $(\frac{1}{2},\frac{1}{2})$ representation we can also expand in other way: $\sum_\lambda \epsilon_\mu(-k,\lambda)a_\lambda(-k) = \sum_\lambda \epsilon_\mu(k,\lambda)a_\lambda(k)$. The field operator will only destroy particles in the second case. **Probably, one should construct the left- and right- parts of the field operator separately each other. In this case the commutation relations may be more complicated.**

Finally, going back to the rest $(S,0)\oplus(0,S)$ objects. **Bogoliubov constructs them introducing the products with δ functions like $\delta(k_0-m)$. Then, he makes the boost of the "spinors" only, and changes the above δ to $\delta(k^2-m^2)$ by hand. Is it possible to make the boost of the $\delta-$ functions consistently in such a way?**

The conclusion is: we still have few questions unsolved in the bases of the quantum field theory, which open a room for generalized theories.

References

[1] P. A. M. Dirac, *Proc. Roy. Soc. A* 114 (1927) 243; *ibid. A* 117 (1928) 610.

[2] J. J. Sakurai, *Advanced Quantum Mechanics.* (Addison-Wesley, 1967).

[3] L. H. Ryder, *Quantum Field Theory.* (Cambridge University Press, Cambridge, 1985).

[4] C. Itzykson and J.-B. Zuber, *Quantum Field Theory.* (McGraw-Hill Book Co., 1980), p. 156.

[5] N. N. Bogoliubov and D. V. Shirkov, *Introduction to the Theory of Quantized Fields.* 3rd Edition. (Nauka, Moscow, 1973), p.28, p, 55.

[6] E. P. Wigner, *Ann. Math.* 40 (1939) 149; S. Weinberg, *Phys. Rev.* 133 (1964) B1318.

[7] E. Majorana, *Nuovo Cimento* 14 (1937) 171.

[8] S. M. Bilenky and B. M. Pontekorvo, *Phys. Repts* 42 (1978) 224.

[9] A. Barut and G. Ziino, *Mod. Phys. Lett. A* 8 (1993) 1099; *Int. J. Mod. Phys. A* 11 (1996) 2081.

[10] D. V. Ahluwalia, *Int. J. Mod. Phys. A* 11 (1996) 1855.

[11] P. Lounesto, *Clifford Algebras and Spinors.* (Cambridge University Press, 2002), Ch. 11 and 12; R. da Rocha and W. Rodrigues, Jr., Preprint math-ph/0506075.

[12] M. Kirchbach, C. Compean and L. Noriega, *Eur. Phys. J. A* 22 (2004) 149.

[13] V. V. Dvoeglazov, *Int. J. Theor. Phys.* 34 (1995) 2467; *Nuovo Cim.* 108A (1995) 1467, *Hadronic J.* 20 (1997) 435; *Acta Phys. Polon. B* 29 (1998) 619.

[14] V. V. Dvoeglazov, *Mod. Phys. Lett. A* 12 (1997) 2741.

[15] M. Markov, *ZhETF* 7 (1937), 579, 603; *Nucl. Phys.* 55 (1964) 130.

[16] I. M. Gelfand and M. L. Tsetlin, *ZhETF* 31 (1956) 1107; G. A. Sokolik, *ZhETF* 33 (1957) 1515.

[17] V. V. Dvoeglazov, *Int. J. Theor. Phys.* 43 (2004) 1287.

[18] B. Nigam and L. L. Foldy, *Phys. Rev.* 102 (1956) 1410.

[19] V. B. Berestetskii, E. M. Lifshitz and L. P. Pitaevskii, *Quantum Electrodynamics.The IV Volume of the Landau Course of Theoretical Physics.* 2nd Edition (Butterworth-Heinemann, 1982).

[20] A. Gersten, *Found. Phys. Lett.* 12 (1999) 291.

[21] V. V. Dvoeglazov, *J. Phys. A* 33 (2000) 5011; *Rev. Mex. Fis. Supl.* 49-S1 (2003) 99.

[22] A. Barut, I. Muzinich and D. N. Williams, *Phys. Rev.* 130 (1963) 442.

[23] A. Sankaranarayanan and Good, jr, *Nuovo Cimento* 36 (1965) 1303.

[24] D. L.Weaver, C. L. Hammer and R. H. Good, jr., *Phys. Rev. B* 135 (1964) 241.

[25] D. V. Ahluwalia, M. B. Johnson and T. Goldman, *Phys. Lett. B* 316 (1993) 102.

[26] R. H. Tucker and C. L. Hammer, *Phys. Rev. D* 3 (1971) 2448.

In: Einstein and Others: Unification
Editor: Valeriy V. Dvoeglazov
ISBN: 978-1-63463-276-8

REPLY TO G. W. BRUHN

*Valeriy V. Dvoeglazov**
UAF, Universidad de Zacatecas, Zacatecas, México

Abstract

I show that the Bruhn critics of my article is based on his misunderstandings. The major part of the critics of the Evans articles has been first given in my works and in private communications to them.

In ref. [1] G. W. Bruhn claims that he found errors in my articles [2, 3]. For instance, the following statements have been given by Bruhn: "Since $\mathbf{B}^{(3)}$ is merely the longitudinal component of a field $\mathbf{B}$ that has an additional transversal component $\mathbf{B}_\perp$ Dvoeglazov's result contradicts the well-known Lorentz transform of the electromagnetic field where the longitudinal component remains unchanged." "This result proves that V. V. Dvoeglazov Equ. (11b) cannot be true".

However, in fact, I called frequently the $\mathbf{B}^{(3)}$ to be "the so-called magnetic field", just see the inverted commas in the words "magnetic field" (the first line, the page 228 of [3]). Next (see the 5th line of the page 230 of [3], just before the equation (11b), which Bruhn doubts) I state explicitly "The 3-vector $\mathbf{B}^{(3)}$ (which is defined by (1) [by Evans indeed]) may *not* be the entry of the antisymmetric tensor field; it is ... the entry of some 4-vector provided that the Evans' definitions for circularly polarized radiation are used." I hope that Bruhn knows that the Lorentz transformation laws are different for an antisymmetric tensor field and a 4-vector field. The equation (11b) of the cited paper is precisely the transformation law for the 3-part of a 4-vector. Its parity properties have been dicussed in the discussion with Comay and Evans, cf. with ref. [4a].

Moreover, in ref. [2] (published previously than [3]), see Eqs. (9), I proved the statement by explicite mathematical calculations. I again explicitely claimed: "... $B^{(0)}$ transforms as zero-component of the 4-vector and $\mathbf{B}^{(3)}$ as the space components of the 4-vector...".

Finally, a quite inaccurate statement is given in the end of the Bruhn paper. The $SO(3)$ group is the subgroup of the Lorentz group. So, it is obvious that $SO(3)$ symmetry (but the different one from that given by M. W. Evans) is compatible with the Lorentz covariance.

*E-mail address: vdvoeglazov@yahoo.com.mx

Thus, in my opinion, the Bruhn paper is some sort of defamation.[1] While I acknowledge the trivial errors in the papers by M. W. Evans, E. Comay and G. W. Bruhn, I continue to state that there are NO any calculational errors in my papers [2, 3, 4].

References

[1] G. W. Bruhn, http://www.mathematik.tu-darmstadt.de/~bruhn/EM-Lorentz-Transform.html. In the similar articles published in the journals Bruhn somehow removed the references to my works. However, it is still on his website. I was not aware about the Bruhn critics existence for a long time.

[2] V. V. Dvoeglazov, Found. Phys. Lett., **10**, No. 4, pp.383-391 (1997), physics/9611009.

[3] V. V. Dvoeglazov, Apeiron, **6**, No. 3-4, pp. 227-232 (1999), physics/9801024.

[4] V. V. Dvoeglazov, Found. Phys. Lett., **13**, No. 4, pp. 387-393 (2000), physics/9907048; V. V. Dvoeglazov and J. L. Quintanar Gonzalez, Found. Phys. Lett., **19**, No. 2, 195-200 (2006), physics/0410169.

[1]Unfortunately, the same sins can be found in the papers and the correspondence by M. W. Evans and E. Comay. So, I am **not** going to enter in discussions with them in the future.

INDEX

E

F

G

H

I

J

L

M

N

O

P

U

V

W

Y